Food Safety
in China
A COMPREHENSIVE REVIEW

Food Safety in China

A COMPREHENSIVE REVIEW

Linhai Wu • Dian Zhu

CRC Press
Taylor & Francis Group
Boca Raton London New York

CRC Press is an imprint of the
Taylor & Francis Group, an **informa** business

MATLAB® is a trademark of The MathWorks, Inc. and is used with permission. The MathWorks does not warrant the accuracy of the text or exercises in this book. This book's use or discussion of MATLAB® software or related products does not constitute endorsement or sponsorship by The MathWorks of a particular pedagogical approach or particular use of the MATLAB® software.

CRC Press
Taylor & Francis Group
6000 Broken Sound Parkway NW, Suite 300
Boca Raton, FL 33487-2742

First issued in paperback 2019

© 2015 by Taylor & Francis Group, LLC
CRC Press is an imprint of Taylor & Francis Group, an Informa business

No claim to original U.S. Government works

ISBN-13: 978-1-4822-1833-6 (hbk)
ISBN-13: 978-0-367-37845-5 (pbk)

Library of Congress Cataloging-in-Publication Data

Wu, Linhai.
    Food safety in China : a comprehensive review / Linhai Wu and Dian Zhu.
        pages cm
    Includes bibliographical references and index.
    ISBN 978-1-4822-1833-6 (hardback)
    1. Food adulteration and inspection--China. 2. Food--Safety measures--Government policy--China. 3. Food contamination--China. I. Zhu, Dian. II. Title.

TX531.W968 2014
363.19'260951--dc23   2014003434

Visit the Taylor & Francis Web site at
http://www.taylorandfrancis.com

and the CRC Press Web site at
http://www.crcpress.com

Contents

Foreword

In recent years, frequent occurrences of food safety incidents have caused loss of human lives and properties worldwide. Therefore, food safety has become a major concern. As China is in a profound state of social transition, food safety incidents have occurred more frequently. Moreover, the media indiscriminately spreads related news and even intentionally exaggerates the problems in such a way that food safety incidents become the focus of widespread concern. People are even shouting, "What on Earth is safe to eat?" Food safety is increasingly becoming a major problem associated with the national economy, the livelihood of the people, and social stability in China. Therefore, a true, objective, and comprehensive description of China's food safety status is of great significance.

Jiangsu Provincial Food Safety Research Base of Jiangnan University is an important specialized institution for the research of food safety management in China. In recent years, with a strong commitment to social responsibility, a group of young- and middle-aged researchers led by Professor Linhai Wu, chief specialist of the Base, and Associate Professor Dian Zhu have carried out a painstaking research. Their years of work culminated in the publication of *Food Safety in China: A Comprehensive Review* in the United States. Part of the relevant findings was published in China and highly regarded by the major Chinese media outlets and the general public. I believe that the publication of the book in the United States will facilitate a comprehensive and objective understanding of China's current food safety status in the international community.

The basic feature of this book is to profoundly reveal the principal contradictions influencing food safety in China, by grasping the basic characteristics of dynamic evolution of food safety, focusing on the entire food supply chain system, and including food producers and operators, consumers, and the government into an organic system. Starting from the production of edible agricultural products, it investigates the changing track of food safety (including the safety of imported and exported food) in key stages, such as production, circulation, and consumption, from the perspective of management, based on field investigations and comprehensive use of various statistical data. An objective assessment of the actual state and future trends of the food safety risks in China is also provided. I believe that the objective assessment of food safety status in China provided in this book will help the world understand the real situation in China, as well as the efforts made by the Chinese government and all sectors of the society to improve food safety.

As China's food safety problems have very complex causes, it is impossible to obtain complete solutions in any single research. Moreover, due to various objective limitations, this book has some inevitable limitations. For example, interdisciplinary discussion is still weak, and the investigation of the effects of food safety standards and food science and technology on food safety is not in-depth enough. I look forward to further research in the relevant fields by Professor Linhai Wu and his fellows.

The publication of this book also provides valuable academic information to academic colleagues concerned about the global and Chinese food safety status and offers a valuable sample for further investigation and research of the experiences in global food safety management. The authors of the book and I also look forward to close cooperation with international colleagues for researching global and Chinese food safety management issues.

Baoguo Sun
Beijing Technology and Business University

Preface

According to a 2010 World Trade Organization report, China is the world's top producer of agricultural products by value, with total production of about 4078 billion yuan (US$536 billion). Unfortunately, China has been the focus of many food safety issues over the years, which has caused fear in the international community. Whether they are aware of it or not, every day, people around the world consume food that was produced in China. Food safety in China has been the subject of numerous news reports from global media agencies for quite some time.

This book will provide a comprehensive review of the efforts put forth by China to prevent food safety issues. This subject matter is of great importance to policymakers around the globe, as well as manufacturers that import and export goods to/from China. Further, global health agencies closely monitor food safety incidents and have a great interest in understanding the steps that China is taking to preventing further food safety issues.

We have striven to provide a comprehensive, objective, and impartial assessment and description of China's food safety situation. Our related preliminary results have been published in China and have been widely recognized by society. However, we know that there are still many problems, as well as many improvements to be made. In particular, the problem of how to provide a more comprehensive, real, and objective description of China's food safety situation has plagued us. We urge researchers worldwide to offer recommendations and criticisms on this book, in order to provide a reference for further improving the quality of our research on food safety issues in China in the future. We very much look forward to close cooperation with international colleagues for future research opportunities regarding global and Chinese food safety management issues.

MATLAB® is a registered trademark of the MathWorks, Inc. For product information, please contact:

The MathWorks, Inc.
3 Apple Hill Drive
Natick, MA 01760-2098 USA
Tel: +1 508 647 7000
Fax: +1 508 647 7001
E-mail: info@mathworks.com
Web: www.mathworks.com

Linhai Wu
Jiangnan University

Dian Zhu
Suzhou University

Acknowledgments

The book *Food Safety in China: A Comprehensive Review* is the research achievement made jointly by the research team led by Professor Linhai Wu, a chief specialist of Jiangsu Provincial Food Safety Research Base of Jiangnan University, and Associate Professor Dian Zhu from the Business School at the Suzhou University. This book integrates the most important research findings of the research team during 2010–2013. Today, it is published in the United States with the support of Taylor & Francis. We thank all colleagues who have supported the publication of this book.

Key researchers from various Chinese colleges, universities, and research institutions who participated in the preliminary and integrated research of this book include (in alphabetical order of first name) the following: Chunhui Liu, China National Institute of Standardization; Fanhua Kong, South China Normal University; He Qian, Jiangnan University; Hongsha Wang, Jiangnan University; Ji Chen, State Administration for Industry & Commerce of the People's Republic of China; Jianhua Wang, Jiangnan University; Liangyun Niu, Anyang Normal University; Lijie Shan, Jiangnan University; Lingling Xu, Jiangnan University; Liqing Xu, Jiangnan University; Peng Liu, Renmin University of China; Qiuling Chen, Shanghai University; Qiuqin Zhang, Jiangnan University; Shijiu Yin, Qufu Normal University; Shuxian Wang, Jiangnan University; Wei Hong, Jiangnan University; Xiaochun Tang, Renmin University of China; Xiaojuan Hong, Nanjing University of Posts and Telecommunication; Xiaoli Wang, Jiangnan University; Xiaomei Zhou, Zhejiang Gongshang University; Xiaowei Wen, South China Agricultural University; Xia Tong, Nantong University; Xujin Pu, Jiangnan University; Yanjing Jiang, China National Food Industry Association; Yanwen Tan, South China Agricultural University; Yaping Huo, Renmin University of China; Yingqi Zhong, Jiangnan University; Yinjun Xu, Qufu Normal University; Zhemin Li, Chinese Academy of Agricultural Sciences; Zhongyi Zhu, Suzhou University; and many others. All these researchers were involved in the research work in varying degrees, provided data, assisted in collating information, or participated in proofreading or graphic production.

The survey reported in Chapter 5 was carried out with the assistance of more than 200 undergraduate students from Jiangnan University from January to March in 2012.

Qiuqin Zhang, our important colleague from Jiangnan University, made great efforts in the publication of this book in the United States by contributing to the proofreading and graphic production of the entire book and also participating in the research.

Professor Wuyang Hu (department of agricultural economics at the University of Kentucky, co-editor of *Canadian Journal of Agricultural Economics*) provided useful recommendations and guidance on the preparation of this book as an international participator.

As authors in charge of this book, we gratefully thank all members of the research team and related personnel for their fruitful efforts!

The large number of works cited in our research should also be noted. We thank all the authors of works cited.

The research of this book was led by Wu and Zhu. They were primarily responsible for the overall design of the book, revision of the research outline, establishing research priorities, and coordinating key issues in the research process. They also made final modifications of all of the research results, while completing their research tasks.

The authors are grateful for the financial support from the Social Science Fund on academic translation project of Jiangsu Province and from the construction project of Jiangsu Province for outstanding innovation team of philosophy and social sciences in the University.

We thank Mr Rock Lau (General Manager), Miss Dawn Hu (senior translator), and Dr. Chris Carter (senior copyeditor) from New Bridge Translation Co. Ltd., Guangzhou, China, for their excellent services including publication consultancy, translating, and copyediting with academic integrity.

Research Introduction
and Main Inclusions

As the introduction of this book, this chapter provides a brief description of the key concepts involved, research thread, methods, content, and main conclusions to give a panoramic description of the topics covered in this book.

0.1 Research Thread and Perspective

In recent years, frequent occurrences of food safety incidents have caused loss of human lives and properties worldwide. Therefore, food safety has become a major concern. As China is in a profound state of social transition, food safety incidents have occurred more frequently. Moreover, the media indiscriminately spreads related news and even intentionally exaggerates the problems in such a way that food safety incidents become the focus of widespread concern. People are even shouting, "What on Earth is safe to eat?" During the 21st Meeting of the Standing Committee of the 11th National People's Congress on June 29, 2011, a proposal to include food safety was made, along with financial security, grain security, energy security, and ecological safety, in the "national security" system (Sina 2011). This demonstrated that food safety risk had become an extremely serious problem at the national level.

China's food safety problems are quite complicated. The basic feature of this book is to provide a full, true, and objective investigation and analysis of the food safety problems affecting China from the professional perspective of a scholar in an impartial manner. To accomplish this, the authors must decide upon the standpoint, perspective, and sequence of thought in which this book is to be presented. In other words, a research thread must be selected, and improper selection may affect the objectivity, accuracy, and scientificity of the research findings. This is a fundamental issue that intrinsically directs the research approach and main content of this text.

0.1.1 Research Thread

In any country, food safety problems can occur at different stages and levels throughout the food supply chain system. In particular, irregularities in the following stages are more likely to induce food safety risks: (1) production of primary agricultural products and other raw materials of food; (2) food production and processing; (3) food transportation; (4) food consumption environment and consumers' awareness of food safety; (5) efforts and technical means of the government in food regulation; (6) social responsibility of food producers and traders, and moral traits and professional qualities of employees; and (7) scientificity, rationality, effectiveness, and operability of the technical specifications in production, processing, circulation, and consumption. In China, the above-mentioned stages mainly involve the government, producers and traders, and consumers. They also involve both technical and management issues. Management issues involve the enterprises, the government regulatory system, and the personal problems of consumers. Moreover, risks can be caused by both natural and human factors. Ultimately, these complex issues span the entire food supply chain system.

Food supply chain refers to the integration of stakeholders involved in each stage, from agricultural production to consumption, ranging from primary food producers and traders to consumers (including the supplier of means of production at the front and back-end regulator at the rear, i.e., the government) (Den Ouden et al. 1996). Although the food supply chain system concept has been constantly enriched and developed in practice, a basic problem has been revealed by the above-mentioned definition, which has been widely accepted throughout the world. In accordance with this definition, producers and operators in the Chinese food supply chain system mainly include agricultural producers (including different forms of entities, such as small private farmers, large-scale farmers, cooperatives, agricultural enterprises, and livestock producers) and entities involved in food production, processing, packaging, logistics, distribution (wholesale and retail), and other relevant stages. They jointly constitute the entity that must prevent and take risks in food production and marketing.* Food safety risk factors can exist in each stage of the food supply chain system due to technical limitations, mismanagement, and other problems of the agricultural producers and entities involved in food production, processing, logistics, and distribution. Because

---

*Agricultural producers and entities involved in food production, processing, logistics, and distribution in the food supply chain system are collectively referred to as "food producers and operators" in this book, in order to distinguish them from the consumers and the government in the food supply chain system.

all stages are closely linked with one another and affected by one another, the guarantee of food safety cannot simply depend on a single manufacturer, but is the common mission of all entities in the supply chain. The relationship between food safety and the food supply chain system has become a research topic of human society development in a new historical period. Therefore, in this text, the thread of Chinese food safety analysis and research is to analyze the quality and safety of Chinese edible agricultural products and food in production, processing, circulation, consumption, and import and export based on the entire food supply chain system and to present the development of Chinese food safety legal system, management system, and standard system, in order to provide an outline description of Chinese food safety to concerned persons throughout the world.

0.1.2 Research Perspective

Numerous pioneering studies have been conducted on the correlation between food safety and the food supply chain system from multiple angles, such as macro and micro perspectives, technology, system, government, market, producers and operators, and consumers (Liu 2012a). However, according to the main characteristics of current food safety risks and the basic nature and causes of major food safety incidents in China, the level of food science and technology is not a major bottleneck restricting and affecting the level of food safety. In fact, the technical deficiencies, environmental pollution, and other factors also have an impact on food safety, for example, photooxidation of milk (Kerkaert et al. 2012) and nitrite peak formation in fresh vegetables (Yan et al. 2006). However, when considering the entire food supply chain system, most of the food safety problems in China are caused by human factors, such as misconduct, noncompliance, or partial compliance to existing food technical specifications and standard systems. This is the authors' distinct point of view. Therefore, at this stage, comprehensive management integrating technical, regulatory, legal, institutional, and political measures must be carried out to effectively prevent food safety risks and protect food safety in China. Moreover, emphasis should be placed on deepening the reform of the supervision system to strengthen food supervision and regulate the behaviors of food producers and operators. This is not only the current problem, but also the future focus in food safety management in China. The reform of China's food safety management system, implemented by the State Council in March 2013, has laid the foundation for the prevention of food safety risks from segmented supervision at the institutional level. However, it will still be difficult to prevent food safety risks if the human factors of food producers and operators are not properly addressed. A preliminary analysis and investigation regarding this topic will be given in Chapter 5. Based on the above-mentioned analysis, the research perspective of this book is to provide a systematic in-depth analysis at the management level.

To sum up, in order to profoundly reveal the principal contradictions influencing food safety in China, this book focuses on the entire food supply chain system and includes food producers and operators, consumers, and the government into an organic system. Starting from the production of edible agricultural products, it investigates the changing track of food safety (including the safety of imported and exported food) in key stages, such as production, circulation, and consumption, from the perspective of management based on field investigations and comprehensive use of various statistical data. An assessment of the actual state and future trends of the food safety risks in China is also provided. In addition, the progresses and the main problems in the construction of the major food safety support systems are analyzed in a selective and targeted manner. In short, this book attempts to comprehensively and accurately describe the overall changes in food safety in China in recent years based on the above-mentioned research thread and perspective to provide adequate information to people concerned with the situation of Chinese food safety to the greatest extent possible.

0.2 Definitions of Major Concepts

Food, edible agricultural products, food safety, and food safety risks are the most important basic concepts in this book. As China's national conditions differ from other countries, the above-mentioned concepts have unique characteristics in China and thus have a somewhat different definition from those in other countries. By drawing on the findings of relevant research (Wu and Xu 2009), this book will provide scientific definitions of these concepts based on realities in China to ensure the scientificity of our research.

0.2.1 Food, Agricultural Products, and Their Relationship

In simple terms, food is any substance consumed by humans. However, it is not easy to further provide an accurate and scientific definition and classification, as these require a comprehensive consideration of various opinions, the realities in China, and the research background of this book.

0.2.1.1 Definition and Classification of Food In China, generally speaking, the simplest definition of "food" is material that can be consumed by humans, including natural foods and processed foods. Natural foods are those that grow in nature; they are unprocessed and can be directly consumed by humans. Processed foods are finished products that are produced or processed with a certain technology for the purpose of human consumption. Processed foods include, among other things, rice, wheat flour, and fruit juice drinks. Additionally, "food" generally does not include products used solely for medical purposes.

According to Article 54 of Chapter 9 Supplementary Provisions in the *Food Hygiene Law of the People's Republic of China* (generally referred to as the *Food Hygiene Law* in the book) implemented on October 30, 1995, "'food' refers to any finished product or raw material intended for people to eat or drink, as well as any product that has traditionally served as both food and medication, with the exception of products used solely for medical purposes." In Article 2.1 in the national standard GB/T 15091-1994, *Basic Terminology in Food Industry*, implemented on December 1, 1994, "food" is defined as any substance that can be eaten or drunk by humans, whether processed, semiprocessed, or raw, but does not include tobacco or substances used only as drugs. Article 99 of Chapter 10 Supplementary Provisions in the *Food Safety Law of the People's Republic of China* (generally referred to as the *Food Safety Law* in this book) implemented on June 1, 2009, provides exactly the same definition of "food" as the GB/T 15091-1994. According to the *General Standard for the Labeling of Prepackaged Foods* (CODEX STAN 1-1985) published by the Codex Alimentarius Commission (CAC), "food" means any substance, whether processed, semiprocessed, or raw, which is intended for human consumption, and includes drinks, chewing gum, and any substance that has been used in the manufacture, preparation, or treatment of "food" but does not include cosmetics, tobacco, or substances used only as drugs.

The wide variety of foods can be classified by different methods according to different standards. The national standard GB/T 7635.1-2002, *Categories and Codes of Major Products in China*, classifies food into two categories: (1) agriculture, forestry, animal husbandry, and fishery products and (2) processed food, beverages, and tobacco (Administration of Quality Supervision, Inspection and Quarantine [AQSIQ] 2002). The agriculture, forestry, animal husbandry, and fishery products include three subcategories: (1) farming products, (2) live animals and animal products, and (3) fish and other fishery products. The processed food, beverages, and tobacco category includes five subcategories: (1) meat, aquatic products, fruits, vegetables, and oils and other similar processed products; (2) dairy products; (3) ground grain products, starches and starch products, soy products, other foods and food additives, and processed feed and feed additives; (4) drinks; and (5) tobacco products.

According to the *Labeling Methods for 28 Product Categories and License Application Units* issued by the AQSIQ (Website of Zhongshan Municipal Bureau of Quality and Technical Supervision of the People's Republic of China 2008), in terms of application for food production licenses, foods are divided into 28 categories, including grain products, edible oils, fats and fat products, condiments, meat products, dairy products, drinks, convenience foods, biscuits, canned foods, frozen drinks, fast-frozen foods, tubers and puffed foods, confectionery products, tea and related products, alcoholic beverages, vegetable products, fruit products, roasted seeds and nuts and nut products, egg products, cocoa and roasted coffee products, sugar, aquatic products, starches and starch products, cakes, soy products, bee products, and special dietary foods.

The *National Food Safety Standard—Standards for Uses of Food Additives* (GB 2760-2011) (Ministry of Health of the PRC 2011) classifies food into sixteen categories: (1) milk and dairy products; (2) fats, oils, and fat emulsions; (3) frozen drinks; (4) fruits, vegetables (including roots and tubers), beans, edible fungus, algae, nuts, and seeds; (5) cocoa products, chocolates and chocolate products (including chocolates and chocolate products with cocoa butter alternatives), and candy; (6) cereals and cereal products; (7) baked goods; (8) meat and meat products; (9) aquatic animals and their products; (10) eggs and egg products; (11) sweeteners; (12) condiments; (13) special dietary foods; (14) drinks; (15) alcoholic beverages; and (16) other foods.

Food concept is a highly specialized issue and is not the research focus of this book. Unless otherwise noted, the definition of food provided by the *Food Safety Law* is used for the purpose of this book.

0.2.1.2 Agricultural Products and Edible Agricultural Products   These are very important concepts in this book. According to the *Law of the People's Republic of China on Agricultural Product Quality and Safety* (generally referred to as *Law on Agricultural Product Quality and Safety* in the book) adopted by the 21st Meeting of the Standing Committee of the 10th National People's Congress on April 29, 2006, the term "agricultural products" is defined as "primary products sourced from agriculture, that is, plants, animals, microbes, and their by-products obtained from agricultural activities." In fact, the concept of agricultural products can be defined in both broad and narrow senses. In a broad sense, the term "agricultural products" refers to products produced by the agricultural sector, including the farming, forestry, animal husbandry, and fishery sectors. In a narrow sense, the term "agricultural products" refers only to staple foods. The broad concept of agricultural products is basically the same as that defined in the *Law on Agricultural Product Quality and Safety*.

Different international organizations and countries have different, often very different, standards for classification of agricultural products. Relevant departments of the Ministry of Agriculture of the People's Republic of China classify agricultural products into five categories: (1) grain and oil, (2) vegetables, (3) fruits, (4) aquatic products, and (5) livestock. The process of manufacturing rough and fine finished products and semifinished products from agricultural products with different processing techniques and methods according to their texture characteristics, chemical composition, and physicochemical properties is called processing of agricultural products. According to the International Standard Industrial Classification (ISIC) of the United Nations (UN), agricultural product processing industry is divided into five categories: (1) food, beverages, and tobacco; (2) textiles, apparel, and leather; (3) wood products and furniture; (4) paper products and printing; and (5) rubber and plastic products. According to the classification by the China National Bureau of Statistics, the agricultural product processing industry includes 12 industries: (1) food processing (including grain and feed processing); (2) food manufacturing (including confectionery, dairy products, canned foods, fermentation products, condiments, and other foods); (3) beverage manufacturing (including alcohol and alcoholic beverages, soft drinks, and tea); (4) tobacco processing; (5) textile; (6) apparel and other fiber products manufacturing; (7) leather, fur, feathers, and their products; (8) wood processing, bamboo, rattan, palm and straw products manufacturing; (9) furniture manufacturing; (10) paper making and paper products industry; (11) printing and record medium reproduction; and (12) rubber manufacturing.

Since agricultural products are a main source of food and an important source of industrial raw materials, they can be divided into edible and nonedible agricultural products. According to the document *A Notice on Pilot Chain Operation of Agricultural Products* issued by the Ministry of Commerce, Ministry of Finance, and State Administration of Taxation of the People's Republic of China in April 2005, edible agricultural products include plants, livestock and fishery products, and their primary processed products intended for consumption. Similarly, the concept of agricultural products is a highly specialized issue and not the research focus of this book. Unless otherwise noted, the definitions of agricultural products and edible agricultural products provided by the *Law on Agricultural Product Quality and Safety* and relevant definitions provided by the Ministry of Commerce, Ministry of Finance, and State Administration of Taxation of the People's Republic of China are used for the purpose of this book.

0.2.1.3 Relationship between Agricultural Products and Food   This seems to be very simple, but that is far from the case. In fact, food is included in agricultural products in some countries, whereas in other countries, agricultural products are included in food. For example, the *Uruguay Round Agreement on Agriculture* and the *Canada Agricultural Products Act* include "food" in the definition of "agricultural products." Although agricultural products are included in food in some countries, the "processing and production" of food is stressed. Nevertheless, no matter how they are defined and classified, there is a clear legal relationship between agricultural products and food in the same country. Similarly, in China, the *Food Safety Law* and the *Law on Agricultural Product Quality and Safety* provide a clear definition of food and agricultural products, respectively, which provides a clear legal relationship between them.

Both a necessary connection and a certain difference exist between agricultural products and food. Agricultural products are primary products derived from agriculture, including agricultural products for direct consumption, food materials, and nonedible agricultural products. Most agricultural products need to be processed to become food. Therefore, food is the extension and development of agricultural products. This is the natural connection between agricultural products and food. The connection between them is also reflected in quality and safety. Most quality and safety problems of agricultural products come from agricultural production; for example, the use of pesticides and fertilizers will often compromise the quality and safety of agricultural products. The quality and safety of food depend primarily on the safety of agricultural products. Furthermore, agricultural products are derived directly from agricultural production activities and thus are the products of primary industry. Foods, especially processed foods, are mainly food products from industrialized processing of agricultural products and thus are the products of second industry. Processed foods have typical industrial characteristics, including a short production cycle, quantity production, delicate packaging, an extended shelf life, and a reduced loss during transportation, storage, and distribution. This is the main difference between agricultural products and food. A simple illustration of the interrelationship between food and agricultural products is given in Figure 0.1.

Currently, a very strict distinction is not made between agricultural products, edible agricultural products, and food in the general discussion of food

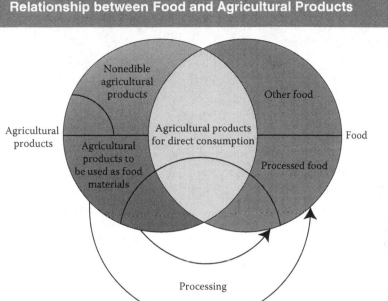

FIGURE 0.1

Relationship between Food and Agricultural Products

safety in the Chinese academic circles. On the contrary, the three concepts cross one another, and agricultural products and edible agricultural products are often included in the meaning of food. Except Chapters 1 and 2, in which the safety of edible agricultural products and food quality and safety in production and processing are analyzed, respectively, and where otherwise indicated, a very strict distinction is also not made between edible agricultural products and food in this book.

0.2.2 Connotation of Food Safety

Food safety is involved throughout the entire development of human society, and it is the material basis and necessary guarantee for a country's economic development and social stability. Therefore, with regard to food safety issues, great concern and attention are directed toward the strategic level of national security in most countries, including developed countries.

0.2.2.1 Quantitative and Qualitative Food Safety At least in China, the connotation of food safety includes both quantitative and qualitative aspects. Quantitative food safety emphasizes the quantity of food, that is, an adequate supply of food, and is the quantitative representation of consumer food demand. The addressing of quantitative food safety is, at all times, a top priority in every country, especially in developing countries. Currently, the problem of quantitative food safety has been solved in all countries in the world, with the exception of a few countries in Africa. Food supply is no longer a principal contradiction. Qualitative food safety concerns the quality of food. Qualitative food safety status is the degree of influence of food hazards on consumer health in a country or region and is characterized by guarantee of food hygiene and rational nutrition structure. Therefore, qualitative food safety stresses the need to ensure that food consumption has no direct or potential adverse effects on human health.

Quantitative and qualitative food safety are the two interrelated and essential aspects in the connotation of food safety. In China, food safety currently concerns the qualitative more so than the quantitative safety.

0.2.2.2 Understanding of Food Safety   In China, some general consensuses have been reached about the understanding of food safety, which were given as follows:

Food safety is dynamic. According to Article 99 of Chapter 10 Supplementary Provisions in the *Food Safety Law*, the term "food safety" means that the food is nontoxic and harmless, satisfies the necessary nutritional requirements, and does not cause any acute, subacute, or chronic hazards to human health. A general review of the history of food safety management in China reveals that the nutritional requirements and the definitions of nontoxicity, harmlessness, and noncontribution to acute, subacute, or chronic hazards vary at different times. Different standards correspond to different levels of food safety. Therefore, food safety is a dynamic concept.

Food safety has legal standards. Since the 1980s, some countries and relevant international organizations have begun to replace legislations for hygiene, quality, nutrition, and other elements with integrated legislations for food safety from the perspective of social systems engineering. The United Kingdom passed the *Food Safety Act* in 1990, the European Union published the *White Paper on Food Safety* in 2000, and Japan enacted the *Food Safety Basic Law* in 2003. Some developing countries have also formulated a food safety law. The replacement of elemental laws, such as food hygiene laws, food quality laws, and food nutrition laws, with integrated food safety laws is generally done to meet the requirements of the development of times.

It also demonstrates the inherent requirements for establishing legal standards for food safety in a country.

Food safety is characterized by social governance. Unlike the concepts of hygiene, nutrition, and quality, the concept of food safety is related to social governance. Different countries have different outstanding problems and governance requirements related to food safety in different historical periods. In developed countries, concerns over food safety are mainly based on problems caused by the development of science and technology, such as the effects of genetically modified foods on human health. In developing countries, the focus of food safety is problems caused by immature market economies, such as counterfeiting and illegal production of toxic and hazardous food at this stage. In China, the food safety problems include all of the above-mentioned concerns.

Food safety is political. It is the most basic social responsibility and a necessary commitment for enterprises and the government to ensure food safety in both developed and developing countries. Food safety is closely linked with the right to life and is unique and mandatory; it is thus a matter of government guarantee or enforcement. Moreover, food safety is often associated with the right to development and characterized by hierarchy and selectivity; therefore, it is also a matter of commercial selection or government advocacy. In recent years, the international community has gradually replaced the concepts of food hygiene and food quality with the concept of food safety, giving more prominence to the political responsibility for food safety.

Based on the above understandings, the full concept of food safety can be expressed in the following way: food (or agricultural products) meets the national mandatory standards and requirements in terms of planting, breeding, processing, packaging, storage, transportation, distribution, and consumption and does not contain any toxic and hazardous substances that may lead to harm or threat to human health, cause illness or death of consumers, or endanger the health of consumers and their descendants. This concept of food safety indicates that food safety covers the safety in both production and operation, in both results and processing, and in both present and future. The definition of food safety provided by the *Food Safety Law* is used for the purpose of this book, with a focus on the "qualitative food safety," and analysis of the overall level of food quality and safety in China is presented on this basis and based on the existing Chinese national standards.

0.2.3 Food Safety, Food Hygiene, and Staple Food Security

In China, the main concepts related to food safety include food hygiene and staple food security. To facilitate understanding of the relevant aspects, a special explanation is provided in Sections 0.2.3.1 through 0.2.3.3.

0.2.3.1 Food Safety and Food Hygiene   As an important concept, the term "food hygiene" appears more than once in this book, especially in the discussion of the Chinese food safety legal system construction and the evolution and reform of the Chinese food safety management system. In fact, this concept is inevitably involved in the study of Chinese food safety problems. According to the Chinese national standard GB/T 15091-1994 *Basic Terminology in Food Industry*, "food hygiene" is defined as "the measures taken to prevent contamination of food by harmful substances during production, harvesting, processing, transportation, storage, distribution, and other stages for the purpose to benefit human health." Food hygiene has the essential characteristics of food safety, including safety in results (nontoxic, harmless, and satisfying the necessary nutritional requirements) and in processing, that is, the conditions and environment that guarantee the safety in results. The differences between food safety and hygiene are in twofold. First, coverage: food safety covers planting, breeding, processing, packaging, storage, transportation, distribution, and consumption, but food hygiene usually does not include planting and breeding. Second, emphasis: food safety is the integration of safety in results and in processing, while food hygiene focuses more on safety in processing, although it contains both aspects.

0.2.3.2 Food Safety and Staple Food Security   One important duty of the Chinese government is to ensure staple food security. In China, it is generally believed that staple food security is the assurance that any person can get enough food to survive and maintain health at any time. Food safety concerns quality requirements, while staple food security concerns an adequate supply. Three main differences exist between food safety and staple food security. First, in concept: staple food is rice, wheat, maize, sorghum, millet, and other grains, as well as tubers and beans, while food has a broader connotation than staple food. Second, in industry range: staple food is mainly produced by planting, while food is produced by planting, breeding, forestry, and other industries. Third, in evaluation indicators: staple food security is mainly to ensure the supply–demand balance, and the evaluation indicators are production levels, inventory levels, and sufficiency of food for the poor; food safety is mainly to provide nontoxic, harmless, healthy, and nutritive food, and the evaluation indicators are physical, chemical, biological, and nutrition indicators.

It should be noted that the meaning of China in this book is the Chinese mainland except Hong Kong, Macao, and Taiwan, if not otherwise indicated. This book is a professional research monograph, rather than a political monograph. If you find any mistake, it is our negligence.

0.2.3.3 Interrelationship between Food Safety and Food Hygiene In summary, the concepts of food safety and food hygiene are neither parallel nor crossed in China. The concept of food safety includes food hygiene. The inclusion of food hygiene in the concept of food safety is not to deny or abolish the concept of food hygiene. On the contrary, it is to look at food hygiene in a more scientific system and in a more macro perspective, which demonstrates China's progress on food safety management. For example, the coordination and integration of food standards under the concept of food safety effectively prevent the overlaps and duplications among food hygiene, quality, and nutrition standards.

0.2.4 Food Safety Risks

Risk is the product of the probability and consequence of the occurrence of risk events (Gratt 1987). The UN Chemical Safety Project defines "risk" as the probability of harmful effects on organizations, systems, or populations (or subpopulations) under certain conditions after the exposure of a particular factor (Shi 2010). Because of different risk characteristics, all risks cannot be described by one definition, and a targeted definition should be used according to specific objects and properties.

With regard to food safety risk, three international expert consultations were held by the Food and Agriculture Organization (FAO) and World Health Organization (WHO) during 1995–1999. CAC defines "food safety risk" as the likelihood and severity of adverse effects, which are caused by a hazard in food, on human health or the environment (WHO/FAO 1997). According to the International Life Sciences Institute (ILSI), "food safety risk" mainly refers to a biological, chemical, or physical factor that will potentially damage or endanger food safety and quality (ILSI 1997). "Biological hazard" mainly refers to bacteria, viruses, fungi, and other microbes that can produce toxins, while "chemical hazard" mainly refers to pesticide and veterinary drug residues, growth promoters, contaminants, and illegal use of additives. "Physical hazard" mainly refers to metal, debris, and other foreign matter. Physical hazards have a smaller effect relative to biological and chemical hazards (Valeeva et al. 2004). Food safety risk varies by country because of differences in technological and economic development. Therefore, a new method should be developed for identifying food safety risks to pool the resources to address key risks, in order to prevent the development of potential risks into real risks and the consequent food safety incidents (Kleter and Marvin 2009). With regard to food risk assessment, a widely accepted definition has been provided by FAO: the scientific assessment of the possible adverse effects of biological, chemical, and physical hazards in food and food additives on human health, including hazard identification, hazard characterization, exposure assessment, and risk characterization. For the purpose of this book, "food safety risk" is defined as the likelihood and severity of adverse effects on human health or the environment due to hazards in food.

Other important concepts and terms, such as food additives, chemical pesticides, and pesticide residues, are also discussed in this book, which are not detailed here due to space limitations. They can be found in relevant chapters.

0.3 Research Period and Methods

0.3.1 Research Period

This book focuses on reviewing the changes in food safety in recent years in China. Taking into account the availability of data, the research period is 2006–2011, but the review starts from 2005 in some chapters. For example, the review of market supply and quality and safety of major edible agricultural products in China starts from 2005. The review even starts from 1949 in some chapters, for example, Chapter 8. A large span of time allows a more clear presentation of the evolution of the Chinese food safety management system. In addition, the review ends in 2011 in most chapters, but ends in 2012 in some chapters, for example, when discussing the assessment of Chinese food safety risks. Due to difficulties in data collection, the research period is not completely unified in this book; however, this will not affect the comprehensive presentation of the current situation and trends of food safety in China to readers.

0.3.2 Research Methods

This book uses a multidisciplinary combination of four research methods, which will be discussed in Sections 0.3.2.1 through 0.3.2.4.

0.3.2.1 Survey Research   A questionnaire survey was carried out among 4289 respondents in 12 provinces in mainland China to investigate the food safety evaluation and concerns of Chinese urban and rural residents. With a wide coverage, numerous survey sites, and a large sample size, this was the largest face-to-face field survey on Chinese food safety management conducted directly in urban and rural residents in China in recent years. The survey had a positive effect in better understanding the food safety evaluation, awareness of the main causes of food safety risks, and top concerns about food safety risks of urban and rural residents, as well as their evaluation of food safety regulation and law enforcement by the government, and in investigating solutions of food safety risks. Relevant details will be presented in Chapter 6. Different survey samples and related analyses will also be presented in other chapters. With the purpose of reflecting public concerns, these surveys provide convincing evidence and ensure the distinct research characteristics of this book.

0.3.2.2 Econometric Modeling   The use of econometric modeling and other modeling methods is generally avoided to enhance readability. However, to ensure the scientificity, accuracy, and stringency of the research, indispensable model analysis methods are still used in some chapters. For example, in Chapter 5, catastrophe models were used to conduct risk assessments on Chinese food safety in three subsystems, production, circulation, and consumption, during 2006–2012, estimate the overall food safety risk level in the corresponding year, and predict the future trend of food safety risks in China.

0.3.2.3 Comparative Analysis   Taking into account the dynamic evolution of food safety, the development of Chinese food safety at different stages was investigated by comparative analysis. For example, the quality and safety of major edible agricultural products in China during 2006–2011 were compared based on monitoring data; the food quality and safety in production and processing during 2005–2011 were analyzed based on pass rates in national food quality checks. In addition, a comparative study was carried out on the composition of operators in food circulation during 2006–2011 based on statistical data to analyze the main problems of food quality and safety supervision in circulation.

0.3.2.4 Literature Search   This is the most basic research method in this book. The book contains numerous literature references to ensure that our research is based on previous studies.

0.3.3 Data Sources

Data from different years were used in this book to provide a panoramic, extensive, and detailed description of the basic situation of food quality and safety in recent years in China. Most data were derived from statistical data at the national level or provided directly by the relevant government departments, while some were obtained from actual surveys. However, some data were obtained from public reports or publications on government websites, some were from existing research literature, and a limited amount of data were collected from ordinary websites. Although the reliability of critical data and main conclusions can be guaranteed, it is difficult to ensure the authority and accuracy of all data. The sources of all data used in this book are provided whenever possible to allow objective judgment by readers.

0.4 Main Research Content

Based on the above-mentioned research thread and perspective, this book devotes four chapters to analyzing the dynamic changes in Chinese food safety during the period of 2006–2011 from four different aspects, that is, market supply and quality and safety of major edible agricultural products (Chapter 1), food industry development and food quality in production (Chapter 2), food safety supervision in circulation and food quality and safety (Chapter 3), and quality and safety of imported and exported food (Chapter 4). Chapter 5 presents a macroscopic analysis of Chinese food safety risks during 2006–2012, from the perspective of management, using econometric models. Chapter 6 investigates food safety evaluation, awareness of the main causes of food safety risks, and top concerns about food safety risks of urban and rural residents through a special survey in 4289 urban and rural residents in 96 sites in 12 provinces, autonomous regions, and municipalities (referred to as provinces and regions). Chapters 7 through 10 objectively describe dynamic changes in the Chinese food safety support systems, efforts of the government regulators in optimizing food safety, and existing problems by investigating the evolutionary development of the Chinese food safety legal system; evolution and reform of the food safety management system; construction and development of a food safety standard system; and food safety information disclosure, respectively. Chapter 11 reveals again the complexity of improving Chinese food safety by analyzing and reviewing the food safety incidents of concern that occurred in China in 2011, as well as the major controversies on food safety standards. Chapter 12 demonstrates the complexity of transformation of the agricultural production model and the difficulty in prevention of food safety risk in China by analyzing vegetable farmers' willingness to adopt biopesticides and the major affecting factors. As the last chapter of this book, Chapter 13 summarizes the main research conclusions and research prospects.

The main research content of each chapter is as follows:

Chapter 1: Market supply and quality and safety of major edible agricultural products. China has a wide variety of edible agricultural products. Based on a brief analysis of the production and market supply of major edible agricultural products that are closely related with people's daily lives, including staple foods, vegetables, fruits, and livestock and aquatic products, as well as the relevant monitoring data at the national level, the quality and safety of major Chinese edible agricultural products were analyzed and the construction and development of the quality and safety supervision system for edible agricultural products were summarized. On this basis, the relationship between the quality and safety of edible agricultural products and the transformation of the agricultural production model were examined, the main problems in the quality and safety of edible agricultural products were analyzed, and the development direction of agricultural product quality and safety supervision was investigated.

Chapter 2: Food industry development and food quality in the production chain. Food manufacturing and processing is the most foundational stage in safe food production and occupies an extremely important position in the food industry chain. This chapter focuses on the basic situation of food industry development in China during 2005–2011 and analyzes the major problems in food quality and safety in the stage of production and processing based on an analysis of the pass rates in national food quality checks during 2009–2011.

Chapter 3: Food safety supervision in circulation and food quality and safety. Food circulation is the final hurdle toward entry into millions of homes. It is a stage most closely associated with consumers. In March 2013, before the reform of the Chinese food safety supervision system, food safety in circulation was managed by the State Administration for Industry and Commerce. In this chapter, based on a brief summary of the progress in constructing the Chinese food safety supervision system in terms of circulation, the special law enforcement inspection of food safety, safety incident management, and the situation of food quality and safety supervision in circulation in 2011 were described. Additionally, consumers' evaluation of food safety in the market was revealed via empirical investigation. Furthermore, based on the food complaints received by the China Consumers' Association during 2007–2011, the food consumption complaint channel accessibility was evaluated, and the major problems in Chinese food quality and safety supervision in circulation were thereby analyzed.

Chapter 4: Quality and safety of import and export food. The global food industry continues to develop in a multifield, whole-chain, deep, and sustainable manner and has an unprecedented profound impact on the Chinese food industry and food safety and quality. This chapter focuses on the quality and safety of imported and exported Chinese food and changes during 2006–2011 based on an analysis of the Chinese scale of international food trade during 1991–2011 and the differences in the imported and exported food category structure between 2006 and 2011. On this basis, the major categories of imported and exported Chinese food that were detained/recalled in 2011 and the causes were analyzed, and the major problems in the quality of imported and exported Chinese food were investigated.

Chapter 5: Food safety risk assessment and risk characteristic analysis at the macrolevel. In this chapter, Chinese food safety risks during 2006–2012 were assessed at the national macrolevel from the perspective of management using econometric models based on the scientificity and availability of data. The actual state of food safety risks was analyzed on this basis to provide a panoramic description of the real changes, main characteristics, and trends of food quality and safety in China. Furthermore, this chapter aims to reveal the main characteristics of food safety in terms of root causes and to investigate principal contradictions in food safety risk prevention in China.

Chapter 6: Food safety evaluation and concerns of urban and rural residents. This chapter presents a special survey of 4289 urban and rural residents in 96 sites in 12 provinces, autonomous regions, and municipalities, including Fujian, Guizhou, Henan, Hubei, Jilin, Jiangsu, Jiangxi, Shandong, Shaanxi, Shanghai, Sichuan, and Xinjiang. Food safety evaluation, awareness of the main causes of food safety risks, and top concerns about food safety risks of urban and rural residents, as well as their evaluation of food safety regulation and law enforcement by the government, were summarized and analyzed based on this survey to better combine the wisdom of the people and lay the foundation for investigating solutions for Chinese food safety risks.

Chapter 7: Evolutionary development of the Chinese food safety legal system. The implementation of the *Food Safety Law* on June 1, 2009, marked the entry into a new stage of the Chinese food safety legal system construction. With the promulgation and implementation of the *Food Safety Law* as a logical starting point, this chapter presents a brief review of the developmental history of the Chinese food safety legal system, analyzes the effects of the current food safety legal system, reveals the major problems, and provides preliminary insights in improving the Chinese food safety legal system.

Chapter 8: Evolution and reform of the Chinese food safety management system. Prior to March 2013, the Chinese food safety management system had always been very controversial. This chapter briefly describes the historical changes of the Chinese food safety management system during 1949–2012 and analyzes the significance of the Chinese food safety management system reform in March 2013, as well as the potential problems the system is currently faced with. In view of the fact that the reformed food safety management system continues to emphasize the local management model of "local governments being fully responsible for food safety," research was carried out on the features of the food safety management system reform in Shenzhen to assess whether local management systems were compatible with local reality.

Chapter 9: Construction and development of a food safety standard system. Food safety standards are an important tool for transforming food science and technology into productivity and effectively guaranteeing food safety. They are a basic technology system for improving food safety. However, contradictions, duplications, overlaps, and extended service of food safety standards have caused great difficulties for food safety supervision in China. This chapter aims to summarize the construction and development of the Chinese food safety standard system, with an emphasis on the new progress after the implementation of the *Food Safety Law* on June 1, 2009.

Chapter 10: Research report on food safety information disclosure. To ensure the scientificity, accuracy, and timeliness of the information, timely release of authoritative information, and objective and accurate responses to public concerns on hot issues concerning food safety are priorities for food safety supervision. This chapter presents an objective description of the implementation of food safety information disclosure systems by investigating the voluntary disclosure of food safety information by the food safety supervision and management departments from June 2012 to July 2013 and related systems.

Chapter 11: Food safety incidents of concern in 2011 in Chinese mainland and a brief review. Every year numerous consumers throughout the world are faced with various food safety risks. In China, as intricate factors are intertwined, food safety risk has become a particularly prominent problem in recent years. Frequent food safety incidents are becoming a major social risk affecting social stability and consumer health. Taking into account the integrity and reliability of data, this chapter analyzes the food safety incidents of concern that occurred in China in 2011, as well as the major controversies on food safety standards. Moreover, a brief review is presented to analyze the complexity of improving Chinese food safety.

Chapter 12: Research on agricultural production mode transformation: Vegetable farmers' willingness to adopt biopesticides in Cangshan County, China. How to prevent food safety risk is a problem of common concern in the international community. Due to the space limitation of this book, this chapter only analyzes vegetable production, which indicates that it is a long and arduous journey to prevent and control food safety risks in China. Vegetable consumption occupies an extremely important position in the food consumption of Chinese residents. However, the use of chemical pesticides has seriously affected the quality and safety of vegetables. Therefore, it is imperative to promote biopesticides. This chapter sheds light on the many difficulties in improving the safety and quality of agricultural products in China via a case study on vegetable farmers' willingness to adopt biopesticides in Cangshan County, Shandong Province, China, and concludes that strong policies must be implemented to motivate farmers to adopt biopesticides and transform the agricultural production model.

Chapter 13: Main conclusions and research prospects. This chapter summarizes the research conclusions presented in this book and gives a preliminary indication of future research emphasis.

Authors

Linhai Wu, PhD, was born in a peasant family in September 1962 in Jiangyin, Jiangsu Province, China. In July 1986, he graduated with a bachelor of science degree from the department of chemistry of Jilin University, which is one of the top educational facilities with respect to the discipline of chemistry in China. In July 2000, he graduated with a doctor of management degree in agricultural and forestry economics from Nanjing Agricultural University. Subsequently, he took a postdoctoral fellowship position at Nanjing University and Renmin University of China. Dr. Wu has been working in the government system for government policy-making consultation since 1987. Despite his long-term service as an important leader in the civil service system, Dr. Wu has not given up academic research and has been involved in the management research of Jiangnan University since 2001. Since September 2012, he has worked as a full-time researcher in food safety management at Jiangnan University.

In view of the importance of food safety management in China, Dr. Wu gave up the technology management research at which he excelled and began food safety management research in June 2008. For more than five years, with great concern regarding food safety risk management in China and throughout the world, Dr. Wu personally went to the fields and visited food enterprises and farmers' markets along with the research team to investigate safe food production by farmers and producers and consumer's perception of food safety risks, while paying close attention to international research fronts. These empirical investigations have provided extremely valuable data and information for the research of China's food safety problems and yielded fruitful results. Since 2009, he has published more than 10 manuscripts in international journals, such as the *Canadian Journal of Agricultural Economics, China Agricultural Economic Review, Food and Agricultural Immunology, International Journal of Food Science & Technology, Journal of the Science of Food and Agriculture, Food Control*, and *Appetite*, and more than 40 excellent manuscripts in famous Chinese journals. Dr. Wu has also published several books in China. Moreover, Dr. Wu continues to actively voice his opinion through the Internet and news media and make recommendations directly to relevant government departments. Dr. Wu has been doing his best to present the reality of the food safety situation in China and call for comanagement of food safety risks by the government, enterprises, and the general public to optimize the prevention of food safety risks.

Dr. Wu has been appointed as a professor and a doctoral tutor in Jiangnan University, the research team convenor of the *Food Safety Development Report of China* (one of the social development reports funded by the Ministry of Education of the People's Republic of China), the chief expert of Jiangsu Provincial Food Safety Research Base, and the convenor of one of the first excellent innovation teams of Humanities and Social Sciences of the colleges and universities of Jiangsu province (Food Safety Risk Prevention Research of China). In addition, he is a member of Expert Committee of the National Sustainable Development Experimental Zone of China. He won two second prizes from the People's Government of Jiangsu Province. The research team led by Dr. Wu won the 2011 First Prize of Science and Technology of the China General Chamber of Commerce for research on key points in measurement and reduction of carbon emissions by the food industry. Dr. Wu also won the Fifth China Agricultural Development Research Paper Award Nomination of 2012 for a manuscript on the main factors affecting consumers' willingness to pay a premium for traceable food and acceptable level of premium published in issue 4 of *Chinese Rural Economy* in 2010. In July 2013, as an expert in management, Dr. Wu was selected as a leading expert in the field of science and technology of Jiangsu Province and became one of the few well-known management scholars in this field.

Dian Zhu, PhD, was born in an ancient and beautiful city in China—Suzhou, Jiangsu Province. Prior to the birth of his lovely daughter, he was engaged in teaching and research in the field of economics at Suzhou University, with a focus on the theory of industrial organization. Between August and September 2008, when the Chinese melamine milk powder incident occurred, his daughter was being fed with milk powder. Like many Chinese fathers, he was angry and afraid. Fortunately, in 2010, he met Professor Linhai Wu from the Jiangnan University and began postdoctoral research in the area of food safety management. For more than three years, he has conducted a number of studies on food safety management based on China's realities in cooperation with Wu, with the purpose of gaining a better understanding of the food safety situation in China. Moreover, he hopes to identify the causes of food safety deterioration using his knowledge of economics and thereby make policy recommendations on preventing food safety risks. Although his ultimate research purpose is to ensure safer food consumption for his daughter, as Adam Smith says, "It is not from the benevolence of the butcher, the brewer, or the baker that we expect our dinner, but from their regard to their own interest." Despite the short period of his food safety research, he is passionate about his research out of love for his daughter.

With a bachelor's degree in economics, a master's degree in agricultural economics, and a doctorate in economics from Shanghai University of Finance and Economics, Dr. Zhu is currently an associate professor at Suzhou University and was selected as one of the most popular teachers by the students. Moreover, he is now engaged in postdoctoral research of food safety management at the School of Food Science at Jiangnan University. He has published numerous manuscripts on food safety research in prestigious Chinese journals, as well as a paper on traceable food in the *Canadian Journal of Agricultural Economics* in cooperation with Wu. He also won first prize in a classroom teaching contest in Suzhou University and received a first prize in the soft science quality engineering of Jiangsu Province.

1

Market Supply and Quality and Safety of Major Edible Agricultural Products

Edible agricultural products and foods, as well as their relationships, are defined and described in the Introduction of this book.* Although China has a wide variety of edible agricultural products, staple foods, vegetables, fruits, and livestock and aquatic products are the major products whose quantity and quality impact the safety of the edible agricultural product industry. This chapter analyzes the production, market supply, and structural changes of edible agricultural products, including staple foods, vegetables, fruits, and livestock and aquatic products, as well as the development of the emerging edible agricultural product market. This chapter also investigates the quality and safety of these major edible agricultural products based on monitoring data in China.

1.1 Production and Market Supply of Major Edible Agricultural Products

In this section, the overall production and market supply of the major edible agricultural products, including staple foods, vegetables, fruits, and livestock and aquatic products, are analyzed.

---

*It should be noted that the agricultural products discussed in this book generally refer to edible agricultural products. Edible agricultural products and agricultural products are not strictly differentiated in this book.

1.1.1 Staple Foods

In the history of China's staple food production, 2011 was a remarkable year. The annual production of staple foods hits a record high and climbed to a new level of 550 billion kg, reaching 571.2 billion kg (571.21 million tons); this climb represented an increase of 24.75 billion kg over 2010. The eight-year cumulative increase was 140.5 billion kg, with an average annual increase of 17.5 billion kg. It was the largest period of staple food production growth since the founding of New China. In addition, the annual staple food yield per unit area hits a record high of 344.4 kg in 2011, representing an increase of 12.8 kg over 2010. The eight-year cumulative increase was 55.6 kg, with an average annual increase of 7 kg. This period of per unit area yield growth was one of the fastest observed since the founding of New China. Furthermore, the national staple food production in China achieved its "four first times" in 2011: (1) the first time reaching 550 billion kg; (2) the first time maintaining a yield of more than 500 billion kg for five consecutive years; (3) the first time in half a century achieving an increase for eight consecutive years; and (4) the first time the per capita availability of staple foods reaching 425 kg. On the whole, China has seen rapid development in its staple food production and has made a historic leap from a staple food supply shortage to an overall balanced supply, and even a surplus in good years. The national staple food reserves were at a record high in 2012 (Figure 1.1).

Economic and social developments, as well as improvements in living standards, have induced significant changes in the food consumption structure in China. Through gradual movement from cereal-based to animal protein-based consumption, this shift has had a great impact on the staple food production structure. Animal protein production requires large amounts of cereal, which has directly stimulated a substantial increase in cereal production in China. Cereal production increased from 427.760 million tons in 2005 to 519.395 million tons in 2011, representing an increase of 21.4% and an average annual growth rate of 3.3% (Table 1.1).* At the same time, the production of root crops and legumes has been declining. The national production of root crops and legumes, respectively, decreased from 34.685 and 21.577 million tons in 2005 to 32.730 and 19.085 million tons in 2011, representing a decrease of 5.6% and 11.5% and an average annual decrease of 1.0% and 2.0%.

---

*Unless otherwise noted, data listed in this section are derived from the *China Statistical Yearbook* and *China National Economic and Social Development Statistics Bulletin* for the corresponding years, published by the National Bureau of Statistics of China, and reproduced by author calculations.

FIGURE 1.1
Annual Production of Staple Foods and Changes in Growth Rate during 2006–2011 in China

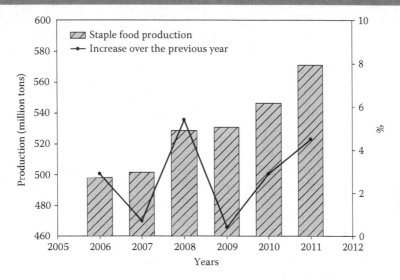

Source. Adapted from *China National Economic and Social Development Statistics Bulletin,* 2007–2012, National Bureau of Statistics of China.

TABLE 1.1
Production of Staple Food Crops and Growth Rates during 2005–2011 in China

Food Crops (million tons)	2005	2006	2007	2008	2009	2010	2011	Average Annual Growth Rate (%)
Staple food	484.022	498.042	501.603	528.709	530.821	546.477	571.210	2.8
Cereals	427.760	450.992	456.324	478.474	481.563	496.371	519.395	3.3
Rice	180.588	181.718	186.034	191.896	195.103	195.761	201.000	1.8
Wheat	97.445	108.466	109.298	112.464	115.115	115.181	117.400	3.2
Corn	139.365	151.603	152.300	165.914	163.974	177.245	192.780	5.6
Root crops	34.685	27.013	28.078	29.802	29.955	31.141	32.730	−1.0
Legumes	21.577	20.037	17.201	20.433	19.303	18.965	19.085	−2.0

Source. Adapted from *China Statistical Yearbook* (2006–2012), National Bureau of Statistics of China.

1.1.2 Vegetables and Fruits

China is the largest consumer and producer of vegetables and fruits in the world (Lu et al. 2010; Pan and Zhang 2011). The production of vegetables and fruits, respectively, was 677 and 227.682 million tons in 2011, each accounting for 30% and 14% of the world's total production. Along with the enhancement of agricultural productivity, improvement in living standards, and adjustments in the planting structure of agricultural products, the vegetable and fruit planting areas, respectively, in China increased from 10,491 and 8,553 thousand hectares in 1996 to 19,639.2 and 11,830.6 thousand hectares in 2011, representing an increase of 87.2% and 38.3%. The rapid growth of fruit and vegetable production ensures that market supply meets multilevel consumer demands (Figure 1.2).

The major fruit species cultivated in China include apples, oranges, pears, grapes, and bananas. The production and supply of apples and oranges have risen sharply since 2002. The annual production of apples and oranges reached 35.985 and 29.440 million tons, respectively, in 2011. The production

FIGURE 1.2
Vegetable and Fruit Planting Areas during 1996–2011 in China

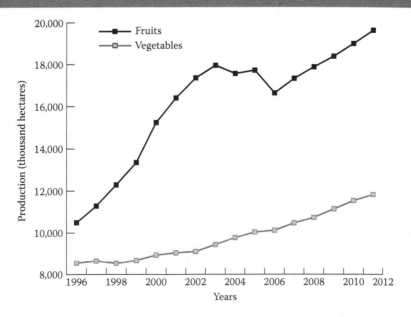

Source. Adapted from *China Statistical Yearbook*, 1997–2012, National Bureau of Statistics of China.

FIGURE 1.3
Production of Major Fruits during 1996–2011

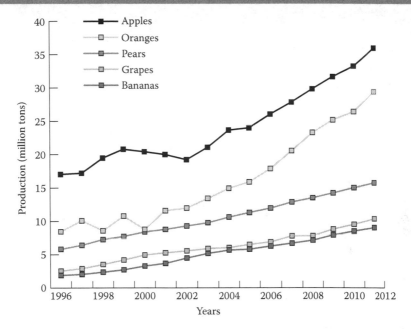

Source. Adapted from *China Statistical Yearbook*, 1997–2012, National Bureau of Statistics of China.

of pears, grapes, and bananas has increased to some extent, but the total production remains relatively small, and was 15.795, 9.067, and 10.4 million tons, respectively, in 2011. The production of major fruit species during 1996–2011 in China is shown in Figure 1.3.

1.1.3 Livestock Products

In China, the production of livestock products such as meat, milk (mainly from cows), and eggs was significantly increased during 1996–2011. However, the increase in the production of assorted livestock products varied for numerous reasons, for example, differences in market growth. Specifically, egg production was stable, while meat and milk production experienced rapid growth (Figure 1.4). The production of meat, milk, and eggs, respectively, increased to 79.57, 36.56, and 28.11 million tons in 2011 from 45.84, 7.358, and 19.65 million tons in 1996, representing an increase of 73.6%, 396.9%, and 43.0%.

As shown in Table 1.2, the growth rate of pork, beef, mutton, cow milk, and egg production remained synchronized with the gross domestic

FIGURE 1.4
Production of Meat, Milk, and Eggs during 1996–2011 in China

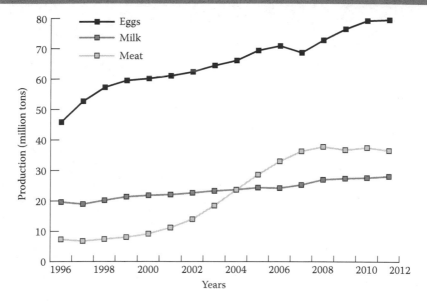

Source. Adapted from *China Statistical Yearbook*, 1997–2011, and *China National Economic and Social Development Statistics Bulletin 2012*, National Bureau of Statistics of China.

TABLE 1.2
Production of Major Livestock Products in the "11th Five-Year" Period in China

Major Livestock Products (million tons)	2006	2007	2008	2009	2010	Growth Rate (%)	Annual Growth Rate (%)
Meat	70.890	68.657	72.787	76.497	79.258	11.8	2.8
Pork	46.505	42.878	46.205	48.908	50.172	9.0	2.2
Beef	5.767	6.134	6.132	6.355	6.531	13.2	3.2
Mutton	3.638	3.826	3.803	3.894	3.989	9.6	2.3
Cow milk	31.934	35.252	35.558	35.188	35.756	12.0	2.9
Eggs	24.240	25.290	27.022	27.425	27.627	14.0	3.3

Source. Adapted from *China Statistical Yearbook* (2007–2011), National Bureau of Statistics of China. Data were reproduced by author calculation.

product (GDP) growth in the "11th Five-Year" period. The production of pork, beef, and mutton, respectively, increased from 46.505, 5.767, and 3.638 million tons in 2006 to 50.712, 6.531, and 3.989 million tons in 2010, representing an increase of 9.0%, 13.2%, and 9.6%. The average growth rate of beef production was higher than the average growth rate of meat production (11.8%), and the growth rates of pork and mutton production were lower than the average growth rate of meat production. Cow milk production increased from 31.934 million tons in 2006 to 35.756 million tons in 2010, representing an increase of 12%, and egg production increased from 24.240 million tons in 2006 to 27.627 million tons in 2010, representing an increase of 14%.

1.1.4 Aquatic Products

Aquatic products contain abundant, easily digestible proteins, as well as a variety of vitamins and minerals necessary for the human body. With the improvement of living standards, aquatic products are increasingly favored by urban and rural residents in China and constitute an increasingly high portion of the diet of Chinese citizens. Thus, the production of aquatic products, driven by consumer demand, has continued to increase substantially in China. The national production of aquatic products increased from 32.881 million tons in 1996 to 56.032 million tons in 2011, representing a cumulative increase of 70%. The per capita production of aquatic products increased from 39.4 kg in 2000 to 41.6 kg in 2011.

National fishing produced 15.7995 million tons of aquatic products in 2011, representing a year-on-year growth of 2.32%, including marine fishing production of 12.4194 million tons (excluding deep-sea fishing production, which produced 1.1478 million tons in 2011). However, because of the growing concern of the international community regarding the environment and fishery resources, countries have increased efforts to protect and compete for these resources. To build a good international environment and achieve sustainable use of resources, China has also begun reforming the development mode of the fishing industry and adjusting the development strategies of marine resources. As a result, fishing production growth has slowed. Aquaculture is of growing importance in China's aquatic production, and both aquacultural area and production continue to increase. The national aquacultural area reached 7835.00 thousand hectares in 2011, representing an increase of 189.73 thousand hectares (2.5%) over 2010. The maricultural area was 2106.40 thousand hectares,

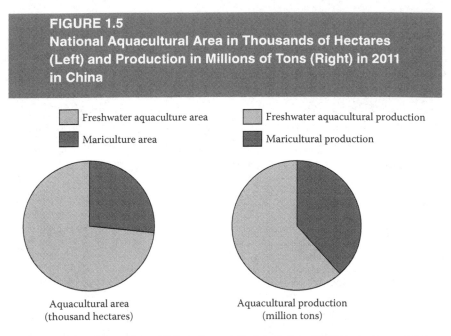

FIGURE 1.5
National Aquacultural Area in Thousands of Hectares (Left) and Production in Millions of Tons (Right) in 2011 in China

Freshwater aquaculture area   Freshwater aquacultural production

Mariculture area   Maricultural production

Aquacultural area
(thousand hectares)

Aquacultural production
(million tons)

Source. Adapted from *National Fisheries Economic Statistics Bulletin 2011*, Fisheries Bureau, Ministry of Agriculture of China.

accounting for 26.9% of the total area of aquaculture, representing an increase of 25.50 thousand hectares (1.23%) over 2010; the freshwater aquacultural area was 5728.6 thousand hectares, accounting for 73.1% of the total area of aquaculture, representing an increase of 164.2 thousand hectares (2.95%) over 2010 (Figure 1.5). In 2011, the national aquacultural production reached 40.233 million tons, representing a year-on-year growth of 5.1%, with its proportion of the total production of aquatic products increasing to 71.8% as well. It included maricultural production of 15.513 million tons and freshwater aquacultural production of 24.719 million tons (Figure 1.5).

Of the wide varieties of available aquatic products in China, fish is the most important; other major varieties include crustaceans, shellfish, and algae. In general, most fish in China come from freshwater aquaculture, while crustaceans come from both freshwater aquaculture and sea fishing, and shellfish and algae generally come from mariculture. The national fish production reached 33.041 million tons in 2011, which was 1.6 times the total production of crustaceans, shellfish, and algae in the same year (Table 1.3).

TABLE 1.3 Production of Aquatic Products in 2011 in China				
Aquatic Products	Fish (million tons)	Crustaceans (million tons)	Shellfish (million tons)	Algae (million tons)
Mariculture	0.964	1.127	11.544	1.602
Freshwater aquaculture	21.854	216.400	0.252	–
Sea fishing	8.640	2.091	0.584	2.7
Freshwater fishing	1.583	0.324	0.287	0.440
Total production	33.041	5.707	12.667	2.06927

Source. *National Fisheries Economic Statistics Bulletin 2011*, Fisheries Bureau, Ministry of Agriculture of China.

1.2 Structural Changes of Major Edible Agricultural Products

Along with the rapid economic growth and improvement in living standards, the diet of urban and rural residents in China continues to undergo profound changes. The relative importance of various agricultural products in Chinese residents' food consumption has changed accordingly, representing a shift from staple food-based to diversified diets. These diets are characterized by increasing proportions of meat, fruits, and vegetables, which are even larger than the proportions of rice and rice-based foods. In addition, along with increased concerns over the Chinese domestic market with regard to food safety, markets for emerging agricultural products have made considerable progress. The major edible agricultural products have undergone a series of internal structure adjustments because of multiple factors, such as national policies for supporting and benefiting agriculture, market demand, and scientific and technological progress; however, the market for these products has maintained a substantial growth on the whole. This changing trend will likely persist for a long period of time.

1.2.1 Planting Structure

As shown in Table 1.4, as the consumer demand for staple foods, such as rice and wheat, has declined relatively since 1995, the planting area of rice and other staple foods has constituted a decreasing proportion of the total

TABLE 1.4
Proportions of Planting Area of Major Crops in Total Cultivated Area

Crops	1995	2000	2005	2008	2009	2010
Rice (%)	20.5	19.2	18.6	18.7	18.7	18.6
Wheat (%)	19.3	17.1	14.7	15.1	15.3	15.1
Corn (%)	15.2	14.8	17.0	19.1	19.7	20.2
Legumes (%)	7.5	8.1	8.3	7.8	7.5	7.0
Root crops (%)	6.4	6.7	6.1	5.4	5.4	5.4
Oil crops (%)	8.7	9.9	9.2	8.2	8.6	8.6
Peanut (%)	2.5	3.1	3.0	2.7	2.8	2.8
Rapeseed (%)	4.6	4.8	4.7	4.2	4.6	4.6
Vegetables (%)	6.3	9.7	11.4	11.4	11.6	11.8

Source.   *China Statistical Yearbook*, 1997–2011, National Bureau of Statistics of China.

cultivated area. As corn becomes an increasingly important industrial raw material and livestock feed, the corn planting area has constituted an increasing proportion of the total cultivated area as a result of the increased demand. The total corn planting area increased from 15.2% to 20.2% of the total cultivated area between 1995 and 2010. As industrial demand continues to increase, the proportion of corn planting area in the total cultivated area will also likely continue to increase.

Legumes are an important raw material for making edible oils. The proportion of the legume planting area began to increase in 1995, as Chinese residents' consumption of edible oil began to increase. However, the proportion of the bean planting area in the total cultivated area has been declining since 2005, mainly because the imported soybean market has achieved rapid growth. Data published by the General Administration of Customs of the People's Republic of China show that China imported a total of 52.64 million tons of soybeans in 2011, and soybean imports were expected to rise to 56.00 million tons in 2012. On the contrary, the proportions of planting area of other raw materials for the creation of edible oils, such as peanut and rapeseed, remain essentially unchanged.

The proportion of yam planting area in the total cultivated area has been declining over recent years, but not rapidly. Chinese residents' diets have been adjusted as income levels increase, which is the main reason for the decline of the yam planting area. For similar reasons, the proportion of vegetable planting area was dramatically increased from 6.3% in 1995 to 11.8% in 2010.

	TABLE 1.5						
	Proportions of Agricultural Products in Key Years						
Year	Rice (%)	Wheat (%)	Corn (%)	Legumes (%)	Root Crops (%)	Meat (%)	Milk (%)	Eggs (%)
1996	0.350	0.198	0.228	0.032	0.063	0.082	0.035	0.011
2000	0.348	0.184	0.196	0.037	0.068	0.111	0.040	0.015
2005	0.304	0.164	0.234	0.036	0.058	0.117	0.041	0.046
2008	0.292	0.171	0.252	0.031	0.045	0.111	0.041	0.057
2009	0.294	0.173	0.247	0.029	0.045	0.115	0.041	0.055
2010	0.287	0.169	0.260	0.028	0.046	0.116	0.040	0.055
2011	0.284	0.166	0.273	0.027	0.046	0.113	0.040	0.052

Source. *China Statistical Yearbook*, 2007–2011; *China National Economic and Social Development Statistics Bulletin 2012*, National Bureau of Statistics of China.

1.2.2 Production Structure

The proportions of production of various major edible agricultural products, with regard to the total production, have undergone a major adjustment since 1996 in China, representing a relatively stable trend (Table 1.5). Proportions of the production of rice, wheat, legumes, and root crops have declined. On the contrary, the proportion of the production of corn continues to rise, while the proportions of the production of meat, milk, and eggs increased rapidly at first and have remained stable since 2005.

These changing trends of the agricultural production structure are basically consistent with the changes in the food consumption structure of Chinese residents: increased consumption of meat, milk, and eggs and decreased consumption of rice, wheat, and other staple foods. This change represents the continuous improvement in the structure of food consumption, as well as Chinese residents' attitude/opinion regarding consumption.

1.3 Production Mode Transition of Edible Agricultural Products

The changes in the consumption structure and concept of China's urban and rural residents not only induced changes in the structure of edible agricultural products, but also stressed the need to transform the agricultural production mode and development mode.* China's agricultural productivity

---

* See the 12th Five-Year Plan for the National Economic and Social Development of China.

has been improving since the late 1990s to meet the supply of agricultural products and guarantee quantitative safety. Moreover, the production mode of edible agricultural products has also undergone profound changes and has gradually transformed into a specialized, standardized, large-scale, intensive mode. The mode is characterized by several aspects, which are described in Sections 1.3.1 through 1.3.4.

1.3.1 Edible Agricultural Product Quality and Safety and Chemical Input

In China, agricultural production has gradually turned into a "petroleum-based industrial agriculture" mode characterized by the high input of chemicals over many years. It has achieved great results in ameliorating agricultural poverty and increasing the supply of agricultural products. However, at the same time, it has caused increasing ecological environmental problems (Wang and Wang 2006). It also seriously affects the quality and safety of edible agricultural products and causes varying degrees of harm to human health (Yang et al. 2010). Since the quantitative safety of edible agricultural products has been, for the most part, guaranteed, higher requirements have been set with respect to qualitative safety that take into consideration China's rapid economic growth and continuous improvement in living standards. The quality and safety of edible agricultural products have caused widespread concern in the past. Therefore, enhancement of quality and safety, improvement of the ecological environment, and promotion of sustainable agricultural development have become the central themes of modern agricultural development.

In recent years, the Chinese government has formulated a series of policies and regulations to help reduce chemical inputs, promote sustainable agricultural development, and improve the quality of agricultural products. To this end, the government launched a series of action projects to promote change in the production mode of edible agricultural products. Since 2009, a subsidy program has been in place to increase soil organic matter and has introduced technical measures, such as straw mulching, green manure, and increased organic manure application, to a total of nearly 20 million hectares. Nationally, testing of soil for formulated fertilization has been conducted on more than 0.667 billion hectares. In addition, programs promoting the building of villages utilizing/promoting conservation tillage and healthy and standardized breeding have been implemented. These programs have introduced conservation tillage to more than 35 million hectares, mechanized no-till seeding to 87 million hectares, and straw mulching after mechanical crushing to 167 million hectares. Programs have also been implemented that assess breeding to promote high yield, high quality, and drought-, water-logging-, high temperature-, disease-, and pest-resistant varieties; to increase subsidies for growing improved crop varieties; and to accelerate the integration process of breeding, production, and promotion of improved crop varieties. At present, the improved crop varieties have covered more than 95% of the total cultivated land and have contributed to a 40% increase in the production of staple foods. Special funds have also been established to provide subsidies to promote the application of green manure for the prevention and control of pests to reduce the use of chemical pesticides (Information Office of the State Council of China 2011).

According to the statistics released by the Ministry of Agriculture of China, cumulative sales of biopesticides were approximately 6.2 billion yuan in 2010, accounting for approximately 12% of the total sales of pesticides. In 2011, the Ministry of Agriculture of China proposed the deployment of uniform professional prevention and control of crop pests to strive to reduce the use of chemical pesticides by 20% in the production of major food crops and cash crops such as cotton, vegetables, and fruits at the end of the "12th Five-Year" period (Ministry of Agriculture of China 2011a).

1.3.2 Edible Agricultural Product Quality and Safety and Organizational Level of Production

Hundreds of millions of small private farmers are the basic units of agricultural production in China. This clearly demonstrates the enormous difficulties in fully achieving agricultural product safety supervision in production in a short period of time. Farmers with bounded rationality generally have certain basic demographics, such as low levels of education, lack of safety production knowledge, and low levels of experience with technology, which are the root cause of the quality and safety risks of agricultural products. Because of difficult economic conditions and lack of technical knowledge of many farmers, the application rate of pesticides is generally high, and highly toxic pesticides are relatively frequently used in China. Thus, the threats to agricultural product quality and safety are often caused by farmers' failure to appropriately/safely apply chemicals (Zhou and Hu 2009).

Policy practice has proven that agricultural cooperatives, as a major route for improving the organization of agriculture in China, can not only effectively promote the application of high-tech, practical agricultural technology and improve the market competitiveness of agricultural production, but also make good use of internal supervisory mechanisms to enhance the

quality and safety of agricultural products. For the common good, cooperative members use fertilizers, pesticides, feed, and other agricultural inputs in
accordance with production standards and technical regulations established
by the cooperative, thus achieving honest and trustworthy production habits.
Standardized agricultural production is being popularized by cooperatives,
and the quality and safety of the agricultural products are guaranteed by
these cooperatives. As of the first half of 2011, there were 446,000 registered
cooperatives with a total of 8,571,000 members throughout China. Of them,
more than 40,000 cooperatives have established quality and safety standards
for agricultural production, more than 24,000 have achieved a quality certification for the production of agricultural products, and 25,600 have a
registered trademark. Moreover, some cooperatives have even established a
product quality traceability system. In Guangdong Province, for example,
1519 cooperatives have achieved product quality traceability, accounting for
19.8% of the total cooperatives in the province. These have greatly improved
the quality and safety of agricultural products (Sun 2011c).

1.3.3 Edible Agricultural Product Quality and Safety and Production Scale

Unfortunately, with the improvement of the market economic system, the
production mode of small farmers is not in accordance with the larger market.
A growing contradiction now exists between small-scale agricultural production and urban and rural residents' growing requirements for the quality and
safety of agricultural products. Small-scale production has become one of the
major bottlenecks in the popularization of agricultural science and technology, improvement of agricultural productivity, and enhancement of quality of
agricultural products. The huge risk posed to the quality and safety of agricultural products by decentralized small-scale farmer production has been
demonstrated in many empirical studies (Li et al. 2007; Xing et al. 2009).

In this context, transfer of land use rights has accelerated since the beginning of the new century; the overall situation is stable and orderly and presents three salient features. First, the overall transfer area has increased; as
of the first half of 2011, the total transfer area was 133 million hectares,
accounting for 16.2% of the total area of cultivated land contracted. Second,
there are more diverse forms of transfer; while subcontracting and renting
are the major forms of transfer, various forms, such as joint stock partner-
ship, cooperative management, entrusted work, and entrusted operation,
continue to emerge. Third, transfer objects are diverse and scale operation is
becoming apparent. Land use rights are mainly transferred between farmers,

which promotes the development of professional large farms, family farms, and professional cooperatives. The only way to develop modern agriculture with Chinese characteristics is to speed up the transfer of land use rights; to develop farming experts, professional large farms, and family farms; and to develop various forms of moderate-scale operations (Chen 2011).

1.3.4 Edible Agricultural Product Quality and Safety and Science and Technology Inputs

The fundamental way to achieve sustainable and stable development of agriculture and to ensure the effective supply of agricultural products on a long-term basis lies in science and technology. Since the beginning of the new century, China has begun to shift from an inefficient labor-intensive production mode toward a more modern fund- and technology-intensive production mode. The investment in research and development and the promotion of agricultural science and technology continue to grow (Table 1.6). Funds allocated by the Chinese government for agricultural research, agricultural technology and service systems, and other related aspects increased from a total of 9.13 billion yuan in 2004 to a total of 18.7 billion yuan in 2011. However, the investment intensity in agricultural science and technology (proportion of investment in agricultural science and technology in agricultural GDP) decreased, because the value of agricultural production increased at a higher rate.

During the "11th Five-Year" period, 13,578 advanced and applicable technologies were introduced, conversed, and promoted in a total area of 16,234.6 hectares in 884 pilot counties (cities). By the end of 2010, 49,282

TABLE 1.6
China's Investment in Agricultural Science and Technology during 2004–2011

Items	2004	2005	2006	2007	2008	2009	2010	2011
Investment in agricultural science and technology (billion yuan)	9.13	11.13	12.62	14.82	15.35	16.16	17.38	18.70
Agricultural GDP (billion yuan)		2141.30	2242.00	2404.00	2862.70	3370.20	3547.70	4053.40	4771.20
Investment intensity in agricultural science and technology (%)	0.44	0.49	0.51	0.54	0.46	0.46	0.43	0.39

Source. Ministry of Agriculture of China.

science and technology service platforms in various forms, such as enterprise R&D institutions, demonstration bases for commercialization of research findings, and farmer cooperative economic and technological organizations, were built in the pilot counties (cities). Since that time, they have effectively promoted the application and promotion of agricultural science and technology. The National Agricultural Technology Parks funded the development and introduction of 7987 projects and promoted the use of 6046 new technologies and 9015 new varieties in 2010.* This is a typical pattern of directing and supporting the construction of modern agriculture based on technological innovation to guarantee quality and safety of the agricultural product supply in China. Now that the 2012 No. 1 Document of the Communist Party of China (CPC) Central Committee and State Council, *Suggestions on Accelerating Agricultural Science and Technology Innovation to Persistently Enhance the Capacity to Ensure Adequate Supplies of Agricultural Products*, has been enacted, it can be expected to enhance the ability to popularize agricultural technology based on technological innovation, which supports the construction of modern agriculture.† Agricultural technology innovation will become an important way to transform the agricultural development mode and promote the transition of the agricultural production mode in China.

1.4 Production and Market Development of Emerging Edible Agricultural Products

Since the late twentieth century, China has gradually built a relatively sound food safety certification system, and markets have been developed for various emerging agricultural products with higher quality and safety. Pollution-free, green, organic, and geographically indicated agricultural products play important roles in the emerging agricultural products. Emerging agricultural products use little or no harmful agricultural chemicals in the production process. Compared with conventional agricultural products, they are characterized by healthfulness, safety, ecological protection, and environmental protection, and thus, they are becoming increasingly recognized by consumers.

---

*See the Overview of Implementation of the "11th Five-Year" National Science and Technology Plan, Ministry of Land and Resources of China.

†See the 2012 No. 1 Document of the CPC Central Committee and State Council: *Suggestions on Accelerating Agricultural Science and Technology Innovation to Persistently Enhance the Capacity to Ensure Adequate Supplies of Agricultural Products*.

1.4.1 Pollution-Free Agricultural Products

Pollution-free agricultural products are proposed in the context of an increasingly serious problem in product quality and safety and increasingly severe environmental pollution. Their primary purpose is to ensure the most basic requirement for agricultural products—edible safety. The Agro-food Quality and Safety Center, Ministry of Agriculture of China, responsible for certification of pollution-free agricultural products, was established in December 2002 and opened in April 2003. As the government has made great efforts to promote pollution-free agricultural products, their production has expanded rapidly, with an exponential growth in the number of certified products and enterprises. By 2011, 29,929 pollution-free agricultural products had been certified and more than 12,302 geographical origins of pollution-free agricultural products had been identified throughout China, with a total production of 174 million tons (Table 1.7).

1.4.2 Green Agricultural Products

In November 1992, the China Green Food Development Center (CGFDC) was formally established. In 1996, the Ministry of Agriculture of China enacted the *Green Food Symbol Regulation*, which indicated that green food (agricultural products) development and management had embarked on a legal, standardized track in China. Two classes of green food products (agricultural products) exist in China: Class A and Class AA. Class A allows for the limited use of chemosynthetic means of production, while Class AA prohibits the use of chemical pesticides, fertilizers, veterinary drugs, food

TABLE 1.7

Production and Market Development of Pollution-Free Agricultural Products during 2006–2011 in China

Year	Number of Geographical Origins	Total Number of Products	Total Production (million tons)	Documented Area (million hectares)
2006	9,358	16,932	400	73.95
2007	10,839	17,722	104	93.52
2008	9,848	16,595	77	630.65
2009	13,360	14,750	83	135.11
2010	13,108	25,187	116	84.57
2011	12,302	29,929	174	74.05

Source. Agro-food Quality and Safety Center, Ministry of Agriculture of China.

TABLE 1.8
Development of Green Agricultural Products (Food)
during 2006–2011 in China

Year	Total Number of Enterprises	Total Number of Products	Annual Sales (billion yuan)	Exports[a] (billion US$)	Monitored Area (million hectares)
2006	4,615	12,868	150	1.96	100
2007	5,740	15,238	192.9	2.14	153
2008	6,176	17,512	259.7	2.32	167
2009	6,003	15,707	316.2	2.16	165
2010	6,391	16,748	282.4	2.31	160
2011	6,622	16,825	313.5	2.30	160

Source. China Green Food Development Center, Ministry of Agriculture of China.
[a] Calculated by an exchange rate of RMB 6.8 = US$1.0.

additives, feed additives, and other substances harmful to the environment and health during the production process. Thus, Class AA is basically equivalent to organic food (agricultural products). Actively advocated and promoted by the Ministry of Agriculture of China, green food (agricultural products) presents an overall strong momentum of development. By the end of 2011, 160 million hectares of monitored area was certified for the production of green goods (agricultural products), with 16,825 products and annual sales of 313.5 billion yuan. In contrast, the numbers were 21 million hectares, 892 products, and 24 billion yuan, respectively, by the end of 1997 (Table 1.8).

1.4.3 Organic Agricultural Products

As consumers are becoming increasingly concerned over food safety and the ecological environment, an increasing consumer demand for organic agricultural products exists, which has resulted in the rapid growth of organic production. By the end of 2011, China had issued 9,337 product certifications, the certified organic production area was 2 million hectares, and the organic conversion area was 440,000 hectares.[*] Different from the pollution-free and green agricultural products, organic agricultural products are independently certified by a number of certification bodies. By the end of 2011, there were 23 organic certification bodies in China. As China is an emerging market and there are a large number of certification bodies, no specific authoritative data, such as data on sales of organic agricultural products, are

[*]Website of the Certification and Accreditation Administration of China (CNCA), http://food.cnca.cn/cnca/spncp/sy/xwdt/03/560933.shtml (accessed June 4, 2012) (in Chinese).

TABLE 1.9

Development of Organic Agricultural Products (Food) Certified by COFCC during 2006–2011 in China

Year	Total Number of Enterprises	Total Number of Products	Domestic Sales (billion yuan)	Exports[a] (billion US$)	Certified Area (million hectares)
2006	520	2278	6.190	0.110	37.78
2007	692	3010	–	–	–
2008	868	4083	17.800	0.153	34.24
2009	1003	4955	19.254	0.186	9.48
2010	1202	5598	14.539	0.095	24.50
2011	1366	6379	–	–	–

Source. China Green Food Development Center, Ministry of Agriculture of China.
[a] Calculated by an exchange rate of RMB 6.8 = US$1.0.

available. The China Organic Food Certification Centre (COFCC) is the largest organic food (agricultural products) certification body in China. By the end of 2011, a total of 1366 enterprises were certified by COFCC, with a total of 6379 products (Table 1.9).

1.4.4 Geographically Indicated Agricultural Products

Geographical indication of agricultural products is a historical and cultural heritage developed from China's long-standing traditional agriculture with well-known geographical/ecological advantages. It is a recognized indicator of origin as well as an important quality mark for agricultural products. It is not only an important starting point and carrier of the quality and safety information of agricultural products, but also an important way to promote the industrialized development of characteristic agriculture with geographical ecological advantages.

The Ministry of Agriculture of China issued the *Agro-Product Geographical Indication Regulation* in December 2007; it began accepting applications and issued a number of special marks for agricultural product geographical indications in February 2008. According to the *Agricultural Intellectual Property Strategy Outline (2010–2020)* issued by the Ministry of Agriculture of China in 2010, by the year 2020, 200 demonstration bases and 50 demonstration counties for the production of geographically indicated agricultural products will be built in China, and more than 800 agricultural product geographical indications will be registered for protection. Geographically indicated agricultural products have been developing rapidly (Table 1.10) at a growth rate far exceeding the expectations of relevant departments.

	2008	2009	2010	2011
TABLE 1.10 **Changes in the Total Number of National Agricultural Product Geographical Indications with Registered Protection during 2008–2011**				
Number of agricultural product geographical indications registered for protection	211	420	536	835

Source.   Ministry of Agriculture of China.

According to *Agricultural Product Geographical Indication Registration Information* published in the third issue (and the following issues) of the announcement by the Ministry of Agriculture of China in 2008, 853 agricultural products had been registered for national geographical indication protection by 2011. These agricultural products cover 13 categories, including fruits, vegetables, grain and oil, livestock, aquatic products, Chinese herbal medicine, teas, poultry, beverages, flowers, textile raw materials, tobacco, and condiments. At present, China has basically developed a multi-level, multidimensional, diversified protection system for agricultural product geographical indications.

1.4.5 Quality and Safety of Pollution-Free, Green, Organic, and Geographically Indicated Agricultural Products

Since 2006, the total production of pollution-free, green, organic, and geographically indicated agricultural products in China has continued to increase and the industry standards have been gradually improving. At present, the development of pollution-free, green, organic, and geographically indicated agricultural products has shifted its focus from scale to quality, moving away from brand creation to brand enhancement (Ministry of Agriculture of China 2012a). By 2011, the total production of pollution-free, green, organic, and geographically indicated agricultural products accounted for more than 40% of the total production of edible agricultural products, representing an increase of 10% over 2010 and covering more than 1000 agricultural products and processed foods. More importantly, the specification of industry standards has ensured the reliable quality and safety of pollution-free, green, organic, and geographically indicated agricultural products. In 2011, the overall sampling qualified rates (according to the green food standard) were 99.5%, 99.4%, and 99.2% for pollution-free agricultural products, green food products, and organic food, respectively; the sampling inspection of geographically indicated agricultural products focusing on monitoring pesticide residues and heavy metal pollution has remained at a qualified rate of 100% for many years. The proportion of safe and high-quality brands of edible agricultural products has been increasing in the consumption structure of Chinese urban and rural residents (Ministry of Agriculture of China 2011b). The development of pollution-free, green, organic, and geographically indicated agricultural products has also improved the international competitiveness of China's agricultural products. China has seen a steady growth in its agricultural exports since 2006 along with a gradual improvement in the quality and safety of agricultural products. The international competitiveness of China's agricultural products has significantly improved, and high-quality green agricultural products have been recognized by more than 40 trading nations; brand influence is also growing. Increasingly more Chinese green and organic foods continue to enter the international market. China's agricultural exports reached US$60.13 billion in 2011, representing an increase of 23% over 2010 (Ministry of Commerce of China 2011).

1.5 Changes in Quality and Safety of Major Edible Agricultural Products Based on Monitoring Data

At the end of the twentieth century, the agricultural development of China entered a new stage with regard to quantitative and qualitative safety, setting a national goal to develop high-yield, high-quality, efficient, ecologically friendly, and safe agriculture. To further ensure agricultural product quality and safety, China developed and implemented the *Pollution-Free Food Action Plan*, the *Law on Agricultural Product Quality and Safety*, and the *Food Safety Law* in 2001, 2006, and 2009, respectively, establishing a preliminary legal security system for agricultural product quality and safety supervision. In recent years, agricultural sectors at all levels have fully performed their regulatory duties to persistently strengthen regulatory measures, steadily improving edible agricultural product quality and safety, and maintaining a development trend of overall stability (Ministry of Agriculture of China 2012b). Because the safety of the agricultural products is guaranteed, this provides important support for the sustained and healthy economic and social development of China.

1.5.1 Vegetable Products

Routine monitoring of pesticide residues, such as methamidophos and dimethoate, in vegetables in large- and medium-sized cities in 31 provinces and autonomous regions of China occurred during 2006–2011. The results indicated a significantly increasing sampling qualified rate for pesticide residues in vegetables and a continuously increasing overall qualified rate of vegetable quality and safety (Figure 1.6). The average sampling qualified rate for pesticide residues in vegetables was 93% in 2006; the rate was consistently greater than 95% for the next five years (2007–2011). Moreover, the qualified rate reached a record high of 97.4% in 2011, representing an increase of 0.6% over 2010. The rate of pesticide residues exceeding their proposed limit in vegetable products has been significantly reduced, and the detected levels of pesticide residue continue to drop, indicating that the overall vegetable quality and safety continue to improve in China.

FIGURE 1.6
Average Qualified Rate for Pesticide Residues in Vegetables during 2006–2011 in China

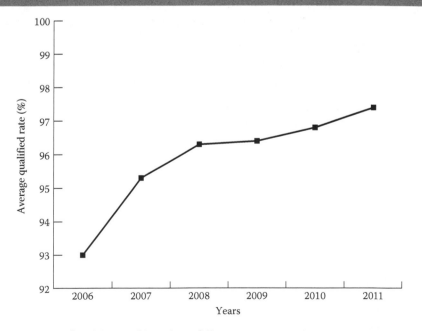

Source.   Data from Ministry of Agriculture of China.

1.5.2 Livestock Products

Monitoring of veterinary drug residues, such as clenbuterol and sulfa drugs, in livestock products in large- and medium-sized cities took place in 31 provinces/autonomous regions of China during 2006–2011. The overall qualified rate of livestock product quality and safety was 98.5% in 2006 and 99.6% in 2011 (the same as in 2010) (Figure 1.7). Thus, the qualified rate of livestock product quality and safety has generally been stable (approximately 99%) and continues to exhibit a slight, yet steady upward trend.

Further analysis indicates that the sampling qualified rate for veterinary drug residues in livestock products has been rapidly increasing over the past few years (Figure 1.8). Monitoring results demonstrated that the sampling qualified rate for veterinary drug residues in livestock products was only 75.0% in 2006, then rapidly increased to >80.0% in 2008, remained at a level of >99.5% for a period of time, and then further increased to 99.6% in 2011.

According to clenbuterol monitoring results in live pigs, the sampling qualified rate for clenbuterol in live pigs was 98.5% in 2006. The qualified rate increased fairly evenly in the subsequent years to 99.5% in 2011, which was 0.2% higher than in 2010 (Figure 1.9).

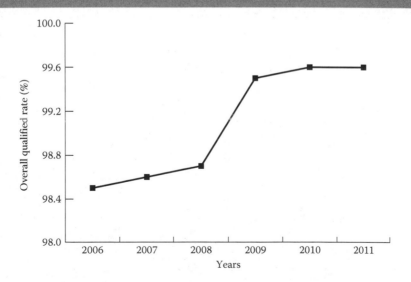

FIGURE 1.7
Overall Qualified Rate of Livestock Product Quality and Safety during 2006–2011 in China

Source. Data from Ministry of Agriculture of China.

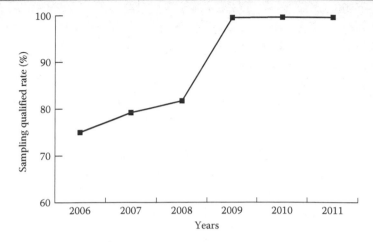

FIGURE 1.8

Sampling Qualified Rate for Veterinary Drug Residues in Livestock Products during 2006–2011 in China

Source.   Data from Ministry of Agriculture of China.

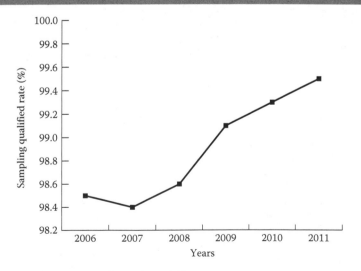

FIGURE 1.9

Sampling Qualified Rate for Clenbuterol in Live Pigs during 2006–2011 in China

Source.   Data from Ministry of Agriculture of China.

1.5.3 Aquatic Products

Monitoring of chlorotoxin, malachite green, and nitrofuran metabolites in aquatic products in large- and medium-sized cities in 31 provinces/autonomous regions of China occurred during 2006–2011. The qualified rate of aquatic product quality and safety displayed a significant upward trend (Figure 1.10). Prior to 2006, the aquatic product quality and safety monitoring only focused on chlorotoxin and the overall sampling qualified rate of aquatic product quality remained at a high level of 98%. As per the Ministry of Agriculture of China, malachite green was added as a routine monitoring object of aquatic products in 2006 for the first time. The sampling qualified rate of aquatic product quality and safety was reduced to 91% in 2006, but significantly increased in the years following. It dropped slightly after a peak of 97.2% in 2009. The qualified rate was 96.8% in 2011, representing an increase of 0.1% over 2010.*

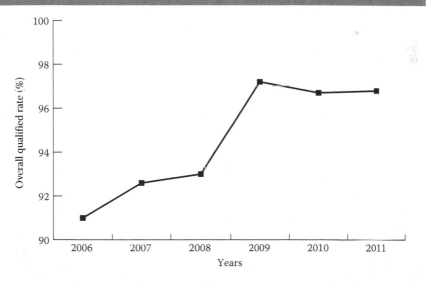

FIGURE 1.10
Overall Qualified Rate of Aquatic Product Quality and Safety during 2006–2011 in China

Source. Data from Ministry of Agriculture of China.

---

* Data were obtained from the *Routine Monitoring Bulletin* of the Ministry of Agriculture of China.

1.6 Main Problems in Enhancing Edible Agricultural Product Quality and Safety

The production mode of edible agricultural products in China is undergoing a positive transformation, and the standardization system of edible agricultural products has been constantly improving. A five-in-one working mechanism has been developed that integrates "service, management, supervision, punishment, and emergency" to ensure the quality and safety of agricultural products (Information Office of the Ministry of Agriculture of China 2007). The quality and safety of major edible agricultural products have maintained a trend of steady, gradual improvement. However, edible agricultural product quality and safety are still faced with many problems and constraints in China, and these issues must be solved if the edible agricultural product quality and safety are to be stabilized and further enhanced. Due to their extreme complexity, Sections 1.6.1 through 1.6.4 analyzes the problems surrounding the key link in the edible agricultural product safety, that is, the production origin of edible agricultural products.

1.6.1 Small-Scale and Scattered Production and Operation Mode

Families are the major units for the production and sale of primary edible agricultural products in China. Compared with other countries, China is characterized by a wide distribution of many small and scattered agricultural product producers and operators. By the end of 2010, there were approximately 240 million farmers, 89% of which were scattered farmers. The cultivated land area was only 5 hectares per household and 1.5 hectares per person.[*] To achieve moderate-scale operation and promote the transition of the agricultural production mode, numerous levels of the Chinese government have made, and continue to make, great efforts to promote land transfer. However, due to the lack of a proper land transfer mechanism/system, as well as general constraints of the farmers, such as traditional views, the organizational level of agricultural production remains low. Data reveal that there are 379,100 farmers' professional cooperatives at present and that approximately 29 million farmers have joined these cooperatives, accounting for only 12% of the total number of farmers in China (Zhang 2011a). However, these professional cooperatives are not becoming the main body of the agricultural production, circulation, and processing. At present, the small scale and decentralization

---

[*] Data were obtained from the *China Statistical Yearbook* (2011).

of agricultural product producers and operators and the poorly organized production and operation modes have posed serious challenges to external governmental supervision and internal supervision among operators, resulting in great difficulties in supervision, high costs, lack of motivation, and inefficiency. In terms of specific practices, it is not conducive to control the quantity and quality of chemical inputs, thus making it difficult to guarantee the quality and safety of edible agricultural products. Due to poor organizational level of the production mode, it is difficult to effectively introduce scientific fertilization techniques to reduce fertilizer inputs; pesticide spraying techniques to reduce pesticide residues; and scientific, practical, and efficient techniques and processes to the production and processing of agricultural products. This further restricts the improvement of agricultural product quality and safety (Sun 2008).

1.6.2 Serious Pollution in Producing Areas

In terms of production conditions, several problems regarding substandard quality, counterfeit and shoddy products, and improper use and abuse of chemical substances exist in many types of agricultural inputs. It is difficult to solve the potential safety risks of agricultural inputs in the short term, which poses a serious threat to agricultural product quality and safety. The outstanding problems are summarized as follows:

1. Excessive and inefficient uses of chemical fertilizers are outstanding problems. The application rate of chemical fertilizers in agricultural production in China ranks first in the world. The application of chemical fertilizers increased from 42.538 million tons in 2001 to 55.617 million tons in 2010, representing an increase of 30% over 10 years (Figure 1.11).[*] However, the efficiency of fertilizer use is low in China. For example, with regard to the quantity of application, with less than 10% of the global arable land, China accounts for one-third of the global nitrogen fertilizer application amount (Li 2009); the average nitrogen use efficiency is approximately 45% in China, well below the 60% of the developed countries.[†] Excessive and inefficient applications of chemical fertilizers have destroyed the agricultural ecological environment, along

---

[*] Data were obtained from the *China Statistical Yearbook* (2011) and reproduced by author calculation.

[†] Research progress on nitrogen use efficiency in the soil–crop system. Central China Agricultural Information, http://ccain.hzau.edu.cn/nyjs/trfl/trzs/201009/t20100921_5199.htm (accessed June 6, 2012) (in Chinese).

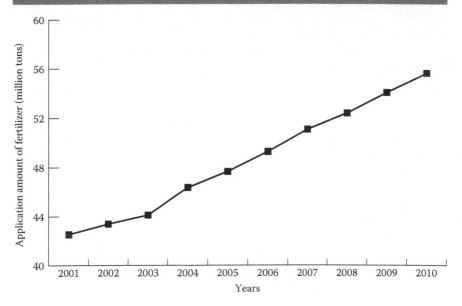

Source. Adapted from *China Statistical Yearbook*, National Bureau of Statistics of China.

with sustainability. As a result, residues of nitrates, nitrites, heavy metals, and other harmful substances in edible agricultural products far exceed the limit, causing harm to human health. Due to the habitual dependency on chemical fertilizers, as well as the limited supply of organic fertilizers and many other constraints, it is still difficult for agricultural producers to effectively reduce the use of chemical fertilizers and adopt scientific fertilization techniques on a large scale in the short term.

2. Abuse of chemical pesticides worsens the ecological environment and severely affects the safety and quality of edible agricultural products. As the Chinese government continues to increase subsidies to support agriculture, agricultural producers have been greatly motivated; however, chemical pesticide application has also substantially increased (Figure 1.12). The Chinese government has banned the use of highly toxic, highly residual, carcinogenic, teratogenic, and/or mutagenic pesticides and has required rigid adherence to safe application standards and rational application guidelines of pesticides. However, banned/prohibited pesticides continue to be used in actual agricultural

FIGURE 1.12
Application Amount of Chemical Pesticides in 2001–2009 in China

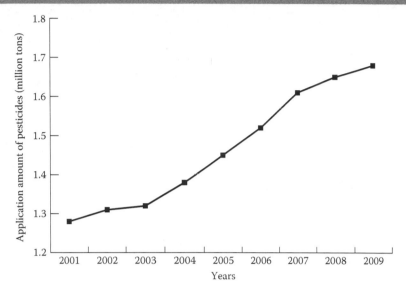

Source. Adapted from *China Rural Statistical Yearbook.*

production, and pesticide abuse violating the safe application standards and rational application guidelines of pesticides remains widespread. In the "toxic cowpeas with isocarbophos" scandal in Hainan Province in 2010, only 21 of the 59 cowpea samples were qualified, while 38 were unqualified, 12 of which contained highly toxic pesticide residues (Lin and Li 2009). Empirical studies have shown that most of the pesticides sprayed in the agricultural environment leach/travel into water, soil, and air. This not only poses great potential safety risks to the production and living environment, but also easily results in a large number of residues in the edible agricultural products, potentially leading to acute or chronic toxicity in humans. In addition, at present, more than 600 pesticides have been registered for use in agriculture, and more than 1500 veterinary drugs have been included in the *Veterinary Pharmacopoeia of the People's Republic of China*. There are more than 2400 pesticide manufacturers and more than 600,000 pesticide traders in China, and most of them are run on a small-scale, self-employed basis. Due to a wide distribution of many small pesticide traders, the management of pesticide traders is also facing difficulties. Therefore, a long period of time will be required to strengthen the supervision of pesticide manufacturers and traders, to guide/educate farmers on the use of pesticides, and to effectively solve the problem of pesticide abuse.

3. Soil heavy metal pollution is a serious threat to the quality and safety of agricultural products. The soil is the material basis for the survival of plants and an important medium for pollutant accumulation. With the rapid development of industrialization, heavy metal pollution of the soil in China is increasing both in severity and area, resulting in incalculable losses to the growth of crops and damage to the quality and safety of agricultural products. It is estimated that an average of 150 heavy metal pollution incidents occur every year in China. According to a national soil pollution survey conducted by the Ministry of Environmental Protection of China in 2011, the heavy metal contaminated soil area has reached 100 million hectares in China (*Guangzhou Daily* 2011), accounting for 8% of the total cultivated area. However, in October 11, 2011, *New Express* (a daily newspaper in China) reported that approximately 200 million hectares of cultivated land was threatened by heavy metal pollution, accounting for one-sixth of the total area of farmland in China. Data published by the Ministry of Land and Resources of China indicate that up to 12 million tons of food is polluted by heavy metals every year, resulting in direct economic losses of more than 20 billion yuan (Beneficiation Technology Network 2012). State Council has officially approved the *12th Five-Year Plan of Heavy Metal Pollution Prevention and Control* and identified 14 provinces/autonomous regions, including Inner Mongolia, Jiangsu, Zhejiang, Jiangxi, Henan, Hubei, Hunan, Guangdong, Guangxi, Sichuan, Yunnan, Shaanxi, Gansu, and Qinghai, as key provinces/autonomous regions in heavy metal pollution management. In recent years, various types of heavy metal pollution incidents have frequently occurred in China, posing a serious threat to the quality and safety of agricultural products. The over standard rates of zinc, cadmium, and chromium were more than 60.0% in all kinds of vegetables in Ningbo, Zhejiang; cadmium had the highest rate (85.0%), followed by chromium (72.3%). Rice, tea, roses, and other agricultural products were polluted by lead in Changxing County, Zhejiang, which resulted in a large area of dead/unusable rice (Liu and Yu 2010). Testing of agricultural products in three agricultural production areas in the Shenyang suburbs revealed that, in the Zhangshi irrigated area, the over standard rate of lead was 100% in Chinese cabbage, the highest lead level in rice exceeded the limit by 0.8 times, and the over standard rates of cadmium, zinc,

and chromium were 20%, 20%, and 10%, respectively, in rice (Zhang 2001). Seventeen agricultural products, including rice and spinach, in an area around a smelting plant in the Taihu Lake region were tested for heavy metal pollution; the results revealed that all agricultural products exceeded the limits of cadmium and lead and that the leaf and stem vegetables exceeded the limits of all heavy metal elements tested (Liu et al. 2006). According to the *Maximum Levels of Contaminants in Food* (GB 2762-2005), the over standard rates of arsenic, cadmium, nickel, and lead were determined to be 95.8%, 68.8%, 10.4%, and 95.8%, respectively, in vegetables along the middle and lower reaches of the Xiangjiang River between Hengyang and Changsha in 2011 (China Low-Carbon Economy Network 2012). Lead pollution was serious in vegetables in the suburbs of Xi'an, Chongqing, and Jingzhou (Teng et al. 2010). Cadmium and lead pollution was (and continues to be) an outstanding problem in aquatic products; for example, cadmium and lead levels were determined to be 200% and 160%, respectively, higher than the standard formulated by the Codex Alimentarius Commission (CAC) in some marine products in Fujian Province (Tang et al. 2010b). Heavy metal pollution in teas is also very serious. In 124 commercially available tea samples tested in the seventh food hygiene inspection by the Ministry of Health in 2003, 12 were unqualified, 11 of which exceeded the upper limit of the standard lead level (Jiang and Gong 2004). More seriously, heavy metal pollution causes long-term or even irreversible effects. It poses a serious and long-term threat to the ecological environment, the quality and safety of edible agricultural products, and the agricultural sustainability of China.

1.6.3 Illegal Use of Chemical Additives Causing Agricultural Product Safety Incidents

The risk of agricultural product quality and safety comes not only from farming, but also from nonfarming processes, for example, production, processing, and circulation. In general, food safety issues in the nonfarming processes are mainly caused by illegal use of food additives. Currently, more than 2,000 types in 23 categories of food additives have been approved for use (Zou 2010), and there are more than 10,000 foods containing additives in China. The Chinese government has clearly defined the applicable range and quantity of food additives. However, some money-oriented, poorly self-disciplined agricultural production enterprises illegally and excessively use chemical additives during production, processing, packaging, and distribution to reap higher economic benefits. This, in turn, severely reduces the quality of the agricultural products. For example, the "clenbuterol scandal," "melamine scandal," and other serious agricultural product quality and safety incidents occurring in China in recent years are generally associated with the abuse of additives.

1.6.4 Principal Contradiction Influencing Agricultural Product Quality and Safety

In this section, the major problems in enhancing the quality and safety of edible agricultural products in China in terms of production origin are focused (Figure 1.13). The "conduction" between agricultural production and edible agricultural product quality and safety can be divided into "direct conduction" and "indirect conduction." Direct conduction refers to illegal and improper conduct in agricultural production directly leading to edible agricultural product problems/issues. Edible agricultural product safety incidents, such as the Hainan "toxic cowpeas" scandal, are generally caused by direct conduction. Indirect conduction between agricultural production and edible agricultural product quality and safety refers to a situation in which improper agricultural production activities initially affect the agricultural

FIGURE 1.13
Main Factors Influencing the Improvement of Edible Agricultural Product Safety and Quality

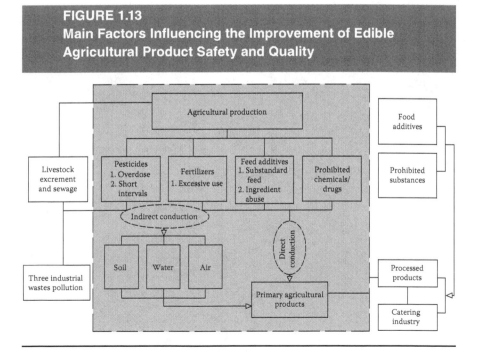

production environment and then affect food safety via the environment. Improper use of agricultural inputs not only directly affects edible agricultural product quality and safety, but also pollutes the environment, which can be indirectly correlated with food safety issues. Of course, rapid industrialization and urbanization have caused significant pollution to the agricultural production environment due to improper management of the relationship between economic development and ecological environment protection.

The direct and indirect conduction of illegal and improper activities in agricultural production and serious production environment pollution are fundamental problems affecting the present and future edible agricultural product quality and safety. The underlying reasons for this problem are the small-scale and decentralized agricultural production and operation mode. As shown in Table 1.11, the production behavior of agricultural producers can be divided into three stages: input obtainment, input used in the production process, and ripening and storage of primary agricultural products. The nine behaviors of agricultural producers that may lead to safety risks in edible agricultural products is listed in Table 1.11 exist in reality. To some extent, small private farmers are the main body of agricultural production in China. In the face of hundreds of millions of small private farmers, it is difficult to make a difference through the limited regulatory power. Therefore, the effective solution to the edible agricultural product

TABLE 1.11
Agricultural Producer Behaviors That May Cause Risks to Edible Agricultural Product Quality and Safety

Stage	Improper or Illegal Behaviors
Input obtainment	1. Purchase of prohibited substances (substances prohibited for production and sales or for use)
	1.1 Because of inability to identify
	1.2 Driven by interest to purchase knowingly
Input use	2. Improper or illegal use in production
	2.1 Because of inability to identify
	2.2 Driven by interest to abuse pesticides, veterinary drugs, and additives
	2.3 Driven by interest to illegally use substances prohibited for production and sales
	2.4 Driven by interest to illegally use substances prohibited for use in this way
Ripening and storage of primary agricultural products	3. Illegal use and improper storage after production
	3.1 Illegal use of prohibited substances in storage after production
	3.2 Improper use of ripeners and preservatives of fruits and vegetables
	3.3 Improper storage resulting in mildew

quality and safety problem lies at the institutional level. Specifically, it is critical to uncover the factors that affect the behavior of agricultural producers, as well as the relationship between the factors, and to ultimately identify the key factors. Furthermore, it is important to identify the agricultural producer set for the prevention and control of edible agricultural product quality and safety risks and incorporate government regulation, market supervision, and contractual governance.

2

Food Industry Development and Food Quality in the Production Chain

The food industry is responsible for producing and providing safe and healthy food for 1.3 billion Chinese people. This basic industry plays an extremely important role in the Chinese national economy. Thus, this chapter focuses on the development of the food industry in China since 2005 based on an analysis of the pass rates from national quality checks of relevant foods. Furthermore, this chapter analyzes the quality and safety of food produced in China, as well as changes in the food industry from the aspects of production and processing.

2.1 Food Industry Development and Quantitative Security

Quantitative security refers to the ability to produce or provide adequate food for survival, and it is paramount in food safety (Wu and Xu 2009). China's food industry has undergone continuous and rapid development during the 30 years of reform and opening up. During this time, the industry scale has expanded rapidly and the production of major foods has increased significantly. In China, the produced amounts of rice, wheat flour, edible vegetable oil, fresh frozen meat, biscuits, juices and juice drinks, beer, instant noodles, and other foods rank in the top of the world's supply. The food industry has become one of the most important economic pillar industries for the Chinese national economy and guarantees the quantity of food. The food industry development in China since 2005 is summarized in Sections 2.1.2 through 2.1.5.

2.1.1 Quantitative Food Safety

Driven by market demand and macroeconomic policy, the production of major foods in China has continued to increase since 2005, and sales rate has reached approximately 97%. As shown in Table 2.1, the production of all major foods in China in 2011 has experienced substantial growth compared with 2005. At present, China's production of rice, wheat flour, instant noodles, edible vegetable oil, refined sugar, meat, and beer continues to rank first in the world. Currently, the abundant types and supply of food in the Chinese market can fully satisfy the growing demand for food and guarantee the quantity of food.

2.1.2 Category Structure of the Food Industry

The food industry structure has been further adjusted in China. As shown in Figures 2.1 and 2.2, changes that occurred in the four major food categories during 2005–2011 include a slight decrease in the proportion of beverage manufacturing sector and a significant decrease in the proportion

TABLE 2.1
Production of Major Foods in China in 2005 and 2011

Products	2005	2011	Cumulative Growth (%)	Average Annual Growth (%)
Rice (million tons)	17.66	88.40	400.5	30.8
Wheat flour (million tons)	39.92	116.78	192.5	19.6
Edible vegetable oil (million tons)	16.12	43.32	168.7	17.9
Refined sugar (million tons)	9.12	11.69	28.1	4.2
Meats (million tons)	77.00	79.57	3.3	0.41
Dairy products (million tons)	12.04	23.88	98.2	12.1
Cans (million tons)	5.00	9.73	94.4	11.7
Soft drinks (million kiloliters)	33.80	117.62	247.9	23.1
Beer (million kiloliters)	31.26	48.99	56.7	7.78
Refined teas (million kiloliters)	0.52	1.77	237.2	22.5

Source.   China Statistical Yearbook, 2006; Overview of 2011 Food Industry Economic Performance in China; and China National Economic and Social Development Statistics Bulletin, 2005 and 2011, National Bureau of Statistics of China.

FIGURE 2.1
Gross Outputs of Four Major Sectors in the Food Industry in 2005 and 2011

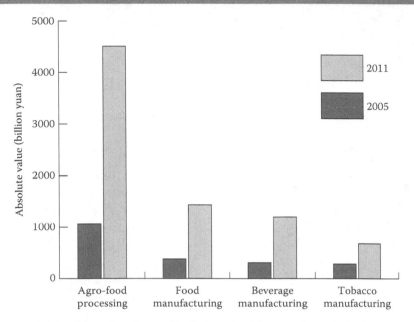

Source. Adapted from *China Statistical Yearbook,* 2006 and *Overview of 2011 Food Industry Economic Performance in China.*

of tobacco manufacturing sector. Furthermore, the agro-food processing sector saw the most rapid growth, representing a cumulative growth of 324.4% in 2011 and an average annual increase of 27.2% over 2005; the absolute value reached 4501.5 billion yuan, and its proportion in the food industry increased from 52.2% in 2005 to 57.7% in 2011. The food manufacturing sector also saw an average annual increase of 24.8%; the absolute value reached 1428.8 billion yuan, and its proportion in the food industry slightly increased to 18.3% over 2005.

In addition, the regional distribution of the food industry has continued to develop to a balanced and coordinated distribution. Eastern China has continued to play a leading and dominant role, while Central China has made efforts to convert its agricultural resources into industrial advantages and achieved further development in the food industry. If the food industry output in Western China is set as 1, the Eastern to Central to Western China ratio was 3.13:1.24:1 in 2005, and it was adjusted to 2.18:1.25:1 in

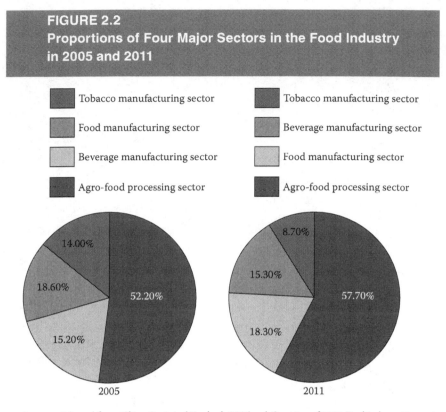

FIGURE 2.2
Proportions of Four Major Sectors in the Food Industry in 2005 and 2011

Tobacco manufacturing sector

Food manufacturing sector

Beverage manufacturing sector

Agro-food processing sector

Tobacco manufacturing sector

Beverage manufacturing sector

Food manufacturing sector

Agro-food processing sector

2005

2011

Source. Adapted from *China Statistical Yearbook*, 2006 and *Overview of 2011 Food Industry Economic Performance in China.*

2011, representing a better balance among the three regions (Figure 2.3). The distribution of the top 10 provinces in terms of food industry output also changed as follows: the Eastern to Central to Western China ratio was 7:1:2 in 2005, while this ratio changed to 6:3:1 in 2011, representing a rapid development of the provinces of the Central China (Figure 2.4).

2.1.3 *The Position of the Food Industry in the Chinese National Economy*

The food industry plays an increasingly prominent role in the Chinese national economy. As shown in Table 2.2, the gross output of the Chinese food industry was 7807.8 billion yuan in 2011, representing a cumulative growth of 284.2% and an average annual increase of 25.2% over 2032.4 billion yuan in 2005. The proportions of food industry output in the gross domestic product and the total industrial output, respectively, increased from 11.09% and 8.08% in

FIGURE 2.3
Proportions of Food Industry Output in Eastern,
Central, and Western China in 2005 and 2011

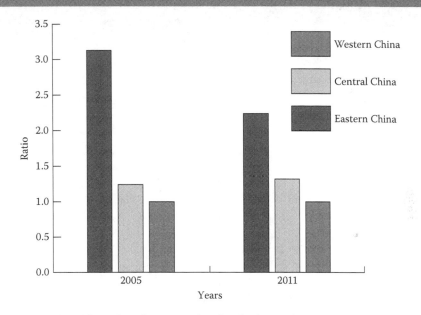

Source. Adapted from *The 12th Five-Year Plan of Food Industry* and *Overview of 2011 Food Industry Economic Performance in China.*

2005 to 16.56% and 9.10% in 2011. The ratio of food industry output to total agricultural output also increased from 51.52% in 2005 to 101.53% in 2011.* These data fully demonstrate that, as a pillar industry in the national economy, the role of the food industry has been further consolidated. It has played an important role not only in promoting the development of the industrial economy, but also in effectively driving the development of related industries, such as agriculture, food packaging, machinery manufacturing, and service.

2.1.4 Contributions to Economic and Social Development

The Chinese food industry has gradually improved in economic efficiency while at the same time experienced rapid development. The total profits and taxes of the Chinese food industry increased to 1214 billion yuan

---

*As of July 10, 2012, the Chinese Government has not yet published data on the total agricultural output of 2011. The present data are estimated based on historical data.

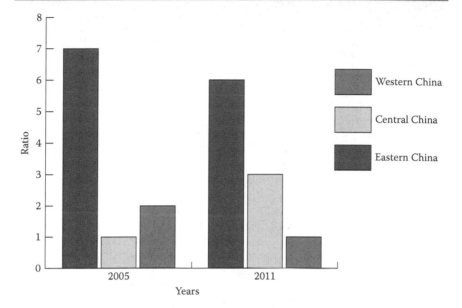

FIGURE 2.4

Distribution of the Top 10 Provinces in Eastern, Central, and Western China in Terms of Food Industry Output in 2005 and 2011

Source. Adapted from *China Statistical Yearbook*, 2006 and *Overview of 2011 Food Industry Economic Performance in China.*

TABLE 2.2

Food Industry Output, Gross Domestic Product, Industrial Output, and Agricultural Output during 2005–2011

Year	Food Industry Output (billion yuan)	Gross Domestic Product (billion yuan)	Proportion (%)	Industrial Output (billion yuan)	Proportion (%)	Agricultural Output (billion yuan)	Proportion (%)
2005	2032.4	18321.7	11.09	25162.0	8.08	3945.1	51.52
2006	2480.1	21192.4	11.70	31658.9	7.83	4081.1	60.77
2007	3242.6	25730.6	12.60	40517.7	8.00	4889.3	66.32
2008	4237.3	30067.0	14.09	50744.8	8.35	5800.2	73.05
2009	4967.8	34140.1	14.55	54831.1	9.06	6036.1	82.30
2010	6127.8	40326.0	15.20	69859.0	8.77	6932.0	88.40
2011	7807.8	47156.4	16.56	85800.3	9.10	7690.5	101.53

Source. *China Statistical Yearbook*, 2006–2012.

TABLE 2.3
Growth of Major Economic Indicators in the Food
Industry during 2005–2011

Year	Output (billion yuan)	Sales Revenue (billion yuan)	Profits and Taxes (billion yuan)	Employee (million people)
2005	2032.4	1993.8	259.0	4.52
2006	2480.1	2439.6	317.4	4.78
2007	3242.5	3171.6	422.9	5.20
2008	4237.3	4142.7	525.9	6.03
2009	4967.8	4729.4	733.6	6.39
2010	6127.8	6006.3	1065.9	6.96
2011	7807.8	7654.0	1214.0	6.83
2005–2011 cumulative growth rate (%)	284.2	283.9	368.7	51.10
2005–2011 average annual growth rate (%)	25.2	25.1	29.4	7.12

Source. *China Statistical Yearbook*, 2006–2010 and *Overview of 2011 Food Industry Economic Performance in China.*

in 2011 from 259 billion yuan in 2005, representing a cumulative growth of 368.7% and an average annual growth of 29.4% (Table 2.3). The number of employees in the Chinese food industry has also increased steadily to 6.83 million employees in 2011, representing a cumulative growth of 51.1% and an average annual growth of 7.12% over 2005 (Table 2.3).

2.1.5 Sustainability

Overall, 70% of the food producers in China are small businesses, and thus, the production levels still lag behind. Furthermore, energy conservation and emission reduction are challenging tasks. With the joint efforts of the entire industry, the sustainability of the food industry has improved since the beginning of the new century. Between 2005 and 2010, the total energy consumption by the Chinese food industry increased from 43.211 to 55.121 million tons of standard coal. The rapid growth of the food industry (over 20.0%) was supported by an average annual growth rate of total energy consumption of 3.27%. As an unfortunate consequence, emissions of the "three wastes" by the Chinese food industry continued to rise during 2005–2010 (Figure 2.5). Emissions of wastewater, SO_2, and solid waste, respectively, increased from 2.080 billion tons, 0.370 million tons, and 0.030 billion tons to 2.758 billion tons, 0.4067 million tons, and 0.037 billion tons (Table 2.4).

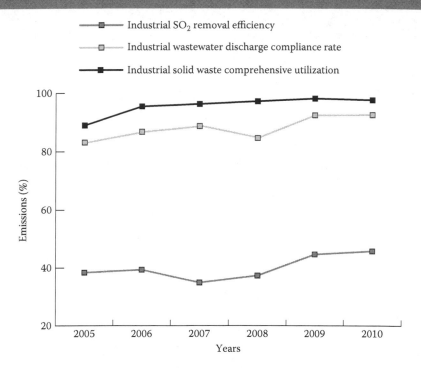

FIGURE 2.5

Emission Reduction Curves of the "Three Wastes" by the Food Industry during 2005–2010

Source. Adapted from *China Statistical Yearbook*, 2006–2011.

Note. Industrial wastewater discharge compliance rate = (standard-compliant industrial wastewater discharges ÷ industrial wastewater discharge) × 100%; industrial SO_2 removal efficiency = (industrial SO_2 removed ÷ industrial SO_2 emissions) × 100%; industrial solid waste comprehensive utilization = (industrial solid waste utilized ÷ industrial solid waste produced) × 100%.

TABLE 2.4

Emissions of the "Three Wastes" by the Food Industry and Its Subsectors during 2005–2010

Sector	Item	2005	2006	2007	2008	2009	2010
Agro-food processing	Wastewater (billion tons)	1.190	0.940	1.490	1.580	1.438	1.431
	SO_2 (million tons)	0.156	0.168	0.170	0.163	0.161	0.169
	Solid waste (billion tons)	0.013	0.015	0.017	0.020	0.021	0.021
Food manufacturing	Wastewater (billion tons)	0.428	0.431	0.428	0.480	0.527	0.545
	SO_2 (million tons)	0.094	0.105	0.117	0.121	0.108	0.1166
	Solid waste (billion tons)	0.009	0.011	0.012	0.012	0.005	0.007

(*Continued*)

TABLE 2.4

(Continued) Emissions of the "Three Wastes" by the Food Industry and Its Subsectors during 2005–2010

Sector	Item	2005	2006	2007	2008	2009	2010
Beverage manufacturing	Wastewater (billion tons)	0.430	0.560	0.630	0.710	0.697	0.755
	SO_2 (million tons)	0.107	0.116	0.124	0.112	0.106	0.112
	Solid waste (billion tons)	0.007	0.008	0.008	0.009	0.009	0.009
Tobacco manufacturing	Wastewater (billion tons)	0.028	0.028	0.029	0.029	0.033	0.027
	SO_2 (million tons)	0.013	0.015	0.014	0.016	0.012	0.010
	Solid waste (billion tons)	0.001	0.0004	0.0005	0.0005	0.0004	0.0004
Food industry	Wastewater (billion tons)	2.080	1.960	2.580	2.800	2.695	2.758
	SO_2 (million tons)	0.370	0.404	0.425	0.412	0.386	0.407
	Solid waste (billion tons)	0.030	0.034	0.038	0.042	0.037	0.037

Source.   *China Statistical Yearbook*, 2006–2011.

However, both the industrial wastewater discharge compliance rate and the industrial SO_2 removal efficiency were greater than 90%, with the exception of solid waste comprehensive utilization. The industrial wastewater discharge compliance rate and industrial SO_2 removal efficiency increased by nearly 10%, and the solid waste comprehensive utilization increased by approximately 7%.

2.2 Changes in Food Quality in Production and Processing: Based on Pass Rates in National Quality Checks

In China, national quality inspection is generally implemented by spot checks. Important industrial products that may endanger human health and/or personal and property safety or those that affect the national economy and also products that have had consumer or relevant organization reports of quality issues are subject to spot checks. At present, tens of thousands of foods in 4 major categories, 22 classes, and 57 subclasses are produced by more than 400,000 food producers and processors in China. The food industry is a basic industry protecting the lives and livelihood of the people. When "relatively unlimited regulatory targets" are regulated by "relatively limited regulatory power," strengthening the food (finished product) quality supervision and inspection in production and processing is of great significance to the protection of overall food safety. National food quality checks are an institutional arrangement to supervise food quality by the government, and thus, the pass rates in national quality checks can basically reflect changes in the overall food quality and safety in China.*

2.2.1 Overall Situation of Food Quality in National Quality Checks

Since 2005, national food quality checks by the Administration of Quality Supervision, Inspection and Quarantine (AQSIQ) of the People's Republic of China cover all main food categories, involving thousands of different types of food. The basic features in national food quality checks during 2005–2011 are summarized in Sections 2.2.1.1 through 2.2.1.3.

2.2.1.1 Pass Rates Have Gradually Increased and Remained at a High Level   As shown in Figure 2.6, the overall pass rates in national quality checks increased from 80.1% in 2005 to 96.4% in 2011, representing an increase of 16.3% during the six years. AQSIQ tested more than 115,000 samples in food production and processing in a nationwide inspection in 2011. Among them, 133 risk items, such as 28 foods, 2 food additives, and 4 food-related products, were subject to routine inspection. In total, 20,352 food samples were tested and quality and safety problems were detected in 689 of them; the positive rate was 3.6%, with a decrease of 0.5% over 2010. In addition, food exports inspected in production and processing by AQSIQ in 2011 amounted to US$ 53.82 billion and included 2.043 million batches, representing an increase of 26.8% and 3.5%, respectively, over 2010. The unqualified rate of food exports reported from overseas was 0.08% in 2011. The pass rates in national quality checks have gradually increased and remained at a high level, which is quickly becoming the norm.

2.2.1.2 Pass Rates Varied Greatly between Different Varieties of Foods   Carbonated beverages, instant noodles, wheat flour, milk powder, meat, edible oil, and other products had a pass rate of over 90% in national quality checks in 2011, representing high quality. The pass rates of some foods even reached 100%. However, the pass rates of other products, such as soy products and tea, were lower than 90%.

---

*Finished products are inspected in national quality checks. Pass rates of finished products are a comprehensive evaluation of quality control in production and processing and a method to verify the effectiveness of the production process controls. Pass rates of food (finished product) in national quality checks can approximately measure the food quality and safety in production and processing.

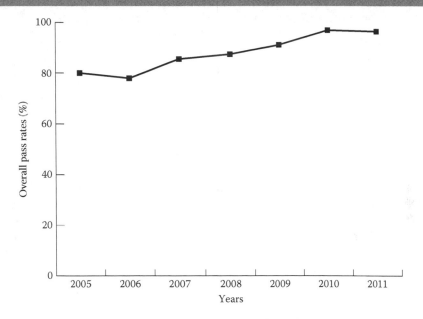

FIGURE 2.6
Changes in Pass Rates in National Quality Checks during 2005–2011

Source.   Data from the official website of China Association for Quality Inspection.

2.2.1.3 The Most Basic Problems in Food Quality Have Not Been Fundamentally Corrected   National food quality checks revealed that some of the major problems in food quality in production and processing in China were corrected during 2005–2011. However, the most basic problems have not been fundamentally addressed. Major problems revealed by the national food quality checks in 2005 included use of food additives beyond the specified amounts and ranges, microbial counts exceeding specified numbers, and nonstandard product labeling. Since 2009, the use of food additives beyond the specified amounts or ranges has become more prominent than the other problems. According to national food quality checks, in 2011, excessive use of food additives, microbial counts exceeding specified numbers, and physical and chemical indicators below standard were still the most basic problems in food quality and safety. In terms of risk monitoring, positive rates of risk items, such as pollutants and quality indicators, were high. Therefore, the use of food additives beyond the specified amounts and ranges is currently the largest risk to food quality and safety in production in China.

2.2.2 Comparison of Pass Rates of Different Varieties in the Same Year

The pass rates varied, to some extent, between different varieties of food in the same national quality checks. As shown in Figure 2.7, data from the AQSIQ spot checks in 2011 revealed that the pass rates of eight food categories that were closely related with people's daily lives were higher than 90.0%, with the exception of soy products (88.7%). The order of pass rates from high to low was carbonated beverages (100.0%), instant noodles (99.0%), wheat flour (99.0%), milk powder (98.0%), meat products (96.4%), sugar (92.7%), and edible vegetable oil (92.5%).

The 2011 national food quality checks found that the main problems in food quality and safety varied between different varieties. In the edible oil,

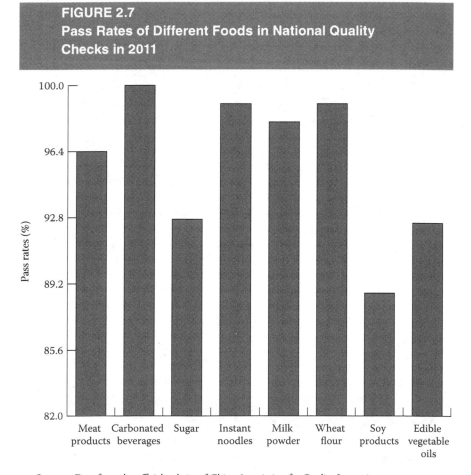

FIGURE 2.7
Pass Rates of Different Foods in National Quality Checks in 2011

Source.   Data from the official website of China Association for Quality Inspection.

characteristics such as aflatoxin B1, acid value, peroxide value, and residual solvents were not up to the standard. In meat products, characteristics such as colony-forming units, coliform, albuterol, nitrite, sunset yellow, carmine, tartrazine, benzoic acid, and acid value were not up to the standard. In soy product, characteristics such as colony-forming units, coliform, benzoic acid, and dehydroacetic acid were not up to the standard. In sugar, sucrose, yeast, reducing sugars, color value, and other indicators in monocrystal rock sugar were not up to the standard.

2.2.3 Comparison of Pass Rates of the Same Varieties in Different Years

Products with a large consumption and a relatively low pass rate in national quality checks, including meat products, wheat flour, edible vegetable oil, puffed foods, and soy products, were selected as examples to describe the changes in pass rates in national food quality checks in recent years and the major food safety problems.

2.2.3.1 Meat Products   As shown in Figure 2.8, the inspection results of nearly 300 meat products by AQSIQ during 2009–2011 revealed that the overall pass rate of meat products increased from 93.0% in 2009 to 96.4% in 2011, and the number of inspected provinces increased from 22 in 2009 to 28 in 2011. The number of inspection items increased from 22 in 2009 to 25 in 2011 which include nitrite, sorbic acid, benzoic acid, allura red, sunset yellow, heavy metals, microbial indicators, acid value, peroxide value, benzopyrene, clenbuterol hydrochloride, ractopamine, salbutamol, volatile basic nitrogen, trimethylamine nitrogen, and so on. National quality checks of meat products cover more provinces and inspection items.

Major noncompliant items identified in the 2009 national quality checks of meat products were coliform out of limits, vegetable protein and starch out of limits, and use of colorants and preservatives beyond the specified amounts or ranges. In the 2010 national quality checks, colony-forming units, coliform, nitrite, and acid values were identified as the major noncompliant items; however, use of allura red and other colorants beyond the specified amounts or ranges was not detected. The pass rate of meat products in the 2011 national quality checks was higher, but more noncompliant items were identified as more provinces and items were inspected. In the 2011 national quality checks, colony-forming units, coliform, albuterol, nitrite, sunset yellow, carmine, tartrazine, benzoic acid, and acid value were identified as the major noncompliant items.

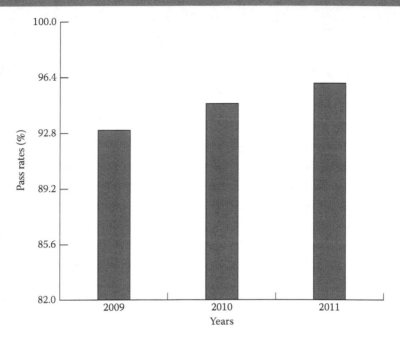

Source.   Data from the official website of China Association for Quality Inspection.

2.2.3.2 Wheat Flour   As shown in Figure 2.9, the inspection results of nearly 100 wheat flour products by AQSIQ during 2009–2011 revealed that the overall pass rate of wheat flour increased from 95.3% in 2009 to 99.0% in 2011. Thirteen items, including ash, fatty acid value, heavy metals, aflatoxin B1, benzoyl peroxide, and sodium formaldehyde sulfoxylate, were tested. The major problem identified in wheat flour in the 2009 national quality checks was that the measured value of benzoyl peroxide did not meet the relevant standards; major noncompliant items identified in 2010 were benzoyl peroxide and ash, and the major noncompliant item identified in 2011 was ash. This indicates that the quality of wheat flour has further improved.

2.2.3.3 Edible Vegetable Oil   As shown in Figure 2.10, the inspection results of more than 200 edible vegetable oil products from nearly 20 provinces by AQSIQ revealed that the pass rate declined from 92.7% in 2009 to 89.8% in 2010. Although the pass rate increased again in 2011, it only reached 92.5%,

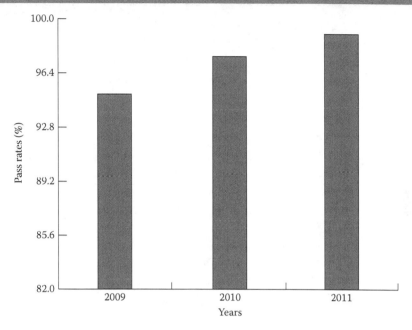

FIGURE 2.9
Pass Rates of Wheat Flour In National Quality Checks during 2009–2011

Source. Data from the official website of China Association for Quality Inspection.

still 0.2% lower than the 92.7% in 2009. The inspection results showed no improvement in reducing the risks in quality and safety of edible vegetable oil produced in China over 2009–2011. Major noncompliant items included aflatoxin B1, acid value, peroxide value, and residual solvents.

2.2.3.4 Puffed Food Puffed food is a new food that emerged in the late 1960s. In a broad sense, it refers to food that is cooked by deep frying, extrusion, hot sand frying, or microwaving and has a significantly increased size after the cooking (Shang and Tang 2007). As shown in Figure 2.11, the inspection results of more than 100 puffed foods from nearly 15 provinces by AQSIQ during 2009–2011 revealed that the pass rates of puffed food in national quality checks were 94.0%, 86.3%, and 100.0%, respectively, in 2009, 2010, and 2011. The pass rate was very unstable, with large fluctuations. Major noncompliant items identified in 2009 were sodium cyclamate and *Escherichia coli*. More noncompliant items were identified in 2010, including colony-forming units, *Escherichia coli*, and lead and residual aluminum, leading to a sharp

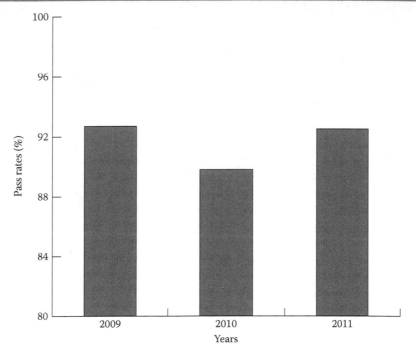

FIGURE 2.10
Pass Rates of Edible Vegetable Oil in National Quality Checks during 2009–2011

Source.   Data from the official website of China Association for Quality Inspection.

decline in the pass rate. On the contrary, the quality of puffed food improved significantly in 2011, with the pass rate increasing to 100%. However, since the objective risks have not been fundamentally eliminated, the foundation for the quality of puffed food to remain stable is still weak.

2.2.3.5 Soy Products   As shown in Figure 2.12, the inspection results of nearly 200 soy products from 15 provinces by AQSIQ during 2009–2011 revealed that the pass rate of soy products was not stable. The pass rate slightly increased from 91.0% in 2009 to 92.9% in 2010, but then dropped significantly to 88.75% in 2011. Among the eight food categories closely related with people's daily lives, soy products had the lowest pass rate in national quality checks. Many noncompliant items were identified in 2009, including colony-forming units, coliform, benzoic acid, dehydroacetic acid, sodium cyclamate, acesulfame, and sorbic acid. Fewer noncompliant items were

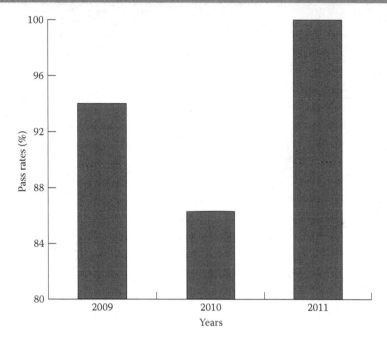

FIGURE 2.11
Pass Rates of Puffed Food in National Quality Checks during 2009–2011

Source. Data from the official website of China Association for Quality Inspection.

identified in 2010, including coliform and colony-forming units; however, the number of noncompliant items identified in 2011 increased again and mainly included colony-forming units, coliform, benzoic acid, sorbic acid, dehydroacetic acid, and sodium cyclamate.

2.3 Major Problems of Food Quality and Safety in Production and Processing

Production and processing are the key processes responsible for guaranteeing food quantity, quality, and safety. The food processing industry has played an important role in China's expansive economic growth and has solved the food supply issue. Unfortunately, the number of food producers and processors in China is deficient compared with that in developed countries, and

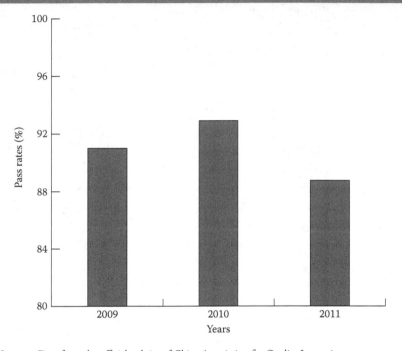

FIGURE 2.12
Pass Rates of Soy Products in National Quality Checks during 2009–2011

Source. Data from the official website of China Association for Quality Inspection.

labor productivity still lags behind that in developed countries as far back as the late 1970s (Chen et al. 2009). Therefore, guarantee of the food supply needs to be further enhanced. Since this book focuses on the analysis of food quality and safety, this chapter analyzes the main problems of food quality and safety in production and processing.

2.3.1 Small-Scale Production and Processing Units

The traditional food industry is an extension of agriculture, and the modern food industry is not only an extension of agriculture, but also a complete food industry chain from "farm to fork." It has developed from simple production and processing of surplus agricultural products to a closely-linked chain integrating marketing, processing, and manufacturing by factories and raw material production by production bases. If the traditional food industry is characterized by workshop-style, small-scale operations, the modern food industry seeks to develop large-scale, intensive, integrated, modern enterprises. Over the years, Chinese food producers and processors have experienced some changes in scale and organizational form. However, the changes are not large enough to meet the needs of modern food industry or the requirements of food quality and safety.

According to the *Quality and Safety of Food in China* published by the Information Office of the State Council of the People's Republic of China in August 2007 and the *12th Five-Year Plan of Food Industry* (National Development and Reform Commission [NDRC] industry No. 3229 [2011]) published by the NDRC and Ministry of Industry and Information Technology of the People's Republic of China in December 2011, there were 448,000 food producers and processors in China in 2006. Of these, 26,000 were above the designated size, 69,000 were below the designated size but had more than 10 employees, and 353,000 had less than 10 employees. As shown in Figure 2.13, of the 448,000 food producers and processors in China in 2006, 94.2% were below the designated size and 78.8% were small workshops with less than 10 employees. The number of food producers and processors below the designated size increased to 31,000 in 2011, which was an increase of 5,000 compared with 2006. However,

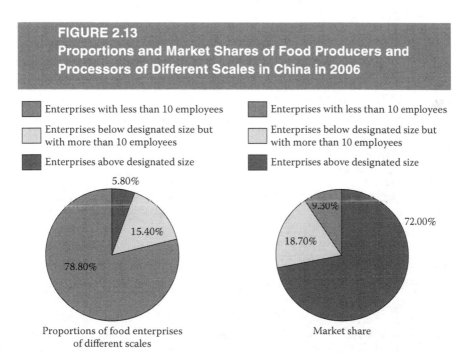

FIGURE 2.13
Proportions and Market Shares of Food Producers and Processors of Different Scales in China in 2006

Enterprises with less than 10 employees

Enterprises below designated size but with more than 10 employees

Enterprises above designated size

Enterprises with less than 10 employees

Enterprises below designated size but with more than 10 employees

Enterprises above designated size

5.80%
15.40%
78.80%

Proportions of food enterprises of different scales

9.30%
72.00%
18.70%

Market share

Source. Adapted from *Quality and Safety of Food in China*, 2007 and Information Office of the State Council of the People's Republic of China.

the "small-scale, decentralized, and nonintensive" model of food producers and processors has not been fundamentally changed. Small, micro, and workshop-style food producers and processors still account for approximately 90% of all food producers in China (China Economic Net 2012).

2.3.2 Diversity in Improper or Illegal Behaviors

In China, with the current market and government failures in food safety supervision and management, human factors, such as improper or illegal production and processing due to aggressive pursuit of economic interests, have become the most important factors causing food safety risks.[*] From the perspective of production and processing, production behaviors of producers and processors can be divided into several stages: the preparation stage, the production and processing stage, and the ready-for-sale stage (Table 2.5). Obviously, all improper or illegal behaviors that may occur at the three stages of food production and processing listed in Table 2.5 exist to varying degrees in objective reality and are even quite common with small/micro food producers and processors. The following findings of various studies provide preliminary confirmation of the possibility of this notion.

Wu and Lin (2005) found that food producers and processors need to register the procurement and usage of raw materials to strengthen management of raw material procurement. By investigating food producers and processors in Xiuzhou District, Jiaxing City, Zhejiang Province, Shi (2006) demonstrated that food producers and processors had no strict management of additive procurement and no consciousness of certificate claiming, with a certificate claiming rate of only 3.88%.

Zhao (2002) and Li (2010) suggested that small-scale and individual food processors had some common problems, such as poor sanitation conditions, use of inferior materials, and nonstandard operation by employees. Yang and Zhang (2011) conducted a questionnaire survey of employees in 29 enterprises providing nutritious meals for students in Chaoyang District, Beijing, on the sanitation conditions of production and processing and knowledge related to food safety; the results revealed the following: only 10.35% of the enterprises had a sound sanitary control system in place, the compliance rate

---

[*] Biological, chemical, and physical factors are natural factors that can directly cause food safety risks. These factors, in a sense, cannot be completely eliminated. In addition to biological, chemical, and physical factors, human misconduct and institutional factors, such as producer issues, information asymmetry, interest, and government regulation, may also lead to food safety risks. This book refers to both human misconduct and institutional factors as human factors. Refer to the Introduction and Chapter 5 in this book.

TABLE 2.5
Behaviors of Producers and Processors That May Lead to Food Safety Risks

Stage	Improper or Illegal Behaviors
Preparation stage prior to production and processing	A1. Purchase of unqualified (illegal) raw materials 　　A1.1 Purchase of substandard raw materials due to deficient testing equipment 　　A1.2 Driven by interest to purchase substandard or illegal raw materials 　　A1.3 Driven by interest to recycle waste 　　A1.4 No detailed record of procurement A2. Improper handling of raw materials 　　A2.1 Contamination of raw materials caused by improper control of storage environment 　　A2.2 No proper purification prior to food processing
Production and processing stage	B1. Irrational or illegal operations in production processes 　　B1.1 Nonstandard operations by operators in production and processing 　　B1.2 Illegal use of food additives and other chemical substances 　　B1.3 Improper operations of production equipments by operators during production 　　B1.4 Absence of real-time quality control during production B2. Substandard sanitation in production 　　B2.1 Environmental health not compliant with relevant provisions 　　B2.2 Substandard operator hygiene 　　B2.3 Wastes are not disposed of in accordance with regulations and reenter the food market
Ready-for-sale stage after production and processing	C1. Improper storage of products prior to sale 　　C1.1 Improper temperature and humidity during storage of products 　　C1.2 Lack of timely packing after completion of processing, causing contamination of food by dust, foreign matter, and microorganisms C2. Improper testing of products prior to sale 　　C2.1 Inadequate testing of quality due to poor operating techniques 　　C2.2 No use of advanced testing equipment for self-checking and evasion of regulatory supervision and inspection to save costs

of employee sanitation conditions was 79.31%, and the compliance rate of sanitation and facilities was as low as 27.59%.

Furthermore, the Beijing Health Inspection Authority inspected 49 food additive producers in Beijing from 2001 to 2003. The results revealed that 83.67% of the producers were equipped with laboratories, and the testing performance of those equipped with laboratories varied; some were only able to determine common indicators and did not determine the content of food additives (Wang 2004). Food additive producers had low product testing capacity. Nearly 80% of food additive producers were not able to test all the items for their own products according to requirements. Some additives contained in the food were undetected by enterprises commissioned to do the testing or the measured values were lower than actual values. Unfortunately, some batches were delivered even if they had failed testing. A survey in all food producers and processors in Heilongjiang Province by Hu and Gan (2009) revealed that, in some food producing and processing facilities, laboratories did not possess the necessary testing equipment and performed practically no function and that all required items were not tested before delivery. Turning post-event monitoring into pre-event prevention was therefore out of the question. A survey conducted in Zhengzhou City by Zhang et al. (2012) found that only 23% of food producers and processors performed regular tests of food additive content in the production process, 67% performed testing only at the last step in the process, and 10% did not even measure food additive content.

In meat processing, some food producers add excessive color fixatives, such as sodium nitrite and potassium nitrate, to give the meat a fresher, more appetizing appearance (Du 2008). Some food producers only label frozen dumplings with brand, fillings, and price but do not mention a production date or a shelf life (Zhao 2009). To increase traffic in order to improve economic efficiency, some transportation departments knowingly transport foods unsuitable for refrigerated transport, thus resulting in food safety problems (Luo and Zhang 2008).

On the contrary, a large number of small/micro food producers and processors have played a positive role in absorbing surplus labor and meeting the diverse needs of the market. However, a considerable number of facts have proved that the small/micro food producers and processors are still prone to food safety problems. As Chinese urban and rural incomes are rising and food consumption demand is escalating, the demand for food quality and safety is higher now than any other time in history. The gap between the growing demand for food safety and decentralized, small-scale food production has become the principal contradiction in the prevention and control of food safety risks in China. Therefore, the ultimate foothold in establishing a food safety risk prevention and control system in China should be optimizing the production and management by food producers and processors at the microlevel. Only by clarifying key factors affecting the behaviors of food producers and processors and implementing a combined system of effective, supporting policies, can a strong barrier against food safety risks be built at the microlevel.

2.3.3 Absence of Processing-Dedicated, Large-Scale Raw Material Sites

The modern food industry is no longer the simple processing of agricultural products, but more like a manufacturing industry. It serves as a connecting link in the industrial chain and has played a positive role in promoting and stimulating the development of agriculture. In particular, to ensure food quality and safety, the modern food industry requires stable, large-scale, specialized raw material sites to produce special varieties of products. The relationship between food production and processing and agricultural development is still at an early stage of supply and demand in China. In essence, the food industry processes whatever is currently agriculturally produced, and no organic connection has been built between the supply of raw food materials and food production and processing. The high quality of food produced and processed by developed countries is guaranteed not only by advanced technology and equipments, but also by processing-dedicated raw material varieties and established raw material sites. A serious long-term shortage of inputs dedicated toward improving the quality of agricultural products, and research remains, and development and production of processing-dedicated agricultural products have led to too many similar varieties of agricultural products and an insufficient supply of high-quality dedicated varieties suitable for processing in China. Therefore, there is an urgent need to establish large-scale, specialized raw food material sites in China.

Furthermore, corresponding to the absence of processing-dedicated large-scale raw material sites, the food industry chain is short in China. Currently, the agro-processing and further agro-processing rates are approximately 90% and 80%, respectively, in developed countries, but are strikingly lower (approximately 45% and 30%, respectively) in China (*The 11th Five-Year Plan of Agro-Processing Industry*). In terms of consumption, unprocessed food plays a primary role in the Chinese consumption structure, while the proportion of processed products is lower than 30%, but is as high as 80% in developed countries (Modern Agriculture of China 2010). More than 30% of crop and aquatic products are processed into food products in developed countries, but the rate is only 2%–3% in China (Ningbo Municipal Statistics Bureau 2005). Furthermore, processed food accounts for approximately 80% of the total food consumed in developed countries, but again the rate is much lower (30%) in China (Li and Li 2007). The food industry is a leading industry in the national economy of developed countries, such as the United States, France, and Japan, as food industry outputs account for more than 20% of the gross national product (GNP) in these countries; however, it is less than 10% in China (*The 12th Five-Year Plan of Food Industry*).

---

*The 12th Five-Year Plan of Food Industry (NDRC industry No. 3229 [2011]), published by NDRC and Ministry of Industry and Information Technology of the People's Republic of China.

2.3.4 Insufficient Science and Technology Inputs in the Food Industry

Based on data availability, data for 2007 were analyzed as an example. As shown in Table 2.6, the overall level of technology and production of the food industry was reflected by five indicators: labor productivity, technical advancement, internationalization, rationality of industrial organization, and sustainability. Investment in research and development accounted for only 0.4% of sales revenue, which was 33.3% of the international advanced level. The internationalization level (proportion of food exports in total national exports) was 3.2%, which was 58.4% of that in developed countries. The average revenue of dominant firms was only 7.4% of the international advanced level, and sustainability was 38.8%, which was 84.2% of the international advanced level. These indicators were weighted and summed to calculate a comprehensive modernization index of the Chinese food industry of 38.3% (the international advanced level was 100.0%). These results revealed a large gap between the Chinese food industry and the international advanced level of food industry.

The investment in research and development accounted for only 0.4% of sales revenue of the Chinese food industry, which was substantially lower than the national industrial average. Insufficient investment in technology has caused many problems, in particular, difficulty in improving key technologies and equipment. For example, in terms of flour processing equipment, performance and quality of flour mills produced in China lag far behind those of the products of Bühler. In terms of edible oil refining equipment, a large gap in mechanical properties and yield exists between separators made in China and those made in developed countries. In addition, spray drying equipment made in China lags far behind the products of Niro in performance and yield. Other direct consequences are low energy efficiency in food production and processing and poor resource conservation, which have resulted in large emissions of the "three wastes" and carbon. For example, beer production in China has 50% lower productivity than that in developed countries, power consumption by equipment made in China is five times than that in developed countries, and water consumption by bottle washers in China is three times than that in developed countries (China Food Machinery and Equipment Network 2006). The amount of wastewater generated in the production process of the food fermentation industry is 4 billion m^3 per year, accounting for 10% of the total amount of national industrial wastewater discharge, and chemical oxygen demand (COD) has reached 5 million tons, accounting for approximately 20% of the total national industrial COD; as a result, these issues have placed great

TABLE 2.6

Comprehensive Modernization Index of the Chinese Food Industry in 2007, US$/Person, Billion US$, %

Index	Subindex	Basic Indicator	Weight	Actual Value	Standard Value	Ratio of Actual to Standard Value	Index Value
Industrial efficiency index	Overall efficiency	Industry-wide labor productivity	30	29021.10	120297.30	24.1	7.2
Industrial structure index	Technical advancement	Investment intensity in research and development	25	0.40	1.20	33.3	8.3
	Internationalization	Proportion of food exports in total national exports	20	3.20	5.48	58.4	11.7
	Rationality of industrial organization	Average revenue of dominant firms	15	4.28	58.09	7.4	1.1
Industrial environment index	Sustainability	Proportion of sewage discharge	10	28.81	34.23	84.2	8.4
	Comprehensive index		100	–	–	–	38.3

Source. Chen, J.G. et al., *China's industrialization report*, Social Sciences Academic Press, Beijing, People's Republic of China, 2009.

pressure on the fragile environment in China (*The 12th Five-Year Plan of Food Industry*).

A more serious consequence of inadequate investment in food industry technology and the low performance of technical equipment is the negative influence on quality and safety of food products. For example, according to surveys in China, 80%–90% of fruits, vegetables, poultry, and aquatic products are transported by ordinary trucks, large amounts of milk and soy products are transported without a cold chain, and approximately 70% of freight vehicles use an open container. Furthermore, less than 10% of trucks/trailers are equipped with a refrigerator and insulated container. Most railway refrigerated transport facilities are also not equipped with standardized insulated refrigerated cars and many of those need improvement/repair. Moreover, China's cold chain system is only a primitive refrigeration equipment market. For these reasons, at present, the cold chain technology owned by China cannot be fully applied to many types of food. China lags far behind the international advanced level in this respect (Zhang 2011b). It is difficult to guarantee food quality when food is transported without a cold chain.

---

* The 12th Five-Year Plan of Food Industry (NDRC industry No. 3229 [2011]), published by NDRC and Ministry of Industry and Information Technology of the People's Republic of China.

3

Food Safety Supervision in Circulation and Food Quality and Safety

Food circulation is not only the first point of contact with consumers, but also the final hurdle toward entry into millions of homes. It is a stage involving the largest number of consumers. According to the *Food Safety Law* and its implementing regulations, industrial and commercial administrative departments are accountable for food safety supervision in circulation in China. Consumer safety in the Chinese food market has been basically guaranteed by strengthening admittance management of food producers and operators, regulating production and operation of food enterprises, striving to control food quality and safety in circulation, and actively preventing and handling food safety incidents in circulation by the industrial and commercial administration systems over the years. However, many problems still remain unsolved.

3.1 Establishment of a System of Food Safety Supervision in Circulation

According to the official document *Decision to Further Strengthen Food Safety* issued by the Chinese State Council in September 2004, a multi-departmental segmented regulatory system was formalized for food safety supervision. It also specified the responsibility of industrial and commercial administrative departments to supervise food safety in circulation. The segmented supervision system has continued to be used under the *Food*

Safety Law passed in 2009. Though partial adjustments have been made to the multidepartmental segmented supervision system, the responsibility of industrial and commercial administrative departments to supervise food safety in circulation has not been changed. After years of efforts, China's industrial and commercial administration systems have established a system of food safety supervision in circulation, as well as related laws and regulations and institutional systems.*

3.1.1 A System of Food Safety Supervision in Circulation

As of the end of 2011, the industrial and commercial bureaus of 27 Chinese provinces (autonomous regions and municipalities) and 11 subprovincial cities set up a functional organization for food safety supervision. The number of full-time and part-time personnel engaged in food safety supervision reached 15,000, and leader accountability for local supervision of food safety in circulation had basically been established nationwide.

3.1.2 Food Safety Regulations and Rules in Circulation

According to the statutory regulatory duties granted by the *Food Safety Law* and *Implementing Regulations of the Food Safety Law*, the State Administration for Industry and Commerce (SAIC) has successively promulgated a series of laws and regulations since 2009. As of 2011, China's industrial and commercial administration systems have basically established a framework of laws and regulations and institutional systems for food safety in circulation. This framework is based on the following systems: food entry system, food procurement inspection system, certificate and invoice claiming system for food safety supervision, food purchase and sale accounting system, food quality commitment system, system of market organizer accountability for food quality, food market inspection system, food safety information disclosure system, substandard food delisting and recall system, and regulatory system of food trader credit classification. Meanwhile, the SAIC has gradually abolished normative documents not complying with the requirements of the *Food Safety Law* and *Implementing*

---

*It should be noted that data used in this chapter, unless otherwise stated, are derived from data published by the Regulation Department for Market Circulation of Food, SAIC of the People's Republic of China, and the book *Instructions for Food Safety Supervision in Circulation* (China Industry and Commerce Associated Press) issued by the Regulation Department for Market Circulation of Food in November 2011. Data of consumer food complaints are derived from statistical information of food complaints accepted by the China Consumers' Association during 2007–2011.

Regulations of the Food Safety Law. The gradual improvement of the legal system of food safety has provided powerful legal and institutional protection to the execution of duties and enhancement of food safety supervision in circulation.

3.2 Special Law Enforcement Inspection of Food Safety and Safety Incident Management in Circulation

Over the years, China's industrial and commercial administration systems have carried out special law enforcement inspections of food safety in circulation by focusing on hot issues that are closely related to consumers and have a strong social impact. Furthermore, substantial efforts have been made to prevent and deal with food safety incidents in circulation and to constantly improve the emergency management of food safety incidents. All of these have effectively safeguarded food market order.

3.2.1 Special Law Enforcement Inspection of Food Safety in Circulation

In 2011, China's industrial and commercial administration systems made efforts for the special law enforcement inspection of food safety in circulation. The five main aspects will be discussed in Sections 3.2.1.1 through 3.2.1.5.

3.2.1.1 Special Rectification of the Rural Food Market   National industrial and commercial systems have given top priority to the protection of consumer safety in rural food markets in food market supervision. Great efforts have been made to crack down on the sale of substandard, expired, and counterfeit food products and other illegal activities in rural food markets. In 2011, China's industrial and commercial systems conducted 9.806 million inspections of rural food business operators and dealt with 23,000 rural food cases to provide basic maintenance of the rural food market order.

3.2.1.2 Special Rectification of the Dairy Market   The quality of dairy products concerns Chinese consumers. To address the concerns, industrial and commercial administrative departments throughout China have conscientiously performed their duties in supervising dairy products in circulation and monitoring the admittance of new players to the dairy market, in particular, the infant formula market. Items of food circulation permission and business scope registration are classified to verify and manage each item individually. Strict control has been imposed on the quality of infant formula milk powder by increasing the number and frequency of sampling inspections. In 2011, China's industrial and commercial systems conducted 5.829 million inspections of dairy retailers and sampled 73,384 sets of dairy products, with 67,985 sets proved to be qualified and a pass rate of 92.64%. Overall, 50,710 sets of infant milk powder were sampled, with 46,809 sets proved to be qualified and a pass rate of 92.31%; 1,362 dairy cases were investigated.

3.2.1.3 Special Rectification of the Illegal Addition of Nonfood Substances and Abuse of Food Additives   Industrial and commercial administrative departments throughout China have severely punished the illegal addition of nonfood substances, abuse of food additives, and illegal sale of food additives in the stage of circulation. In 2011, China's industrial and commercial systems conducted 606,000 inspections of food additive traders, detained 56,000 kg of nonfood substances and food additives and 214,000 kg of food subject to abuse of food additives, and investigated 4,611 relevant cases. Strict control and punishment have been maintained with respect to the illegal addition of nonfood substances and abuse of food additives.

3.2.1.4 Special Rectification of the Edible Oil Market and Illegal Sale of Swill-Cooked Dirty Oil   By focusing on wholesale markets and bazaars, industrial and commercial administrative departments throughout China have investigated the ins and outs of edible oil traders and have gradually standardized the qualification of edible oil traders. The administrative departments have also made close inspection of the procurement sources of edible oil traders, especially bulk edible oil traders, and continue to thoroughly examine certificates of incoming inspection and related invoices, as well as implement bulk edible oil labeling requirements. Furthermore, the departments have enhanced investigation of and punishment for sale of counterfeit and poor-quality edible oils, especially swill-cooked dirty oil, and have cracked down on the illegal sale of swill-cooked dirty oil and edible oil purchased from informal sources. In 2011, China's industrial and commercial systems detained 380,300 kg of swill-cooked dirty oil and edible oil purchased from informal sources, investigated 447 cases of substandard edible oil, and destroyed 216 swill-cooked dirty oil production dens in conjunction with other departments. All of these efforts have effectively safeguarded the edible oil market order.

3.2.1.5 Special Rectification of the Wine Market Industrial and commercial administrative departments throughout China have carried out special rectification of liquor and wine markets and focused on preventing infringement of registered trademarks and counterfeiting of specific names, packaging, and decorations of well-known liquor and wine brands. Great efforts have been devoted to investigating the infringement of famous Chinese liquor trademarks and sale of counterfeit and substandard wines. In 2011, China's industrial and commercial systems detained 730,338.5 kg of counterfeit and poor-quality liquors and wines and investigated 7,014 relevant cases.

3.2.2 Food Safety Incident Management in Circulation

In 2011, China's industrial and commercial administration systems properly managed 18 cases of food safety emergencies, including the "stained steamed bread" incident, and actively responded to public opinion about food safety, thereby basically achieving dynamic controllability of food safety in circulation.

3.2.2.1 Shanghai "Stained Steamed Bread" Incident In April 2011, steamed corn bread stained with colorants and treated with preservatives to prevent mildew was found to be sold in many supermarkets in Shanghai, China. Every day, some 30,000 pieces of stained steamed bread were sold to more than 30 supermarkets, including Lianhua, Hualian, and Dia. After the incident was identified, to control progression of the incident, the Shanghai Industrial and Commercial Bureau quickly activated contingency plans, investigated the incident straight away, and made sure that all stained steamed bread was pulled from the shelves in involved supermarkets at once. On April 13, 2011, the Shanghai Municipal Bureau of Quality and Technical Supervision revoked the food production license of the branch of Shanghai Shenglu Foods Co. Ltd. that produced the stained steamed bread. The company's legal representative and other four suspects were held in criminal detention. Because of this incident, food safety accountability in circulation was further strengthened in Shanghai.

3.2.2.2 Plasticizer Incident On May 24, 2011, the Ministry of Health and Welfare of Taiwan reported to the Administration of Quality Supervision, Inspection and Quarantine (AQSIQ) that Yu Shen Spice Co. Ltd. was found to have been adding di-2-ethylhexyl phthalate (DEHP; a commonly known plasticizer) to food products that may have entered the market of mainland

China. Industrial and commercial administrative departments throughout China rapidly investigated the local market for food containing DEHP and other plasticizers. As of June 12, 2011, plasticizers had been detected in eight samples from four companies in mainland China. In face of the plasticizer incident, China's industrial and commercial systems actively performed their duties, took quick actions and resolute measures, and provided timely and effective management to take control of the situation, protecting the order of the food additive market and the legitimate rights and interests of the consumers.

3.2.2.3 Clenbuterol Incident   In March 2011, it was exposed that the Henan Shuanghui Group had added clenbuterol to pig feed. The SAIC quickly set up a special team to investigate the site of the incident. To further improve the nationwide supervision of pork containing clenbuterol, in mid-March 2011, the SAIC decided to carry out special rectification of clenbuterol from April 2011 to February 2012 by the national industrial and commercial systems. As a result, 510,000 inspections were made at bazaars and wholesale markets and 4,370,000 inspections were made at pork retailers. During these inspections, 3,695 unlicensed pork retailers were eliminated, and 229,000 kg of substandard pork, including 13,000 kg of pork containing clenbuterol, were detained. Furthermore, 1504 relevant cases, with a total value of 6.78 million yuan, were investigated, and 22 cases were transferred to the judicial court.

3.2.2.4 Clean-Up of Swill-Cooked Dirty Oil   At the end of June 2011, the Chinese news media exposed the industry chain of swill-cooked dirty oil in the Beijing–Tianjin–Hebei regions. The investigation revealed a huge industrial chain of swill-cooked dirty oil, with processing dens employing high technology in Tianjin, Hebei, and even Beijing, and that the swill-cooked dirty oil was introduced into the supermarkets in the form of small packages. Since late August 2011, the Ministry of Public Security of the People's Republic of China has directed departments of public security all over China to fight against the crimes of swill-cooked dirty oil. As of November 2011, the departments of public security had detected 128 cases in which swill-cooked dirty oil was produced and sold as edible oil, and they had arrested more than 700 criminal suspects, verified more than 60,000 tons of oil involved, and destroyed 60 criminal networks involving 28 provinces integrating collection, primary processing, reselling, deep processing, wholesale, and sales. The hazards of the crimes of swill-cooked dirty oil to edible oil safety have been initially contained.

3.3 Food Quality and Safety Supervision in Circulation

While strengthening the special law enforcement inspection of food quality and safety in circulation to actively and effectively prevent and manage food safety incidents, Chinese industrial and commercial administrative departments have achieved positive results in enhancing regular supervision of food quality and safety in circulation and in the food market.

3.3.1 Quality Checks and Monitoring in the Food Market

Since 2006, the industrial and commercial administrative departments at all levels have established monitoring systems combining self-inspection by operators, inspection required by consumers, and sampling inspection by industrial and commercial departments, as well as the four-level rapid detection systems comprising provincial, municipal, and county industrial and commercial bureaus and industrial and commercial agencies. Meanwhile, great efforts have been made toward improving and innovating the mechanisms and methods of food sampling inspection in circulation, scientifically identifying the scope, varieties, items, and form of sampling inspection, and improving the capacity to standardize food sampling inspection according to law. At present, the sampling scope and varieties in the Chinese food market have been expanding, more and more efforts have been devoted, and classified supervision systems have been gradually established. As shown in Figure 3.1, Chinese industrial and commercial administration departments sampled 285,200, 208,300, and 332,000 sets of food in 2009, 2010, and 2011, respectively, and observed a general increasing trend in pass rates.

3.3.2 Conduct Regulation of Food Business Operators

By combining food sampling inspection with classified supervision of the food market, the Chinese industrial and commercial administration departments have made great efforts in monitoring qualification, operating conditions, food appearance, employees, sources, packaging and labels, trademarks and advertisements, and market organizer accountability of food business operators, according to the characteristics of different food business patterns and sites. In addition, emphasis has been placed on shopping malls, supermarkets, wholesale markets, and food wholesalers. Basic data regarding conduct

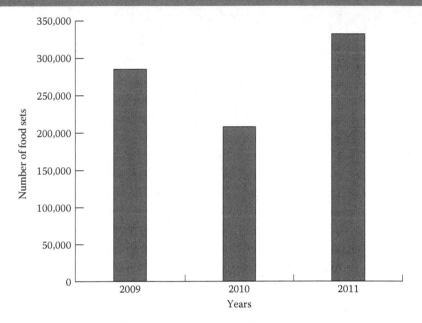

FIGURE 3.1
Number of Food Sets Sampled by the Chinese
Industrial and Commercial Administration
Departments during 2009–2011

Source.   Data from State Administration for Industry and Commerce.

regulation of food business operators and the total number of relevant cases during 2007–2011 are shown in Table 3.1.

3.3.3 Admittance Regulation of Food Market Players

In strict accordance with the *Food Safety Law*, as well as the *Food Circulation Permit Management Regulations* and the *Registration and Management System of Admittance of Food Market Players*, since 2009, China's industrial and commercial administration departments have strictly managed the admittance of food market players. The administrative departments have also standardized the approval and issuance of licenses, adhered to the principle of approval before issuance, and implemented legitimate registration procedures since 2009. In combination with the strict approval and issuance of food circulation permits, business registration management and classified credit supervision of food business operators in circulation have been enhanced to continuously

TABLE 3.1

Basic Data Regarding Regulation of Food Business Operators by Industrial and Commercial Departments during 2007–2011

	Total Number of Cases	Total Value of Cases (million yuan)	Confiscated Value (million yuan)	Penalty (million yuan)
2007	79,000	580.00	–	–
2008	77,556	1122.90	103.72	688.72
2009	85,307	411.81	19.99	318.22
2010	94,854	385.75	36.39	354.48
2011	105,032	511.63	49.27	414.18
Total	441,749	3012.09	209.37	1775.60

Source. State Administration for Industry and Commerce (SAIC).

improve the effectiveness of supervision. The industrial and commercial administration departments issued 0.201, 2.169, and 2.1148 million food circulation permits in 2009, 2010, and 2011, respectively, all over China.

3.3.4 Supervision and Early Warning of Illegal Food Advertising by Food Producers and Operators

Chinese industrial and commercial administrative departments have severely punished illegal food advertising in circulation and disclosed risk warning information, for example, by exposing the illegal food advertising and publishing food safety information. These measures not only give serious warning to law breakers and associated media, but also disclose food safety risk warning information to consumers, thus clearing up market warning signs and protecting the interest of consumers. The SAIC has exposed seriously illegal food advertising published in television, newspapers, and other media all over China since January 2009. Table 3.2 shows the basic information of illegal food advertising exposed by the SAIC during 2009–2011.

3.3.5 Development of a Long-Term Regulatory Mechanism

To meet the demand of new situations and tasks of food safety supervision based on the current reality of food safety supervision in circulation, Chinese industrial and commercial administration departments have taken development and improvement of a long-term regulatory mechanism for food safety in circulation as a fundamental, long-term, important task. Efforts have been devoted to

TABLE 3.2
Basic Information of Illegal Food Advertising Exposed
by the SAIC during 2009–2011

Announcement	Announcement Time	Description	Monitoring Time
SAIC illegal advertising announcement [2012] No. 1	January 16, 2012	Nagqu Snowland Cordyceps, health food advertising; Dongfangzhizi Shaungqi Capsules, health food advertising; JianDu Runtong Capsules, health food advertising	Fourth quarter of 2011
SAIC illegal advertising announcement [2011] No. 5	November 28, 2011	Beauty Seeking Meirongbao Capsules, health food advertising; Zangmi Snowland Cordyceps Capsules, food advertising	Third quarter of 2011
SAIC illegal advertising announcement [2011] No. 4	August 10, 2011	Yixuan Cordyceps and Pine Tea, food advertising; Jinwang Propolis and Balsam Pear Soft Capsules, health food advertising; Guolao Wengan Tea, food advertising; Tongren Restoring Oral Insulin, health food advertising	Second quarter of 2011
SAIC illegal advertising announcement [2011] No. 3	June 13, 2011	Shouruixiang Quansong Tea, food advertising; Guoyan Qianliefang (claiming to treat prostatitis), food advertising; Houde Propolis Soft Capsules, health food advertising	Second quarter of 2011
Illegal advertising exposed by Beijing and Kunming industrial and commercial departments	March 10, 2011	Kuilikang (claiming to enhance sexual potency), food advertising; Cordyceps Health Wine, health food advertising; Kunming: Tongren Tangke, health food advertising; Zhifengtang Propolis, health food advertising	First quarter of 2011
SAIC illegal advertising announcement [2011] No. 1	January 30, 2011	Energetic Antihypertensive Enzyme (Huoli Jiangya Mei), food advertising released on Page A31 of *Zhengzhou Evening News* on December 3; Gaogenguo (health food to enhance sexual potency), food advertising released on Page A13 of *Lanzhou Evening News* on December 2	Fourth quarter of 2010
SAIC illegal advertising announcement [2010] No. 7	November 11, 2010	MAXMAN, food advertising released on Page 11 in *Sanqin Daily* on October 13; Tianmaisu (claiming various effects in health protection), food advertising released on Page 17 of *Taiyuan Evening News* on October 13; Minyuanqing (claiming to boost immunity), health food advertising released on Xinjiang TV on September 3	Third quarter of 2010

(Continued)

TABLE 3.2
(Continued) Basic Information of Illegal Food
Advertising Exposed by the SAIC during 2009–2011

Announcement	Announcement Time	Description	Monitoring Time
SAIC illegal advertising announcement [2010] No. 6	September 21, 2010	Ximu L-carnitine Milk Tea, food advertising; Dongfangzhizi Shaungqi Capsules, food advertising; Jingshiqing Corn Stigma and Coix Seed Tea, food advertising Xueyinghua Natto Compound Capsules, food advertising; Acid Discharging Kidney Care Tea (Paisuan Shen Cha), food advertising	Second quarter of 2010
SAIC illegal advertising announcement [2010] No. 4	May 10, 2010	Tongren Qiangjing Capsules (claiming to enhance sexual potency), food advertising released on Page A10 of *Modern Evening Times* (Heilongjiang) on March 20; Ximo Immune Enhancing Capsules, health food advertising released on Page 09 of *Nanning Evening News* on March 20	First quarter of 2010
SAIC and State Food and Drug Administration illegal advertising announcement [2010] No. 3	February 10, 2010	Pear Flower Antihypertensive Vine Tea, health food advertising released on Page A09 of *Southland Metropolis Daily* (Guangxi) on December 3; Beiqishen Haohanlianglibang Soft Capsules (claiming to enhance sexual potency), food advertising released on Page A32 of *Strait Metropolis Daily* (Fujian) on December 3	Fourth quarter of 2009
SAIC illegal advertising announcement [2009] No. 8	October 27, 2009	Zezheng Duowei Zhikang Capsules (claiming to improve memory), health food advertising released on Page 4 of *Writers Digest* (Beijing) on September 18; Dubang Chaoyingpai Maiqishen Capsules (claiming to enhance immunity and sexual potency), health food advertising released on Page 09 of *Nanning Evening News* on September 17; LiverWise, health food advertising released on Page A31 of *Jinghua Times* (Beijing) on September 17	Third quarter of 2009

(Continued)

TABLE 3.2

(Continued) Basic Information of Illegal Food Advertising Exposed by the SAIC during 2009–2011

Announcement	Announcement Time	Description	Monitoring Time
SAIC illegal advertising announcement [Q2 2009] No. 6	July 29, 2009	Zhifengtang, health food advertising released on *Chutian City Newspaper* (Hubei) on June 11; Mannsih, health food advertising released on *Northern News Daily* (Inner Mongolia) on June 10	Second quarter of 2009
SAIC illegal advertising announcement [Q1 2009] No. 5	May 17, 2009	Vital Protein A (Shengming A Danbai), food advertising released on *Xi'an Evening News* on March 18; Beili Capsules, health food advertising released on *Modern Evening Times* (Heilongjiang) on March 18; Zhongma, food advertising released on *Yanzhao Evening News* (Hebei) on March 16; Shengshou Qiaoqi Capsules, food advertising released on Set No.1 of Qingdao TV on March 26	First quarter of 2009
SAIC illegal advertising announcement [2009] No. 2	February 11, 2009	Aidongli (claiming to enhance sexual potency), health food advertising released on *Peninsula City News* on December 3	Fourth quarter of 2008

Source. Illegal advertising announcement of the SAIC, http://www.saic.gov.cn/zwgk/gggs/wfgggg/.

improving law enforcement mechanisms with emphasis on ensuring legal registration, product quality conformance, and legitimate operation of food business operators. Further efforts have been made to develop operator self-regulatory mechanisms and food safety traceability systems with emphasis on the sourcing, selling, and withdrawal of food, as well as to enhance coordination and cooperation between various departments with emphasis on establishing seamless connection between departments. Development of social supervision mechanisms based on the media, consumers, and hired supervisors has also been a large focus.

3.4 Consumer Food Safety Evaluation and Complaints

Between January and March 2012, the research team of this book conducted a survey of 4289 consumers in 12 provinces, autonomous regions, and municipalities (hereinafter referred to as the respondents). According to household register, the respondents comprised 2143 urban and 2146 rural consumers (Chapter 6). This book analyzes the survey results of consumer food safety evaluation and investigates issues related to consumer food complaints based on relevant data published by the China Consumers' Association (CCA) during 2007–2011.

3.4.1 Consumer Food Safety Evaluation

According to the survey data, this book attempts to describe consumer food safety evaluation from six aspects, that is, ranking of preferred sites to purchase food, ranking of food safety purchased from different sites, freshness of edible agricultural products, food sanitation, authenticity of food quality certification and labeling information, and experience in purchasing unsafe food.

3.4.1.1 Ranking of Preferred Sites to Buy Food   As shown in Figure 3.2, for the 4289 respondents, the order of preference for sites to purchase foods was as follows: supermarkets (73.02%), bazaars (12.92%), grocery stores (10.35%), itinerant vendors (2.05%), and others (1.66%).

As shown in Table 3.3, urban and rural respondents had the same preference for sites to buy food, but the specific proportions were different.

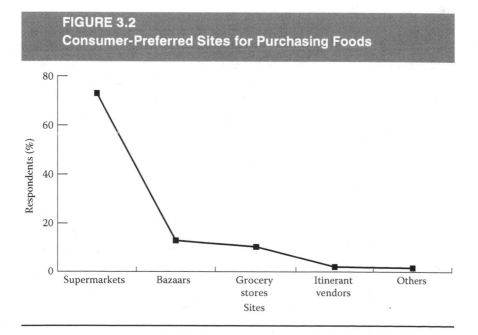

FIGURE 3.2
Consumer-Preferred Sites for Purchasing Foods

TABLE 3.3
Preferred Sites to Buy Food by Urban and Rural Respondents

	Urban Respondents		Rural Respondents	
Site	Sample Size	Proportion (%)	Sample Size	Proportion (%)
Supermarket	1727	80.59	1405	65.47
Bazaar	223	10.41	331	15.42
Grocery store	138	6.44	306	14.26
Itinerant vendors	27	1.26	61	2.84
Others	28	1.31	43	2.00

TABLE 3.4
Safety Evaluation of Food Purchased from Different Sites of Urban and Rural Respondents

	Total		Urban Respondents		Rural Respondents	
Site	Number of Respondents Trusting Food Safety	Proportion (%)	Number of Respondents Trusting Food Safety	Proportion (%)	Number of Respondents Trusting Food Safety	Proportion (%)
Supermarket	4051	94.45	2005	93.56	2046	95.47
Bazaar	2220	51.76	1090	50.86	1130	52.73
Grocery store	1476	34.41	640	29.86	836	39.01
Itinerant vendors	285	6.64	120	5.60	165	7.70

Note. Respondents were asked to evaluate the food safety of supermarkets, bazaars, grocery stores, and itinerant vendors, respectively.

The proportion of rural respondents preferring supermarkets was 15% lower than that of urban respondents. The main reason for the difference may be the lower income levels of rural respondents, which result in an inability to offer the relatively expensive products in supermarkets in rural areas.

3.4.1.2 Ranking of Food Safety Purchased from Different Sites As shown in Table 3.4, food purchased from supermarkets (94.45%) had the highest safety evaluation of respondents, followed by the bazaars (51.76%). Food purchased from grocery stores and itinerant vendors had the lowest safety evaluation, which was 34.41% and 6.64%, respectively. The ranking of food safety evaluation of different sites was also the same between urban and rural respondents. However, the food safety evaluation of grocery stores was nearly

FIGURE 3.3
Evaluation of Freshness of Major Edible Agricultural Products

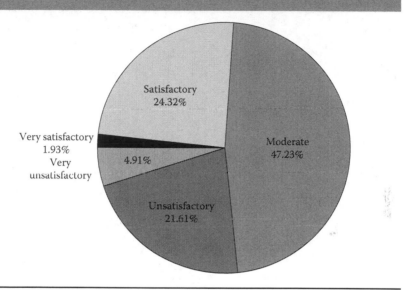

10% higher in rural respondents than in urban respondents. Although the food safety of grocery stores has been questioned, their role in the rural food shop system should not be underestimated.

3.4.1.3 Freshness of Edible Agricultural Products   As shown in Figure 3.3, 47.23% of respondents gave a "moderate" evaluation of the freshness of major edible agricultural products. The proportions of "satisfactory" and "unsatisfactory" products were similar, 24.32% and 21.61%, respectively. The proportions of "very satisfactory" and "very unsatisfactory" were as low as 1.93% and 4.91%, respectively.

As shown in Figure 3.4, the freshness evaluation of major edible agricultural products on the market was roughly the same between rural and urban respondents. However, it was better in urban respondents than in rural respondents. These results not only reflect the improvement of major edible agricultural products on the urban market, but also indicate that rural respondents have higher requirements for freshness of edible agricultural products than urban respondents.

3.4.1.4 Food Sanitation   As shown in Figure 3.5, 45.72% of respondents considered food sanitation to be "moderate." The total proportion of

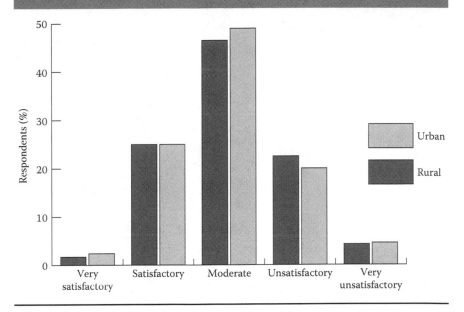

FIGURE 3.4
Evaluation of Freshness of Major Edible Agricultural Products by Rural and Urban Respondents

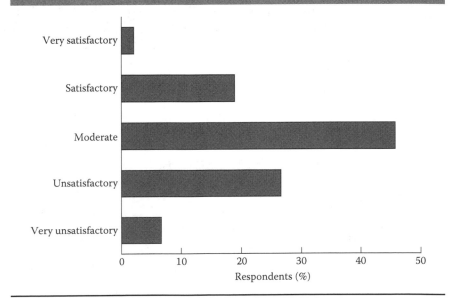

FIGURE 3.5
Overall Evaluation of Food Sanitation on the Market

FIGURE 3.6
Evaluation of Food Sanitation by Rural and Urban Respondents

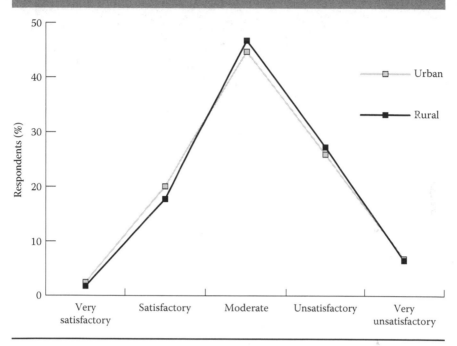

"unsatisfactory" combined with "very unsatisfactory" responses was 33.29%, approximately 12% higher than 20.98%, which was the total proportion of the combination of "satisfactory" and "very satisfactory" responses. The evaluation of food sanitation was essentially the same between rural and urban respondents (Figure 3.6). However, the proportions of both "satisfactory" and "very satisfactory" responses were slightly higher in urban respondents than in rural respondents, while the proportions of both "unsatisfactory" and "moderate" responses were lower in urban respondents than in rural respondents. It is clear that urban respondents gave a better evaluation of food sanitation than rural respondents, indicating that food sanitation remains to be improved in rural markets.

3.4.1.5 Authenticity of Food Quality Certification and Labeling Information   As shown in Figure 3.7, 46.02% of respondents considered the authenticity of food quality certification and labeling information to be "moderate." The total proportion of "unsatisfactory" combined with

FIGURE 3.7
Evaluation of Authenticity of Food Quality Certification
and Labeling Information

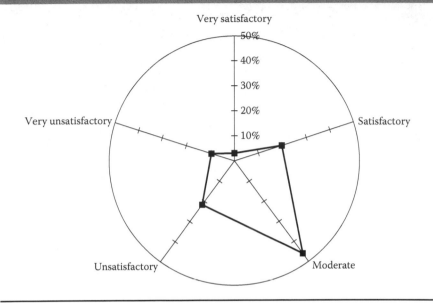

"very unsatisfactory" responses was 31.18%, higher than the 22.80% for the combination of "satisfactory" and "very satisfactory" responses.

As shown in Figure 3.8, the distribution of the evaluation of the authenticity of food quality certification and labeling information was similar between rural and urban respondents. The proportions of "satisfactory" and "very satisfactory" responses were lower in urban respondents than in rural respondents, while the proportion of "very unsatisfactory" responses was higher in urban respondents. This finding is associated with the fact that urban respondents were more concerned about food safety than rural respondents.

3.4.1.6 Experience in Purchasing Unsafe Food   As shown in Figure 3.9, 5.3% of respondents "often" purchased unsafe food (e.g., expired, moldy, or other substandard food), 36.5% "sometimes" purchased unsafe food, and the proportions of "seldom" and "never" responses were 32.3% and 6.8%, respectively. The total proportion of "often" and "sometimes" responses in urban respondents combined (46.1%) was 11.5% higher than that in rural respondents (34.7%).

FIGURE 3.8
Evaluation of Authenticity of Food Quality Certification and Labeling Information by Rural and Urban Respondents

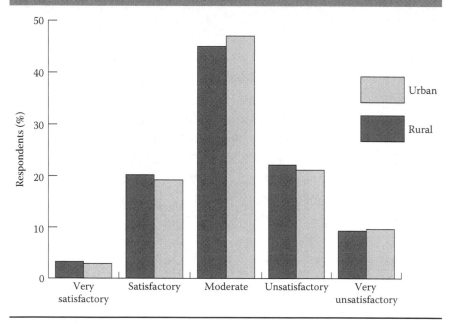

3.4.2 Consumer Food Complaints and Evaluation of Complaint Channel Accessibility

Data of consumer food complaints used in this book are derived from statistical information of food complaints accepted by the CCA during 2007–2011. Data of evaluation of complaint channel performance were collected by a survey conducted among 4289 respondents in 12 provinces, autonomous regions, and municipalities.

3.4.2.1 Changes in Number of Consumer Food Complaints The CCA is a semiofficial nongovernmental organization possessing legal personality. Its main aim is to perform social supervision of goods and services, protect the legitimate rights and interests of consumers, guide consumers in rational, scientific consumption, and promote the healthy development of a market economy. Since 2005, the CCA has been developing a statistical system of consumer complaints. Table 3.5 shows the number of food complaints accepted by the CCA during 2007–2011.

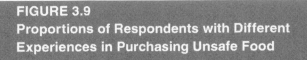

FIGURE 3.9
Proportions of Respondents with Different
Experiences in Purchasing Unsafe Food

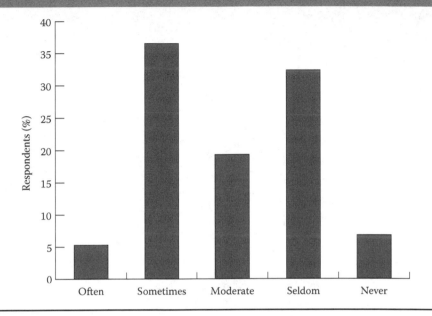

TABLE 3.5
Number of Consumer Food Complaints Accepted by
the CCA during 2007–2011

	2007	2008	2009	2010	2011	Average Annual Growth Rate (%)
Number of complaints	36,815	46,249	36,698	34,789	39,082	1.5
Increase over the previous year (%)	−12.57	25.63	−20.65	−5.20	12.34	−0.09

Source. Statistical information of food complaints accepted by the CCA during 2007–2011.

In 2007, 36,815 consumer food complaints were accepted by the CCA, 12.57% lower than the number of complaints collected in 2006 (42,106). In 2008, when the melamine milk powder incident occurred, the number of consumer food complaints increased sharply, representing an increase of 25.6% over 2007. The number decreased in 2009 and 2010, but increased again in 2011, representing an increase of 12.34% over 2010.

	TABLE 3.6 Composition of Consumer Food Complaints Accepted by the CCA in 2011	
Item	*Number of Complaints*	*Proportion (%)*
Quality	24,860	63.61
Safety	1,362	3.48
Price	2,307	5.90
Quantity	2,549	6.52
Advertising	393	1.01
Counterfeiting	940	2.41
Fake	395	1.01
Marketing contract	984	2.43
Personal dignity	46	0.12
Others	5,246	13.42
Total	39,082	100.00

Source. Statistical information of food complaints accepted by the CCA in 2011.

3.4.2.2 Major Compositions of Consumer Food Complaints The CCA accepted 39,082 consumer food complaints in 2011 (Table 3.6). Overall, 67.09% of the complaints were related to food quality and safety; 24,860 of these were related to food quality (accounting for 63.61%) and 1,362 were related to food safety (accounting for 3.48%). Major food safety problem complaints included incomplete information of labeling; unauthorized alterations of production date (which results in food spoilage within the labeled shelf life); inclusion of foreign bodies; use of inferior materials; excessive use of food additives and sweeteners; use of chemical additives not intended for food processing; illegal use of chemical substances outside the usable range specified by food regulations; improper control of pathogenic microorganisms (which results in frequent food contamination); excessive chemical fertilizer and pesticide residues; heavy metal and other harmful substances in agricultural products; and illegal use of antibiotics, hormones, and other harmful substances.

3.4.2.3 Methods Adopted to Solve Food Problems As shown in Figure 3.10, of the 4,289 respondents from 12 provinces, autonomous regions, and municipalities, 48.81% would "negotiate with the vendor" when having purchased unqualified food. This is a direct and rapid way to address the immediate problem, but is prone to disputes and not conducive to the elimination of such food safety problems. Furthermore, 21.78% of the respondents would "simply treat it as bad luck," while 10.32%, 13.34%, 3.41%, and 2.34% of the

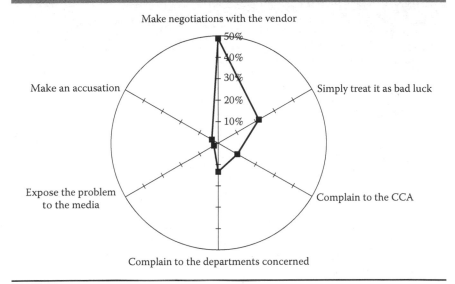

FIGURE 3.10
Major Behaviors Adopted by Consumers to Solve Food Problems

respondents would complain to the CCA, complain to the industrial and commercial departments or other departments concerned, make an accusation, and expose the problem to the media, respectively. Overall, 37.60% of urban respondents would complain to the CCA or the departments concerned, make an accusation, or expose the problem to the media, which was 13.3% more than the proportion of rural respondents. However, in general, "making negotiations with the vendor" and "simply treating it as bad luck" were the major ways adopted by the respondents to solve the food problems, instead of complaining to the CCA.

3.4.2.4 Evaluation of Complaint Channel Accessibility As shown in Figure 3.11, 32.30%, 27.21%, and 25.84% of respondents considered the CCA, the departments concerned, and the court, or the media, to be "moderately accessible," "less accessible," and "poorly accessible," respectively, in terms of complaining of food problems. Only 2.89% and 11.76% of respondents considered them to be "highly accessible" and "accessible."

As shown in Figure 3.12, 32.01% and 32.67% of urban and rural respondents, respectively, considered the departments concerned, including the CCA, to be "moderately accessible." The proportions of "less accessible" and "poorly accessible" were higher than those of "highly accessible" and

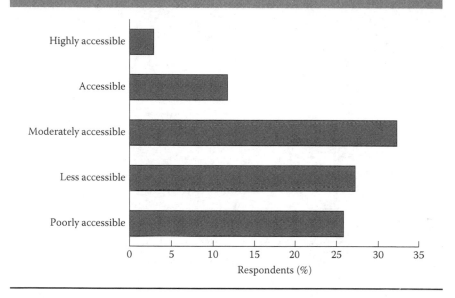

FIGURE 3.11
Evaluation of Complaint Channel Accessibility

FIGURE 3.12
Evaluation of Complaint Channel Accessibility
by Urban and Rural Respondents

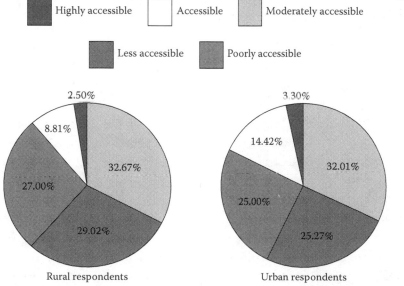

"accessible" in both urban and rural respondents. The major difference was that the total proportion of "highly accessible" and "accessible" was 6.4% higher in urban respondents (17.72%) than in rural respondents (11.31%).

3.5 Main Problems of Food Quality and Safety Supervision in Circulation

As described earlier, remarkable results have been achieved in food safety supervision in circulation, and the overall level of food quality and safety is stable and shows a positive trend. However, deep-seated problems have not been fundamentally resolved. Small-scale, decentralized agricultural production prevails in China, and food safety risks generated in production are often reflected in circulation. Moreover, because of the low requirements of market admittance, a large number of food businesses exist in circulation. The majority of them are small- and medium-sized businesses, and small workshops, street vendors, and small restaurants abound. This small-scale, decentralized, and disordered pattern has become a serious problem. A considerable proportion of operators involved in food circulation lacks the necessary equipment, facilities, and standardized management and thus does not meet the basic requirements for transacting food business. Unfortunately, some operators have a poor sense of integrity and self-discipline and deliberately break the law. As for consumers, there are large differences in living standards between regions, urban and rural areas and groups; low-income groups have a low standard of consumption and low consumer safety awareness and thus are more likely to become victims of substandard food. These problems are deep-seated factors that have long-term effects on Chinese food safety. Finding new ways to effectively monitor, prevent, or eliminate these problems to avoid the harm of substandard food on the health of consumers is an important indicator of the government's ability to govern. As a matter of fact, three main problems of food quality and safety supervision in circulation in China exist. These problems will be discussed in Sections 3.5.1 through 3.5.3.

3.5.1 Complex Composition of Operators in Food Circulation

Operators in food circulation have a very complex composition, which further increases the difficulty in monitoring circulation. The complexity is mainly manifested in a few major aspects described in Sections 3.5.1.1 through 3.5.1.3

3.5.1.1 Complex Food Business Patterns   Operators in food circulation include food stores, supermarkets, and wholesale markets, as well as food enterprises, individual businesses, and farmers' professional cooperatives. These various business patterns result in a complex situation. Circulation connects production and consumption and has an important role in guaranteeing food safety. The complex food business patterns make food safety and quality problems more likely to occur in circulation.

3.5.1.2 The Majority Are Small- and Medium-Sized Businesses   All over China, operators engaged in food circulation comprised 0.516 million enterprises, 4.699 million individual businesses, and 3000 farmers' professional cooperatives, accounting for 9.89%, 90.05%, and 0.06% of total operators, respectively, by the end of 2011 (Table 3.7). Clearly, individual businesses are the absolute majority. As decentralized small businesses, they widely spread in urban and rural areas throughout the country, especially in rural areas and remote mountainous areas. The individual businesses have relatively poor performance in safety management level and a relatively poor sense of integrity and self-discipline, which increases the difficulty in supervision and law enforcement.

3.5.1.3 Rapid Growth of Small- and Medium-Sized Businesses   As shown in Table 3.7, during the five-year period between 2006 and 2011, the number of operators in food circulation increased from 2.885 million in 2006 to 5.218 million in 2011, representing an average annual growth of 12.58%. Despite this rapid growth, the individual businesses-based pattern has not changed, but has been further enhanced. In 2011, individual businesses accounted for 90.05% of the total operators in food circulation, 2% higher than the 88.01%

TABLE 3.7
Composition of Operators in Food Circulation in 2006 and 2011

	Total Number of Operators (million)	Number of Enterprises (million)	Number of Individual Businesses (million)	Number of Farmers' Professional Cooperative (million)
2006	2.885	0.346	2.539	–
2011	5.218	0.516	4.669	0.003
Average annual growth rate (%)	12.580	8.330	13.000	–

Source.   SAIC.

in 2006. The increase in the number and proportion of individual businesses makes it more difficult to supervise food quality and safety in circulation.

3.5.2 Unbalanced Development of Regulatory Capacity

The imbalance in the development of regulatory capacity is mainly reflected in the three aspects described in Sections 3.5.2.1 through 3.5.2.3.

3.5.2.1 Insufficient Law Enforcement Resources Specialized regulatory agencies are insufficient. As of December 2011, there were 31 provincial, 332 municipal, and 2,831 county industrial and commercial bureaus, as well as 238,000 elementary industrial and commercial agencies, in China. However, some provincial, municipal, and county industrial and commercial bureaus have not yet set up independent food safety regulatory agencies. Insufficient law enforcement resources are allocated to food safety supervision. Elementary industrial and commercial agencies are the primary undertaker of food safety regulation and law enforcement in circulation. At present, law enforcement officers from 238,000 elementary industrial and commercial agencies have to undertake regulatory enforcement tasks involving 103 laws, 201 regulations, and 124 rules. In addition to food safety supervision, elementary industrial and commercial agencies have to undertake more than 10 tasks, such as market inspection, annual inspection, comprehensive management of public security, and special rectification. In particular, in remote and mountainous areas, some industrial and commercial agencies have less than 10 employees, but govern an area of hundreds of kilometers and perform inspections and regulatory enforcement by walking and cycling. There is a clear contradiction between the arduous regulatory tasks and staff shortages in the elementary industrial and commercial agencies. Furthermore, a serious lack of professional administrative staff exists. According to incomplete statistics on regulatory capacity in 2010 provided by relevant agencies, dedicated food safety supervision staff in provincial, municipal, and county industrial and commercial bureaus all over China only included 300 food professionals. It is obvious that the talent pool cannot meet the needs of the actual regulatory tasks at hand.

3.5.2.2 Insufficient Funds Vertical management is implemented for industrial and commercial departments below the provincial level. In some provinces, as municipal and county governments provide insufficient financial support to food safety supervision in circulation, it is difficult for industrial and commercial departments to work effectively. The implementation of the

Food Safety Law provides a clear division of responsibilities in food safety supervision. Health departments withdraw completely from the circulation. As a result, the industrial and commercial departments have to undertake more tasks of food safety supervision in circulation. However, the funds provided for food safety supervision have not increased with the increase in supervision tasks, which makes it difficult for the industrial and commercial departments to effectively perform their duties.

3.5.2.3 Insufficient Instruments and Equipment   The industrial and commercial departments are faced with a large funding gap, especially in detection. For example, the industrial and commercial systems all over China were equipped with only 3474 integrated food detectors by the end of 2010. To put this into perspective, if each industrial and commercial agency were to be equipped with one detector, there would be a shortfall of more than 17,000 detectors. According to the official document *A Notice on Further Strengthening the Supervision of Dairy Quality and Safety* issued by the Chinese State Council in 2010, the industrial and commercial departments should strengthen sampling inspection of dairy products and infant formula milk powder. By the end of 2010, there were 1,666,500 dairy traders and 707,300 infant formula milk powder traders in China. Based on the actual demand, an annual fund of 558 million yuan is needed for sampling of dairy products, 433 million yuan for sampling of food containing milk, 1365 million yuan for sampling of infant formula milk powder, and 3367 million yuan for sampling of other foods. In total, an annual fund of 5723 million yuan is needed to fulfill these demands. However, the actual funds appropriated by local governments are insufficient, which has directly resulted in the insufficiency in regulatory instruments and equipment, and the consequent inability to meet the requirements of food safety supervision.

3.5.3 Unbalanced Development of Regulatory Work

The imbalance in the development of regulatory work is mainly represented in Sections 3.5.3.1 through 3.5.3.3.

3.5.3.1 Law Enforcement Needs to Be Improved   The general situation of food safety in circulation is stable and has maintained a positive trend in China. However, the situation remains very difficult. The development of regulatory work is unbalanced among regions. The regulatory work is insufficient in some regions. In particular, with respect to the food safety emergencies occurring in recent years, these incidents occurred in circulation and caused a threat to the health of the consumers, as well as economic losses, though they were not rooted in circulation. This indicates that sourcing management and self-regulation of traders, as well as local regulation and law enforcement, has not been effectively implemented.

3.5.3.2 Supervision of Food Business Operations Needs to Be Enhanced   The industrial and commercial departments have undertaken many tasks for food safety supervision in circulation. However, implementation of the responsibilities and obligations of food business operators remains a weakness in current and future food safety supervision in circulation. Concerted efforts are required to establish and improve a long-term mechanism for food safety supervision in circulation.

3.5.3.3 Food Safety Emergency Response Capacity in Circulation Needs to Be Further Reinforced   Today, public opinion information on food safety spreads over many channels at a very high speed and has a wide range of influence and powerful magnification effect; therefore, it should be taken seriously. However, current food safety emergency plans in circulation are still not perfect. No public opinion monitoring and response system has been established. A large gap exists between the current response to public opinion on food safety and the general requirements of "active prevention and early intervention" and "early discovery, early reporting, and early management." This gap must be closed.

It should be noted that China's food safety supervision system underwent a major reform in March 2013. As a result of the reform, it has become the responsibility of food and drug regulatory authorities to supervise food safety in circulation. The industrial and commercial administrative departments will no longer undertake this function. However, the reform does not mean the resolution of problems in food safety supervision in circulation. The problems still exist objectively, and the food and drug regulatory authorities have to undertake these long-term and arduous tasks. The reform of China's food safety supervision system will be analyzed later in this book.

4

Quality and Safety of Import and Export Food

Along with an increasingly sophisticated division of modern international food industry labor, as well as the further development of international trade, the traditional production mode has been phased out and the food chain consistently extended. However, the probability of food contamination is increasing in the delivery of the extended food chain, leading to increased food safety risks. The high speed, wide scope, and large impact of food contamination spread through international trade on human health have become common problems faced by numerous countries. The food safety of any country is closely related to international food trade, and no country can stand aloof in the context of economic globalization. This chapter focuses on the quality and safety of imported and exported Chinese food and changes during 2006–2011 based on an analysis of the Chinese scale and structure of international food trade.*

4.1 Quality and Safety of Exported Food

Since the reform and opening up, agricultural production and the food industry have experienced rapid development in China, putting a complete end to its long history of food supply shortages (Zhang 2010). Since the 1990s, both the total food exports and pass rates of exported food have continued to increase in China. China plays an indispensable role in regulating the global food supply and demand and in guaranteeing the food supply of the world.

---

*Foods are classified according to the Standard International Trade Classification (SITC) of the United Nations in this chapter and consist of food and live animals (SITC0), beverages and tobacco (SITC1), and animal and vegetable oils, fats, and waxes (SITC4).

4.1.1 Overview of Exported Food

The basic features of food export development in China are discussed in Sections 4.1.1.1 through 4.1.1.3.

4.1.1.1 Total Exports Have Been Expanding   As shown in Table 4.1, food exports from China have increased from US$7.905 billion in 1991 to US$53.341 billion in 2011. This represents an increase of 5.7 times over two decades. Food exports from China have had an average annual growth rate of 14.0% since the beginning of the new century. Because they were severely affected by the 2008 global financial crisis, in 2009, food exports from China

TABLE 4.1

Chinese Exported Food Composition during 1991–2011 According to the Standard International Trade Classification

Year	Food and Live Animals (US$ billion)	Beverages and Tobacco (US$ billion)	Animal and Vegetable Oils, Fats, and Waxes (US$ billion)	Total Export (US$ billion)	Annual Growth Rate (%)
1991	7.226	0.529	0.150	7.905	11.15
1992	8.309	0.720	0.139	9.168	15.98
1993	8.399	0.901	0.205	9.505	3.68
1994	10.015	1.002	0.495	11.512	21.12
1995	9.954	1.370	0.454	11.778	2.31
1996	10.231	1.342	0.376	11.949	1.45
1997	11.075	1.049	0.647	12.771	6.88
1998	10.513	0.975	0.307	11.795	−7.64
1999	10.458	0.771	0.132	11.361	−3.68
2000	12.282	0.745	0.116	13.143	15.69
2001	12.777	0.873	0.111	13.761	4.70
2002	14.621	0.984	0.098	15.703	14.11
2003	17.531	1.019	0.115	18.665	18.86
2004	18.864	1.214	0.148	20.226	8.36
2005	22.480	1.183	0.268	23.931	18.32
2006	25.723	1.193	0.373	27.289	14.03
2007	30.743	1.397	0.303	32.442	18.88
2008	32.762	1.529	0.574	34.865	7.47
2009	32.628	1.641	0.316	34.585	−0.80
2010	41.148	1.906	0.355	43.410	25.52
2011	50.495	2.276	0.570	53.341	22.88

Source.   *China Statistical Yearbook,* 2011; the United Nations COMTRADE database.

have decreased for the first time since 2000. However, along with the gradual recovery of the global economy, food exports from China quickly resumed rapid development, and the average annual growth rate has continued to hit record highs in recent years.

4.1.1.2 Export Structure Has Been Gradually Optimized Analysis of the Chinese food export structure during 2006–2011 revealed that the main exported foods were aquatic products, vegetables, fruits, meat, and meat preparations (Table 4.2). A brief overview of this analysis is provided in the subsequent paragraphs.

The proportion of fruit and vegetable exports has continued to increase, and sugar preparations and honey exports have experienced rapid growth. Exports of fruit and vegetable from China increased from US$8.87 billion in 2006 to US$19.135 billion in 2011, more than doubling in five years. The proportion of these items in food exports increased from 32.50% in 2006 to 35.87% in 2011. In recent years in China, fruits and vegetables have had sustained high yields, and improvement in refrigeration technology, logistics, and the supply chain have supported the rapid growth of these exports (Chen 2008). Furthermore, sugar preparations and honey exports rapidly increased from US$0.713 billion in 2006 to US$1.7 billion in 2011, increasing by US$0.987 billion in five years; their advantage as high value-added exports is becoming increasingly apparent.

Exports of animal fats, feeding stuff for animals, processed animal and vegetable oils, and waxes have sharply increased. The most significant growth was seen in exports of animal oils and fats. The exports of animal oils and fats increased from US$0.025 billion in 2006 to US$0.145 billion in 2011, representing a five-year increase of 4.9 times. Feed exports increased from US$0.535 billion in 2006 to US$2.085 billion in 2011, representing a five-year increase of nearly three times. Exports of processed animal and vegetable oils, fats, and waxes increased from US$0.048 billion in 2006 to US$0.171 billion in 2011, representing a striking five-year increase of 253.35%.

Cereal exports have had a slow growth, and fixed vegetable oils and fats exports have shown negative growth. Cereal exports from China increased slowly from US$1.542 billion in 2006 to US$1.652 billion in 2011. The proportion of cereal exports in total food exports decreased from 5.65% in 2006 to 3.10% in 2011. This pattern is closely related to the large population and scarce land resources of China. Cereal exports from China are unlikely to increase significantly in the future, and cereal imports may increase instead to protect the domestic supply and demand balance. As domestic demand for vegetable oils and fats continues to grow in China,

TABLE 4.2
Chinese Food Exports by Category and Structural Changes in 2006 and 2011

Food category	2006		2011		Increase or Decrease in 2011 Relative to 2006	
	Value of Exports (US$ billion)	Proportion (%)	Value of Exports (US$ billion)	Proportion (%)	Increase or Decrease in Value (US$ billion)	Increase or Decrease in Proportion (%)
Total value of exports	27.290	100.00	53.341	100.00	26.051	95.46
Food and live animals chiefly for food	25.722	94.24	50.495	94.67	24.773	96.30
1. Live animals chiefly for food	0.333	1.22	0.571	1.07	0.238	71.39
2. Meat and preparations	2.031	7.44	2.957	5.54	0.925	45.55
3. Dairy products and eggs	0.189	0.69	0.282	0.53	0.093	49.20
4. Fish, crustaceans, and mollusks, and preparations thereof	8.949	32.79	16.969	31.81	8.020	89.61
5. Cereals and cereal preparations	1.542	5.65	1.652	3.10	0.110	7.13
6. Vegetables and fruit	8.870	32.50	19.135	35.87	10.265	115.72
7. Sugar, sugar preparations, and honey	0.713	2.61	1.700	3.19	0.986	138.27
8. Coffee, tea, cocoa, spices, and manufactures thereof	1.171	4.29	2.431	4.56	1.260	107.60

9. Feedstuff for animals (not including unmilled cereals)	0.535	1.96	2.085	3.91	1.551	289.95
10. Miscellaneous edible products and preparations	1.389	5.09	2.714	5.09	1.325	95.34
Beverages and tobacco	1.194	4.37	2.276	4.27	1.082	90.69
1. Beverages	0.628	2.30	1.135	2.13	0.507	80.82
2. Tobacco and tobacco products	0.566	2.07	1.141	2.14	0.575	101.64
Animal and vegetable oils, fats, and waxes	0.373	1.37	0.570	1.07	0.197	52.92
1. Animal oils and fats	0.025	0.09	0.145	0.27	0.121	492.19
2. Fixed vegetable oils and fats	0.300	1.10	0.254	0.48	−0.046	−15.39
3. Processed animal and vegetable oils, fats, and waxes	0.048	0.18	0.171	0.32	0.123	253.35

Source. The United Nations COMTRADE database.

vegetable oils and fats exports declined from US$0.3 billion in 2006 to US$0.254 billion in 2011, with the proportion in food exports decreasing from 1.10% to 0.48%.

4.1.1.3 Exported Countries Are Basically Stable Major countries/regions receiving food exports from China in 2006, in order of value of exports, were Japan (US$7.545 billion, 27.65%), the European Union (EU; US$3.530 billion, 12.94%), the United States (US$3.369 billion, 12.35%), the Association of Southeast Asian Nations (ASEAN; US$2.761 billion, 10.12%), Hong Kong (US$2.507 billion, 9.19%), Republic of Korea (US$2.487 billion, 9.11%), and Russia (US$0.803 billion, 2.94%). In 2006, food exports from China to these seven regions amounted to US$23.002 billion, accounting for 84.29% of the annual food exports of China, whereas those to Japan, the EU, and the United States accounted for more than 50% of the annual food exports.

Major countries/regions receiving food exports from China in 2011, in order of value of exports, were Japan (US$9.980 billion, 18.71%), ASEAN (US$9.054 billion, 16.97%), the EU (US$6.429 billion, 12.05%), the United States (US$6.159 billion, 11.55%), Hong Kong (US$5.361 billion, 10.05%), Republic of Korea (US$3.590 billion, 6.73%), and Russia (US$1.789 billion, 3.35%). In 2011, food exports from China to these seven regions amounted to US$42.362 billion, accounting for 79.42% of the annual food exports of China. Variations in food exports from China to these major countries/ regions during 2006–2011 are shown in Figure 4.1.

As shown in Figure 4.1, Japan was the largest food export market for China during 2006–2011; ASEAN surpassed the EU and the United States to become the second largest food export market since 2009 and is on a trend to overtake Japan. Chinese food export market shares have remained relatively stable in the EU, the United States, and Hong Kong, and food exports to Republic of Korea and Russia have had a slow growth, with declining export market shares in the two countries.

4.1.2 Quality and Safety of Exported Food

According to the *Quality and Safety of Food in China* published by the Information Office of the State Council of the People's Republic of China in August 2007, the pass rates of Chinese export food held steady at 99% in the years preceding the first half of 2007 (Information Office of the State Council of the People's Republic of China 2007). Relevant data published after that time also demonstrated that the pass rates of Chinese exported

FIGURE 4.1
Food Exports from China to Major Countries during 2006–2011

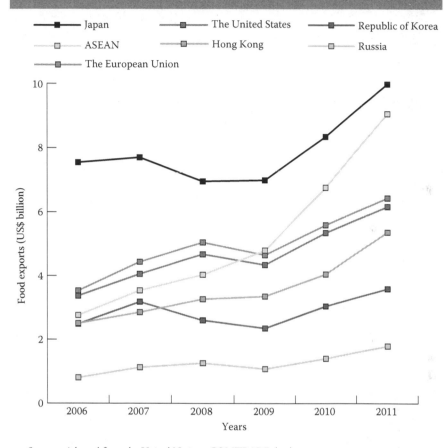

Source. Adapted from the United Nations COMTRADE database.

food have remained above 99%. Although Chinese exported foods have maintained high pass rates, the quality and safety of Chinese exported food is still a concern in the global food market due to differences in inspection standards of food quality and safety in various countries, as well as complex influences of political, economic, and other factors.

4.1.2.1 Major Countries Rejecting Chinese Food Exports Detention/recall of Chinese food exported to the United States, Japan, the EU, Republic of Korea, and Canada since 2006 is summarized in Table 4.3. In 2011, a total of 1628 food batches exported from China to the United States, Japan, the EU,

TABLE 4.3
Detention/Recall of Chinese Exported Food during 2006–2011

Issuing Country (Region)	Issuing Institution (Organizations)	2006 Number of Batches	2006 Proportion (%)	2007 Number of Batches	2007 Proportion (%)	2008 Number of Batches	2008 Proportion (%)	2009 Number of Batches	2009 Proportion (%)	2010 Number of Batches	2010 Proportion (%)	2011 Number of Batches	2011 Proportion (%)	Annual Growth Rate (%)
The United States	Food and Drug Administration (FDA)	765	52.18	870	54.48	707	42.90	1058	46.04	867	46.56	620	38.08	−4.11
Republic of Korea	Republic of Korea's National Veterinary Research and Quarantine Service and Food and Drug Administration	–	–	–	–	262	15.90	609	26.50	366	19.66	378	23.22	12.99
Japan	Ministry of Health, Labour and Welfare (MHLW)	502	34.24	461	28.87	284	17.23	304	13.23	247	13.27	213	13.08	−15.76
The European Union	Food and Feed Committee	199	13.57	266	16.66	395	23.97	226	9.83	270	14.50	332	20.39	10.78
Canada	Food Inspection Agency	–	–	–	–	–	–	101	4.40	112	6.02	85	5.22	−8.26
Total		1466	100.00	1597	100.00	1648	100.00	2298	100.00	1862	100.00	1628	100.00	2.12

Source. Adapted from WTO/TBT-SPS Notification and Enquiry of China.

Republic of Korea, and Canada were detained or recalled by relevant agencies of the respective countries. Of them, 620 batches were detained by the US Food and Drug Administration (FDA). With regard to the other four countries, 332 batches were recalled by the EU Food and Feed Committee, 378 batches were detained by the Republic of Korea's National Veterinary Research and Quarantine Service and Food and Drug Administration, 213 batches were detained by the Japan Ministry of Health, Labour and Welfare (MHLW), and 85 batches were recalled by the Canadian Food Inspection Agency.

4.1.2.2 Major Categories of Detained/Recalled Chinese Exported Food Information regarding exported food categories detained or recalled in 2011 is shown in Table 4.4. In the 1628 detained/recalled batches of food exported to the United States, Republic of Korea, Japan, the EU, and Canada in 2011, the top five categories were aquatic products (433 batches, 26.60%), meat and meat preparations (239 batches, 14.68%), vegetables and vegetable preparations (211 batches, 12.96%), cereals and cereal preparations (139 batches, 8.54%), and dried fruits and nuts (119 batches, 7.31%).

4.1.2.3 Major Causes of Detention/Recall As shown in Table 4.5, the major causes of detention/recall of Chinese exported food in 2011 were unacceptable quality (296 batches, 18.18%), excessive pesticide and veterinary drug residues (261 batches, 16.03%), noncompliance with quarantine regulations (193 batches, 11.86%), unacceptable food additives (154 batches, 9.46%), microbiological contamination (149 batches, 9.15%), unacceptable certificate (114 batches, 7.00%), unacceptable labeling (105 batches, 6.45%), and excessive biotoxin contamination (95 batches, 5.84%).

Specifically, the major causes of detention/recall in recent years, according to the Administration of Quality Supervision, Inspection and Quarantine (AQSIQ) Research Centre for International Inspection and Quarantine Standards and Technical Regulations in 2012, are unacceptable quality and microbial contamination for aquatic products, noncompliance with quarantine regulations for meat and meat preparations, excessive pesticide residues for vegetables and vegetable preparations, and excessive biotoxin contamination for dried fruits and nuts. Use of nonfood additives was a major cause of detention/recall for plant-derived Chinese medicinal materials and fruit products, and inclusion of unapproved genetically modified ingredients and excessive heavy metal content was a major cause for cereal products. In addition, unacceptable food additives were the major cause of detention/recall for baked goods, spices, and fruit juices and drinks. The major causes of detention/recall of Chinese exported food during 2009–2011 are shown in Table 4.5.

TABLE 4.4
Detention/Recall of Chinese Exported Food by Category in 2011

Food Category	Number of Batches	Proportion (%)	Food Variety	Number of Batches	Proportion (%)
Aquatic products	433	26.60	Fish products	197	12.10
			Other aquatic products	80	4.91
			Shrimp products	75	4.61
			Aquatic products	56	3.44
			Crab products	18	1.11
			Shellfish products	6	0.37
			Sea grass and algae	1	0.06
Meat and meat preparations	239	14.68	Other meat and meat preparations	72	4.42
			Poultry and poultry preparations	62	3.81
			Cooked meat	41	2.52
			Beef and beef preparations	27	1.66
			Pork and pork preparations	19	1.17
			Mutton and mutton preparations	13	0.80
			Casings	5	0.31
Vegetables and vegetable preparations	211	12.96	Vegetables and vegetable preparations	184	11.30
			Edible mushrooms	27	1.66
Cereals and cereal preparations	139	8.54	Cereal products	110	6.76
			Cereal-based processed products	22	1.35
			Beans (dry)	5	0.31
			Cereals	2	0.12
Dried fruits and nuts	119	7.31	Dried fruits	87	5.34
			Dried fruits (nuts), roasted seeds, and nuts (cooked)	32	1.97
Fats and oils	62	3.81	Oilseeds	57	3.50
			Vegetable oils	2	0.12
			Animal oils (fats)	2	0.12
			Virgin vegetable oils	1	0.06
Live animals chiefly for food	45	2.76	Bone	27	1.66
			Hair	9	0.55
			Other animal products	3	0.18
			Live animals	3	0.18
			Down	2	0.12
			Skins	1	0.06
Cakes and biscuits	44	2.70	Cakes and biscuits	44	2.70
Sugars	34	2.09	Sugar, candy, chocolate, and other preparations containing cocoa	33	2.03
			Raw sugar	1	0.06

(Continued)

TABLE 4.4
(Continued) Detention/Recall of Chinese Exported Food by Category in 2011

Food Category	Number of Batches	Proportion (%)	Food Variety	Number of Batches	Proportion (%)
Other processed food	33	2.03	Other processed foods	30	1.84
			Other fruit preparations	3	0.18
Eggs and egg preparations	30	1.84	Nonedible fresh eggs	29	1.78
			Egg preparations	1	0.06
Cans	29	1.78	Canned aquatic products	14	0.86
			Canned fruits	7	0.43
			Canned vegetable	5	0.31
			Other cans	1	0.06
			Canned beverages	1	0.06
			Canned nuts and beans	1	0.06
Chinese herbal medicines	29	1.78	Plant-derived Chinese medicinal materials	19	1.17
			Animal-derived Chinese medicinal materials	10	0.61
Feedstuff	29	1.78	Feedstuff	29	1.78
Spices	28	1.72	Spices	28	1.72
Vegetable spices	25	1.54	Vegetable spices	25	1.54
Beverages	25	1.54	Beverages	25	1.54
Foods for special dietary uses	16	0.98	Foods for special dietary uses	11	0.68
			Health food	4	0.25
			Foods containing medicines	1	0.06
Tea	16	0.98	Tea	16	0.98
Bee products	15	0.92	Bee products	15	0.92
Dairy products	12	0.74	Milk and milk products	12	0.74
Candied fruit	6	0.37	Candied fruit	6	0.37
Plant-derived foods	6	0.37	Fruits	4	0.25
			Beans	2	0.12
Other plant derived foods	3	0.18	Other plant-derived foods	3	0.18
Other animal-derived foods	1	0.06	Other animal-derived foods	1	0.06
Total amount	1628	100.00		1628	100.00

Source.   2011 Analysis Report on Detention/Recall of Chinese Exports, AQSIQ Research Centre for International Inspection and Quarantine Standards and Technical Regulations.

TABLE 4.5
Major Causes of Detention/Recall of Chinese Exported Food during 2009–2011

	2009		2010		2011	
Causes of Detention/Recall	*Number of Batches*	*Proportion (%)*	*Number of Batches*	*Proportion (%)*	*Number of Batches*	*Proportion (%)*
Unacceptable quality	172	7.5	362	19.4	296	18.18
Excessive pesticide and veterinary drug residues	233	10.1	223	12.0	261	16.03
Noncompliance with quarantine regulations	20	0.9	101	5.4	193	11.86
Unacceptable food additives	622	27.1	293	15.7	154	9.46
Microbiological contamination	441	19.2	212	11.4	149	9.15
Unacceptable certificate	265	11.5	172	9.2	114	7.00
Unacceptable labeling	62	2.7	103	5.5	105	6.45
Excessive biotoxin contamination	64	2.8	92	4.9	95	5.84
Other unacceptable items	–	–	–	–	78	4.79
Nonfood additives	–	–	–	–	72	4.42
Contaminant	63	2.7	69	3.7	53	3.26
Genetically modified ingredients	20	0.9	42	2.3	29	1.78
Irradiation	–	–	–	–	9	0.55
Unacceptable packaging	20	0.9	56	3.0	6	0.37
Quarantine objects	–	–	–	–	3	0.18
Illegal imports	–	–	–	–	3	0.18
Illegal trade	296	12.9	111	6.0	3	0.18
Pests	–	–	–	–	3	0.18
Threat to humans	14	0.6	16	0.9	1	0.06
Chemical property-related problems	3	0.1	–	–	1	0.06
Noncompliance with storage and transportation regulations	3	0.1	10	0.5	–	–
Total	2298	100.0	1862	100.0	1628	100.00

Source. 2011 Analysis Report on Detention/Recall of Chinese Exports and Analysis of TBT Risks and Early Warning (production), issue 10 of 2011 and issue 9 of 2009, AQSIQ Research Centre for International Inspection and Quarantine Standards and Technical Regulations.

Further analysis revealed that excessive pesticide and veterinary drug residues, unacceptable food additives, and microbial contamination were the most basic causes of detention/recall of Chinese exported food. As shown in Table 4.5, the proportion of food batches detained/recalled for excessive pesticide and veterinary drug residues in the total detained/recalled export food batches was on the rise during 2009–2011. Excessive pesticide and veterinary drug residues are still the most important factor negatively affecting food exports from China. The proportion of food batches detained/recalled for unacceptable food additives in the total detained/recalled exported food batches decreased from 27.1% in 2009, through 15.7% in 2010, to 9.46% in 2011. The proportion of food batches detained/recalled for microbial contamination in the total detained/recalled exported food batches decreased significantly from 19.2% in 2009 to 9.15% in 2011. In general, the problems of unacceptable food additives and microbial contamination in Chinese food exports appear to have been initially controlled.

4.1.2.3.1 Excessive Pesticide and Veterinary Drug Residues   Excessive pesticide residues in vegetables and vegetable preparations and excessive veterinary drug residues in aquatic products, and meat and meat preparations are a particularly prominent problem in food exports from China. Abuse or misuse of pesticides and veterinary drugs in agricultural production and animal husbandry is the major cause of excessive pesticide and veterinary drug residues. At present, 500–600 cases of exported food detention/recall occur annually due to excessive pesticide residues, with resulting economic losses of more than 7 billion yuan (Wang 2008). A total embargo on export of frozen spinach from China to Japan was implemented due to excessive pesticide residues in several batches in 2002. In 2002, British health authorities detected chloramphenicol residues in honey from China, which led to the EU's decision to prohibit all honey product exports from China according to the rapid alert mechanism. This also triggered chain reactions in the United States and Japan. As a result, the leading position of China in the world honey market was swiftly taken over by Argentina. The EU did not lift the embargo on honey product exports from China conditionally until July 2004. Due to the "malachite green" incident in June 2005, Japan has enhanced detection of malachite green in eel products from China, which has severely disrupted eel exports from China. Japan began implementing the positive list system and had set more than 10,000 maximum allowable residue limits for 734 pesticides, veterinary drugs, and feed additives in 2006. This has led to a sharp decline in exports from China to Japan. A total of 261 batches of food exports were detained or recalled because of excessive pesticide and veterinary drug residues in 2011, accounting for 16.03% of the total batches.

4.1.2.3.2 Unacceptable Food Additives   A total of 177 food export batches from China to the United States were reported for unacceptable food additives in 2008, accounting for 23.3% of the total batches detained or recalled in that year. Of the 177 batches, 39 batches of candy and candied fruit were detained for containing unsafe colorants and other additives. In total, 154 batches of food exports were detained or recalled for excessive food additives in 2011, accounting for 9.46% of the total batches being detained or recalled that year. Vegetables and vegetable preparations (41 batches), dried fruits and nuts (20 batches), and sugars (13 batches) were the top three categories detained/recalled for excessive food additives. In general, the exported food safety problems related to food additives mainly occur during food processing and are caused by illegal use of additives.

4.1.2.3.3 Microbial Contamination   Bacterial contamination is the most common type of microbial contamination in the food industry. *Escherichia coli* and *Salmonella* contamination are common forms of bacterial contamination. Raw material contamination, inadequate sterilization, and improper storage or improper temperature control during food processing, storage, and distribution may result in excessive bacteria and pathogens (China Food Safety Strategy Research Group of Development Research Center of the State Council 2005). Unfortunately, this is a very prominent problem in food exports from China. In October 2006, frozen cooked octopus and frozen squid from China were detained by the Kobe Inspection Office and Osaka Inspection Office due to coliform-positive results (Yue 2010). In April 2007, 257 batches of food exported from China were detained by the FDA, of which 137 batches were rejected because of detection of *Salmonella* and other prohibited ingredients. A total of 149 export food batches from China were detained or recalled due to microbial contamination in 2011, accounting for 9.15% of the total detained/recalled batches. Of them, 130 batches were detained or recalled for bacterial contamination, which is the most important cause of microbial contamination. Moreover, aquatic products, meat and meat preparations, and vegetables and vegetable preparations were most frequently detained/recalled for bacterial contamination.

4.1.2.3.4 Excessive Biotoxin Contamination   Aflatoxin, a highly toxic substance, was classified as a Class I carcinogen by the Cancer Research Institution of World Health Organization (WHO) in 1993. As the influences of weather and water are not considered during production, processing, and transportation, dried fruits and nuts exported from China are rejected for excessive aflatoxin contamination upon arrival at the destination after verified as being acceptable in China. Trade frictions in food export caused by aflatoxin contamination have become an increasingly serious problem. Aflatoxin was detected in 63 batches of peanuts exported from China by the EU in 2006. Each returned container caused a direct economic loss of 60,000 yuan, resulting in a loss of more than US$600,000 at one time for a company in Shandong. In 2011, 95 food batches exported from China to the United States, Japan, Republic of Korea, the EU, and Canada were detained or recalled due to excessive biotoxin contamination, accounting for 5.84% of the total batches. Excessive biotoxin contamination is a serious problem, especially in dry fruits and nuts, oils, and fats.

4.1.2.3.5 Heavy Metal Contamination   Serious agricultural environmental pollution is the main cause for heavy metals and other hazards exceeding permissible limits. In recent years, heavy metal contamination has become an outstanding problem in cereal products, vegetables, and aquatic products exported from China. In July 2007, the *Wall Street Journal* reported the problem of heavy metal contamination in vegetables, corn, and other plant-derived foods produced in China, which has raised concerns of the EU, Japan, and other importing countries. In 2011, 15 batches of food exported from China to the United States, Japan, Republic of Korea, the EU, and Canada were rejected or detained for heavy metal contamination, of which 12 batches were cereal products.

4.2 Quality and Safety of Imported Food

Since the reform and opening up, food import to China has developed rapidly, as total food imports have continuously increased, product structures have been constantly adjusted, and the market structure has been optimized. This has played an important role in balancing and regulating the food supply and demand and meeting diverse consumer demand within China. It has also promoted the reconstruction of the Chinese food industry through demonstration effects. However, along with the rapid development and increased total imports, the amount of unacceptable imported food has increased. The causes of nonconformity are diverse and complicated.

4.2.1 Overview of Imported Food

The basic features of food import development in China are discussed in Sections 4.2.1.1 through 4.2.1.3.

4.2.1.1 Rapidly Growing Total Imports   Changes in food imports of China since 1991 are shown in Table 4.6. Food imports of China experienced rapid

TABLE 4.6

Chinese Imported Food Composition during 1991–2011 According to the Standard International Trade Classification

Year	Food and Live Animals (US$ billion)	Beverages and Tobacco (US$ billion)	Animal and Vegetable Oils, Fats, and Waxes (US$ billion)	Total Exports (US$ billion)	Annual Growth Rate (%)
1991	2.799	0.200	0.719	3.718	−16.90
1992	3.146	0.239	0.525	3.910	5.16
1993	2.206	0.245	0.502	2.953	−24.48
1994	3.137	0.068	1.809	5.014	69.79
1995	6.132	0.394	2.605	9.131	82.11
1996	5.672	0.497	1.697	7.866	−13.85
1997	4.304	0.320	1.684	6.308	−19.81
1998	3.788	0.179	1.491	5.458	−13.47
1999	3.619	0.208	1.367	5.194	−4.84
2000	4.758	0.364	0.977	6.099	17.42
2001	4.976	0.412	0.763	6.151	0.85
2002	5.238	0.387	1.625	7.250	17.87
2003	5.960	0.490	3.000	9.450	30.34
2004	9.154	0.548	4.214	13.916	47.26
2005	9.388	0.783	3.370	13.541	−2.69
2006	9.994	1.041	3.936	14.971	10.56
2007	11.500	1.401	7.344	20.245	35.23
2008	14.051	1.920	10.486	26.458	30.69
2009	14.827	1.954	7.639	24.420	−7.70
2010	21.570	2.428	9.017	33.015	35.20
2011	28.771	3.685	11.629	44.084	33.53

Source. *China Statistical Yearbook*, 2011; the United Nations COMTRADE database.

development during 1991–2000. The total food imports increased from US$3.718 billion in 1991 to US$6.099 billion in 2000; this climb represented an increase of 64.04% despite the constant fluctuations during this period. Food imports of China have experienced a faster and more stable development since the beginning of the new century, exceeding US$10 billion for the first time in 2004, then rapidly exceeded US$20 billion in 2007, US$30 billion in 2010, and US$40 billion in 2011. Despite the influences by domestic inflation and the global financial crisis, as well as some degree of fluctuation in particular years, food imports of China increased rapidly on the whole during 2001–2011. The average annual growth rate was 20.71%, higher than the average annual growth rate of exports (14.0%) over the same period and higher than the average annual growth rate of imports during 1991–2000.

4.2.1.2 Trade Structure of Imports The major imported food categories of China are currently vegetable fats and oils, aquatic products, and vegetables and fruits. The basic features of the imported food structure of China during 2006–2011 are discussed in the subsequent paragraphs.

Imports of live animals (chiefly for food), meat and meat preparations, dairy products, and birds' eggs have increased dramatically. Of these, imports of live animals experienced the largest increase, from US$0.063 billion in 2006 to US$0.377 billion in 2011, representing nearly a fivefold increase. Chinese consumers' trust in domestic meat and meat preparations, dairy products, and birds' eggs has plummeted due to frequent safety incidents in the Chinese meat and dairy industries in recent years. This impact has increased China's import of these products. Imports of meat and meat preparations increased from US$0.728 billion in 2006 to US$3.405 billion in 2011, representing an increase of 3.68 times over the five-year period; their proportion in total food imports increased by 2.86%. Imports of dairy products and birds' eggs increased from US$0.566 billion in 2006 to US$2.655 billion in 2011, representing an increase of 3.69 times over the five-year period; their proportion in total food imports increased from 3.78% in 2006 to 6.02% in 2011 (Table 4.7).

Imports of beverages, miscellaneous edible products and preparations, coffee, tea, cocoa, spices, and manufactures thereof also rose significantly. As shown in Table 4.7, beverage imports increased from US$0.577 billion in 2006 to US$2.547 billion in 2011, representing an increase of 3.4 times over the five-year period; their proportion in total food imports increased from 3.86% in 2006 to 5.78% in 2011. Imports of coffee, tea, cocoa, spices, and manufactures thereof increased from US$0.247 billion in 2006 to US$0.888 billion in 2011, representing an increase of 3.4 times over the five-year period. Imports of miscellaneous edible products and preparations increased from US$0.685 billion in 2006 to US$2.441 billion in 2011, representing an increase of 2.56 times over the five-year period; their proportion in total food imports increased by 1.00%. The rapid growth of imports of beverages, miscellaneous edible products and preparations, coffee, tea, cocoa, spices, and manufactures thereof is mainly a result of the continuous increase in domestic consumption and changes in the consumption structure in China.

Aquatic products still account for a large proportion of food imports. As shown in Table 4.7, imports of aquatic products increased from US$3.157 billion in 2006 to US$5.754 billion in 2011, representing a growth rate of 82.27%, while their proportion in total food imports decreased from 21.09% in 2006 to 13.05% in 2011. Although the growth of aquatic product imports has been relatively slow, these products still account for a large proportion

TABLE 4.7
Chinese Food Imports by Category and Structural Changes in 2006 and 2011

Food Category	2006		2011		Increase or Decrease in 2011 Relative to 2006	
	Value of Imports (US$ billion)	Proportion (%)	Value of Imports (US$ billion)	Proportion (%)	Value of Imports (US$ billion)	Proportion (%)
Total import value	14.971	100.00	44.084	100.00	29.113	194.46
Food and live animals chiefly for food	9.994	66.76	28.771	65.26	18.777	187.87
1. Live animals chiefly for food	0.063	0.42	0.377	0.85	0.313	493.46
2. Meat and meat preparations	0.728	4.86	3.405	7.72	2.676	367.59
3. Dairy products and birds' eggs	0.566	3.78	2.655	6.02	2.088	368.65
4. Fish, crustaceans, mollusks, and preparations thereof	3.157	21.09	5.754	13.05	2.597	82.27
5. Cereals and cereal preparations	0.909	6.07	2.367	5.37	1.458	160.38
6. Vegetables and fruits	1.720	11.49	5.527	12.54	3.807	221.31
7. Sugar, sugar preparations, and honey	0.621	4.15	2.150	4.88	1.529	246.40
8. Coffee, tea, cocoa, spices, and manufactures thereof	0.247	1.65	0.888	2.01	0.641	259.54
9. Feedstuff for animals (not including unmilled cereals)	1.298	8.67	3.209	7.28	1.912	147.35
10. Miscellaneous edible products and preparations	0.685	4.58	2.441	5.54	1.755	256.09
Beverages and tobacco	1.041	6.95	3.685	8.36	2.644	254.10
1. Beverages	0.577	3.86	2.547	5.78	1.970	341.15
2. Tobacco and tobacco products	0.463	3.09	1.138	2.58	0.675	145.62
Animal and vegetable oils, fats, and waxes	3.936	26.29	11.629	26.38	7.692	195.15
1. Animal oils and fats	0.173	1.15	0.427	0.97	0.254	147.21
2. Fixed vegetable oils and fats	3.475	23.21	10.258	23.27	6.782	195.15
3. Processed animal and vegetable oils, fats, and waxes	0.288	1.92	0.944	2.14	0.656	227.46

Source. The United Nations COMTRADE database.

of food imports. The main reason for the slow growth is the significantly increased domestic supply capacity.

4.2.1.3 Market Structure of Imports Major countries/regions importing food to China in 2006 were ASEAN (US$4.697 billion, 31.38%), the United States (US$1.529 billion, 10.21%), the EU (US$1.308 billion, 8.74%), Russia (US$1.281 billion, 8.56%), Australia (US$0.728 billion, 4.86%), Peru (US$0.625 billion, 4.17%), and New Zealand (US$0.547 billion, 3.65%). In 2006, food imports to China from these seven countries/regions amounted to US$10.715 billion, accounting for 71.57% of the annual food imports of China (Figure 4.2).

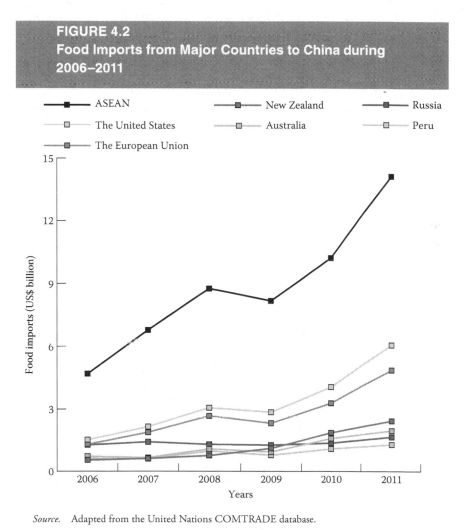

FIGURE 4.2
Food Imports from Major Countries to China during 2006–2011

Source. Adapted from the United Nations COMTRADE database.

Major countries/regions importing food to China in 2011 were ASEAN (US$14.145 billion, 32.09%), the United States (US$6.080 billion, 13.79%), the EU (US$4.883 billion, 11.08%), New Zealand (US$2.447 billion, 5.55%), Australia (US$1.975 billion, 4.48%), Russia (US$1.672 billion, 3.79%), and Peru (US$1.304 billion, 2.96%). In 2011, food imports to China from these seven countries/regions amounted to US$32.506 billion, accounting for 73.74% of the annual food imports of China. The food import market structure of China was changed in 2011, and ASEAN, the United States, and the EU remain the top three countries in the food import market of China, with their market shares steadily growing. New Zealand surpassed Russia and Australia to rank fourth in the food import market of China. The Australian market share first rose, then fell, and finally became stable. Russian and Peruvian market shares exhibited a gradual declining trend (Figure 4.2).

4.2.2 Quality and Safety of Imported Food

Along with its economic and social development, domestic food consumption of China is becoming more diverse, directly leading to the continuous growth of food imports in both scale and food categories. However, the amount of unacceptable imported food has also increased. According to data published by the AQSIQ of the People's Republic of China, 1543, 1753, and 1818 batches of unacceptable imported food were rejected in 2009, 2010, and 2011, respectively. The unacceptable fraction of imported food has increased year by year.

4.2.2.1 Main Origins of Unacceptable Imported Food According to relevant data published by the AQSIQ, the top ten source countries of unacceptable imported food in 2011 were Taiwan, the United States, Malaysia, France, Australia, Spain, Republic of Korea, Thailand, Vietnam, and Japan (Table 4.8). A total of 1231 food batches imported from the ten countries were found unacceptable, accounting for 67.71% of the total unacceptable batches in food imports in 2011. Source distribution of unacceptable import food to China in 2011 is detailed in Table 4.8.

4.2.2.2 Major Causes of Unacceptability of Imported Food According to the *2011 Information of Unacceptable Import Food and Cosmetics* published by the AQSIQ, the major causes of unacceptability of imported food to China in 2011 were microbial contamination, unacceptable food additives, unacceptable quality, unacceptable labeling, inability to provide relevant certificates,

TABLE 4.8
Countries of Source of Unacceptable Imported Food to China in 2011

Countries of Source of Unacceptable Imported Food	Number of Batches	Proportion (%)
Taiwan, China	302	16.61
The United States	213	11.72
Malaysia	146	8.03
France	118	6.49
Australia	91	5.01
Spain	87	4.79
Korea	79	4.35
Thailand	74	4.07
Vietnam	63	3.47
Japan	58	3.19
Italy	53	2.92
India	48	2.64
New Zealand	47	2.59
Germany	45	2.48
The United Kingdom	40	2.20
Canada	36	1.98
Indonesia	30	1.65
Belgium	25	1.38
The Philippines	21	1.16
Norway	21	1.16
Czech	19	1.05
Hong Kong, China	18	0.99
Austria	15	0.83
Sri Lanka	15	0.83
Turkey	12	0.66
Sweden	11	0.61
Brazil	9	0.50
The Netherlands	9	0.50
Slovenia	8	0.35
Mexico	7	0.41
Chile	7	0.39
Ireland	6	0.39
Switzerland	6	0.33
Singapore	6	0.44
Hungary	6	0.33
Argentina	5	0.28
Poland	5	0.28
Denmark	5	0.28
Russia	4	0.22

(Continued)

TABLE 4.8
(Continued) Countries of Source of
Unacceptable Imported Food to China in 2011

Countries of Source of Unacceptable Imported Food	Number of Batches	Proportion (%)
Ecuador	3	0.17
Kenya	3	0.17
South Africa	3	0.17
Portugal	3	0.17
Slovakia	3	0.17
China[a]	3	0.17
Ethiopia	2	0.11
Pakistan	2	0.11
Croatia	2	0.11
Peru	2	0.11
Myanmar	2	0.11
Syria	2	0.11
Israel	2	0.11
Macau, China	2	0.11
Egypt	1	0.06
Belarus	1	0.06
Iceland	1	0.06
Faroe Islands	1	0.06
Finland	1	0.06
Gambia	1	0.06
Ghana	1	0.06
Cambodia	1	0.06
Bangladesh	1	0.06
Nepal	1	0.06
Senegal	1	0.06
Armenia	1	0.06
Iraq	1	0.06
Iran	1	0.06
Total	1818	100.00

Source.   *2011 Information of Unacceptable Import Food and Cosmetics*, AQSIQ Import and Export Food Safety Authority.

[a] Three of the unacceptable imported food batches in 2011 were produced in China; in actuality, they were not imported food but rejected imported food returned to China via the import procedure.

heavy metal contamination, excessive pesticide and veterinary drug residues, excessive biotoxin contamination, nonconformity between cargo and certificates, unacceptable packaging, irradiation, inclusion of unapproved genetically modified ingredients, originating from epidemic area, and failure in sensory testing (Table 4.9). In particular, 522 batches were detained/recalled

TABLE 4.9
Major Causes of Unacceptability of Imported Food to China in 2011

Causes of Imported Food Unacceptability	Number of Batches	Proportion (%)
Microbiological contamination	522	28.71
Unacceptable food additives	406	22.33
Unacceptable quality	211	11.61
Unacceptable labeling	186	10.23
Inability to provide relevant certificates	178	9.79
Heavy metal contamination	74	4.07
Excessive pesticide and veterinary drug residues	62	3.41
Excessive biotoxin contamination	40	2.20
Nonconformity between cargo and certificates	37	2.04
Unacceptable packaging	23	1.27
Irradiation	11	0.61
Unapproved genetically modified ingredients	10	0.55
Originating from epidemic area	10	0.55
Failure in sensory testing	6	0.33
Inclusion of banned or restricted substances	5	0.28
Pests	2	0.11
Others	35	1.93
Total	1818	100.00

Source. *2011 Information of Unacceptable Import Food and Cosmetics*, AQSIQ Import and Export Food Safety Authority.

---

for microbial contamination, accounting for the largest fraction, 28.71%, of all unacceptable batches of imported food in 2011.

4.2.2.2.1 Microbial Contamination WHO estimates that hundreds of millions of people suffer from food-borne diseases worldwide each year, and the cause of such diseases is most commonly consuming food and drinking water contaminated with pathogenic microorganisms (Zhou et al. 2008). According to the WHO, the annual global economic losses caused by food spoilage alone are tens of billions of dollars (Lu 2012). Analysis of the relevant statistical data of unacceptable imported food to China in 2011 revealed that the main cause of unacceptability of imported food was microbial contamination, especially by *Escherichia coli* (Table 4.10).

4.2.2.2.2 Excessive Food Additives or Use of Banned Additives Illegal use of food additives or use of nonfood substances is not only a major problem leading to detention/recall of Chinese food exports but also a major cause of unacceptability of imported food to China. It has become the most

TABLE 4.10
Specific Microbial Contamination-Related
Causes of Unacceptability of Imported Food
to China in 2011

No.	Specific Cause of Imported Food Unacceptability[a]	Number of Batches	Proportion (%)
1	Excessive *Escherichia coli*	218	11.99
2	Excessive colony-forming units	216	11.88
3	Excessive molds	81	4.46
4	Excessive *Vibrio parahaemolyticus*	15	0.83
5	Detection of *Listeria monocytogenes*	11	0.61
6	Excessive *Staphylococcus aureus*	11	0.61
7	Excessive yeast	16	0.88
8	Detection of *Salmonella*	7	0.39
9	Illegal use of *Lactobacillus sporogenes*	4	0.22
10	Detection of *Enterobacter sakazakii*	1	0.06
11	Failure of commercial sterilization	1	0.06

Source. 2011 Information of Unacceptable Import Food and Cosmetics, AQSIQ Import and Export Food Safety Authority.

[a] Sometimes one kind of imported food was unacceptable because of several kinds of microbes, so the sum of the third column is more than 522 in this table.

prominent problem in global food safety. Specific food additive-related causes of unacceptability of food imported to China in 2011 are shown in Table 4.11. As shown in Table 4.11, colorants, preservatives, plasticizers, and color fixatives are most likely to be abused as food additives.

4.2.2.2.3 Heavy Metal Contamination   Heavy metal contaminants are difficult to wash away with water runoff or to decompose by microorganisms in the soil. They can cause damage to the environment and pose a threat to human health through enrichment of organisms in the food chain. At present, heavy metals of special concern are mercury, cadmium, chromium, arsenic, and lead. Specific heavy metal-related causes of unacceptability of food imported to China in 2011 are shown in Table 4.12.

4.2.2.2.4 Excessive Pesticide and Veterinary Drug Residues or Use of Banned Pesticides and Veterinary Drugs   Many countries, especially developed countries, attach great importance to the problem of pesticide and veterinary drug residues and have very strict regulations regarding pesticide and veterinary drug residue limits in agricultural products and animal-derived foods. However, illegal use of pesticides and veterinary drugs and use of prohibited drugs is still a common

TABLE 4.11
Specific Food Additives-Related Causes of Unacceptability of Imported Food to China in 2011

No.	Specific Cause of Imported Food Unacceptability[a]	Number of Batches	Proportion (%)
1	Colorants	110	6.05
2	Preservatives	89	4.90
3	Plasticizers	56	3.08
4	Color fixatives	46	2.53
5	Processing aids for food industry	25	1.38
6	Antioxidants	20	1.10
7	Bleaches	20	1.10
8	Nutrient supplements	18	0.99
9	Sweeteners	11	0.61
10	Bulking agents	11	0.61
11	Humectants	11	0.61
12	Unacceptable food additives	7	0.38
13	Thickening agents	5	0.28
14	Chelating agents and clarifying agents	3	0.17
15	Buffer agents and neutralizing agents	2	0.11
16	Acidity regulators	2	0.11
17	Others	21	1.16

Source.   2011 Information of Unacceptable Import Food and Cosmetics, AQSIQ Import and Export Food Safety Authority.

[a] Sometimes one kind of imported food was unacceptable because of several kinds of food additives, so the sum of the third column is more than 406 in this table.

TABLE 4.12
Specific Heavy Metal-Related Causes of Unacceptability of Imported Food to China in 2011

No.	Specific Cause of Imported Food Unacceptability[a]	Number of Batches	Proportion (%)
1	Excessive copper	39	2.15
2	Excessive arsenic	18	0.99
3	Excessive lead	11	0.61
4	Excessive cadmium	4	0.22
5	Excessive manganese	2	0.11
6	Excessive chromium	2	0.11

Source.   2011 Information of Unacceptable Import Food and Cosmetics, AQSIQ Import and Export Food Safety Authority.

[a] Sometimes one kind of imported food was unacceptable because of several kinds of heavy metal, so the sum of the third column is more than 74 in this table.

TABLE 4.13
Specific Pesticide and Veterinary Drug Residues-Related Causes of Unacceptability of Imported Food to China in 2011

No.	Specific Cause of Imported Food Unacceptability	Number of Batches	Proportion (%)
1	Detection of ractopamine	30	1.65
2	Detection of malachite green	13	0.72
3	Detection of nitrofurazone metabolite	7	0.39
4	Detection of chloramphenicol	6	0.33
5	Excessive phorate	2	0.11
6	Dichloro-diphenyl-trichloroethane	1	0.06
7	Detection of enrofloxacin	1	0.06
8	Excessive triazophos	1	0.06
9	Detection of carbofuran	1	0.06

Source. *2011 Information of Unacceptable Import Food and Cosmetics*, AQSIQ Import and Export Food Safety Authority.

phenomenon. It is also a major cause of unacceptability of food imported to China. Specific pesticide and veterinary drug residue-related causes of unacceptability of food imported to China in 2011 are shown in Table 4.13.

4.2.2.2.5 Unacceptable Labeling   Standardization of Chinese labeling for importing food to China is an important means to ensure food safety and hygiene (Chen 2007). According to the *General Standard for Food Labeling of China*, imported food labels should include basic contents, such as food name, net content, list of ingredients, country of origin, production date, shelf life, and information of the Chinese distributor. Major labeling problems in food imported to China in 2011 were false food name, withholding information of ingredients, inconsistency between label content and product, and missing basic contents. A total of 186 batches were detained/recalled for labeling problems in 2011, accounting for 10.23% of the total unacceptable batches in that year.

4.3 A Brief Comparison of the Quality and Safety between Imported and Exported Food and Basic Conclusions

On the one hand, ensuring the quality and safety of imported food is imperative for protecting the food safety of Chinese urban and rural residents. On the other hand, it is important to ensure the quality and safety of exported food to guarantee the health of people in importing countries, which is also essential for building a responsible image and developing an international reputation. Therefore, it is necessary to compare the quality and safety between imported and exported food.

4.3.1 Similarities and Differences

4.3.1.1 Similarities in Quality and Safety between Imported and Exported Food   As shown in Table 4.14, similarities in major problems in quality and safety between imported and exported food are excessive microbial contamination, unacceptable food additives, excessive pesticide and veterinary drug residues, and heavy metal contamination.

4.3.1.2 Differences in Quality and Safety between Imported and Exported Food   Major problems in quality and safety also have obvious differences between imported and exported food. According to the frequent food safety incidents occurring in China and the major causes of unacceptability of Chinese export food in 2009, 2010, and 2011 (Table 4.5), major problems in quality and safety of Chinese exported food are improper and illegal use of pesticides, veterinary drugs, and food additives, leading to excessive pesticide and veterinary drug residues and unacceptable food additives. Most of the problems occur during the course of production. The direct cause is mismanagement in production and processing. Thus, human and management factors play the leading role.

As shown in Table 4.9, the major causes of unacceptability of food imported to China in 2011 were pesticide and veterinary drug residues,

TABLE 4.14
Comparison of Major Causes of Unacceptability between Chinese Exported and Imported Food in 2011

	Exported Food		Imported Food	
	Number of		Number of	
Major Unacceptable Item	*Unacceptable Batches*	*Proportion (%)*	*Unacceptable Batches*	*Proportion (%)*
Microbiological contamination	149	9.15	522	28.71
Substandard food additives	154	9.46	406	22.33
Excessive pesticide and veterinary drug residues	261	16.03	62	3.41
Heavy metal contamination	21	1.29	74	4.07

Source.   2011 Analysis Report on Detention/Recall of Chinese Exports, AQSIQ Research Centre for International Inspection and Quarantine Standards and Technical Regulations; *2011 Information of Unacceptable Import Food and Cosmetics*, AQSIQ Import and Export Food Safety Authority.

unacceptable food additives, heavy metal contamination, and especially microbial contamination. However, microbial contamination is a possible problem in production, processing, storage, transportation, and marketing and cannot be completely avoided. Therefore, the safety risks of imported food are mainly caused by natural factors.

4.3.2 Basic Conclusions

Based on the preliminary analysis of quality and safety problems in Chinese imported and exported food, several basic conclusions can be drawn. These conclusions are described in Sections 4.3.2.1 through 4.3.2.3.

4.3.2.1 The Quality and Safety of Exported Food Is Adequately Protected   Despite the increasingly strict technical trade measures implemented by other countries, the pass rates of Chinese exported food remained stable at >99%. An overall upward trend was observed in detained/recalled batches of Chinese food exported to the United States, Korea, Japan, the EU, and Canada during 2006–2011, representing an annual growth rate of 4.01%. However, compared with the average annual growth rate of Chinese food exports of 14.0% since the beginning of the new century, the current growth rate of detained/recalled batches is significantly lower and presents an improving trend.

4.3.2.2 It Is Rather Difficult to Stabilize and Enhance the Quality and Safety of Exported Food   Major problems in the quality and safety of Chinese exported food are improper and even illegal use of pesticides, veterinary drugs, and food additives, which leads to excessive pesticide and veterinary drug residues and unacceptable food additives. Most risks lie in the course of agricultural and industrial production and are caused by acts of bad faith of producers due to the pursuit of commercial interests. Unfortunately, it is difficult to completely solve these problems in the short term.

4.3.2.3 Safety Risks of Imported Food May Further Increase   Although the overall quality and safety of imported food are at relatively high levels, food safety risks are increasing with the growing demand for imported food. There were 1543, 1753, and 1818 batches of unacceptable import food in 2009, 2010, and 2011, respectively. The unacceptable fraction of imported food has also increased year by year. Therefore, establishing technical trade measures integrating technology, regulations, and culture against imported food in the World Trade Organization framework by drawing on international experience may need to be considered as an important agenda.

5

Food Safety Risk Assessment and Risk Characteristic Analysis at the Macrolevel

In this chapter, Chinese food safety risks during 2006–2012 were assessed at the national macrolevel from the perspective of management using econometric models based on the scientificity and availability of data. The actual state of food safety risks was analyzed on this basis to provide a panoramic description of the real changes, main characteristics, and trends of food quality and safety in China. Furthermore, this chapter aims to reveal the main characteristics of food safety in terms of root causes and to investigate principal contradictions in food safety risk prevention in China.

5.1 Food Safety Risk Assessment from the Perspective of Management

Food safety risk assessment is an important part of the risk analysis framework. Currently, food safety risks are mainly assessed at the technical level. However, because of the wide variety and different properties of available foods, different technical standards and methods are applied to risk assessment of different foods. In reality, there is no doubt that it is very important to conduct safety risk assessments and to publish these assessment results for a single variety or category of food. However, with the well-developed information network that is available today, food safety incidents regarding pork, vegetables, milk powder, edible vegetable oils, and so on are being exposed one after another. With the large boost in network communication, people are more or less skeptical of food safety and doubt whether it is safe to eat food produced in China. Therefore, appropriate assessment of food safety risks from an overall point of view and response to social concern are important realistic problems that must be addressed.

5.1.1 Brief Review of Existing Research

In China, food safety risk assessment from the perspective of management started at the end of the last century. It was not until recent years that real development in this field was achieved. Li (2004) divided food safety into three dimensions, that is, quantity, quality, and sustainability, based on the connotation of food safety, and proposed a full set of specific indicators that reflected these three dimensions. The result was the development of a comprehensive assessment indicator system for food safety risks and determined weights of these indicators. Zou (2005) proposed specific indicators for food safety monitoring and inspection items, such as the pass rate in food inspection, the pass rate in monitoring of agricultural products used as raw materials, and the over standard ratios of residual pesticides in farm produce. Tang (2005) designed a four-layer alert indicator system of food safety, including 18 alert indicators, and performed a preliminary analysis of warning sign indexes based on the ideas in constructing the alert system of food safety and alert requirements for food safety. Han (2006) presented prediction models for microbial density and shelf life of fresh food on the basis of analyzing the supply chain of fresh food in the supermarket and simulated the model using the Monte Carlo method. The results of that study provided a conceptual model for microbial risk analysis and control of food. Li and Zhang (2007) identified and assessed possible food safety risks during the 2008 Beijing Olympic Games using the standard method of Australia and New Zealand—risk analysis matrix, based on data published by the Beijing Health Inspection and the Beijing Food Contaminant Monitoring Network, along with relevant data reported by the Ministry of Health.

In general, a large number of Chinese researchers have studied food safety risk assessment in China and have provided a number of different perspectives. However, deficiencies in this field remain. Primarily, no systematic and integrated assessments or analyses of food safety risks have been conducted based on different subsystems, including production, circulation, and consumption, throughout the food supply chain. Therefore, this chapter provides a quantitative and comparative analysis of food safety risk level during 2006–2012 in China using catastrophe models from the perspective of the entire food chain according to international experience.

5.1.2 Evaluation Methods from the Perspective of Management

Food safety risks can be assessed from the perspective of management using a variety of methods. For example, Zhou et al. (2009) proposed a Food Safety Comprehensive Index (FSCI) and constructed a comprehensive food safety assessment model using principal component analysis. In that study, food safety risk levels during 2000–2005 in China were assessed using the overall pass rate in food hygiene inspection, the pass rate in chemical testing, and other indicators and concluded that the overall food safety level in China improved steadily but remained at middle and lower levels during this period. However, as the authors pointed out, some indicators cannot be accurately represented by data using the above-mentioned method, and the use of an expert scoring method results in large uncertainties because of the impact of human factors. To overcome these shortcomings of the above-mentioned method, a quantitative analysis of Chinese food safety risks was conducted using catastrophe models in this chapter.

5.1.2.1 Principle of Catastrophe Models   Cusp, swallowtail, and butterfly are the most commonly used catastrophe models in the assessment of sociological and economic issues. Relevant details are shown in Table 5.1.

As shown in Table 5.1, the catastrophe function $f(x)$ represents the potential function of the state variable x in a system, and the coefficients of the state variables, *a, b, c, and d*, represent control variables of the state variable.

5.1.2.2 Catastrophe Progression Method   Catastrophe progression is the combination of a bifurcation equation of catastrophe models and a membership function of fuzzy mathematics. The key of this method lies in the deduction of the normalization equation based on the several varieties of bifurcation equations in R. Thom's elementary catastrophe theory, and the

TABLE 5.1
Three Commonly Used Catastrophe Models

Catastrophe Type	Number of State Variables	Number of Control Variables	Catastrophe Function
Cusp	1	2	$f(x) = x^4 + ax^2 + bx$
Swallowtail	1	3	$f(x) = (1/5)x^5 + (1/3)ax^3 + (1/2)bx^2 + cx$
Butterfly	1	4	$f(x) = (1/6)x^6 + (1/4)ax^4 + (1/3)bx^3 + (1/2)cx^2 + dx$

combination of the normalization equation with the membership function of fuzzy mathematics to deduce the catastrophe fuzzy membership function. By separately calculating the first- and second-order derivatives for the state variable x of the potential function, the following normalization Equations 5.1 through 5.3 could be derived for cusp, swallowtail, and butterfly catastrophe models.

$$x_a = \sqrt{a}; \; x_b = \sqrt[3]{b} \tag{5.1}$$

$$x_a = \sqrt{a}; \; x_b = \sqrt[3]{b}; \; x_c = \sqrt[4]{c} \tag{5.2}$$

$$x_a = \sqrt{a}; \; x_b = \sqrt[3]{b}; \; x_c = \sqrt[4]{c}; \; x_d = \sqrt[5]{d} \tag{5.3}$$

In the equations, x_a, x_b, x_c, and x_d are the corresponding x values of a, b, c, and d. The control variables, in descending order of importance, are ranked as follows: $a > b > c > d$. This method eliminates the subjectivity of researchers in giving "weight" to various indicators. A normalization equation is derived using the bifurcation point set equation in decomposed form and is then used to convert different morphological states of various control variables within the system into the same morphological state, that is, a morphological state represented by the state variable. Comprehensive assessment is then conducted using the normalization equation. Corresponding quantitative values are calculated for each control object of the same object using the normalization equation, and the value of the state variable is determined according to the relationship between corresponding control variables. The relationship between the control variables at the same level are complementary or noncomplementary. These are detailed in Table 5.2.

TABLE 5.2
Two-Dimensional Analysis of Catastrophe Types Based on "Complementary" and "Noncomplementary" Principles

Catastrophe Type Relationship between Control Variables	Cusp	Swallowtail	Butterfly
Complementary relationship	Complementary cusp catastrophe	Complementary swallowtail catastrophe	Complementary butterfly catastrophe
Noncomplementary relationship	Noncomplementary cusp catastrophe	Noncomplementary swallowtail catastrophe	Noncomplementary butterfly catastrophe

If complementary relationships exist among control variables, which compensate for the shortfall of one another, the value of the state variable should be the average of the state variables corresponding to each control variable. In noncomplementary relationships, control variables can neither replace each other, nor compensate for each other; thus, the minimum control variable becomes a "bottleneck," and the value of the state variable should be the value of the state variable corresponding to the minimum control variable (Chen et al. 2010).

5.1.2.3 Indicator System of Food Safety Risk Assessment Because of the complexity of food safety issues, numerically measurable indicators and factors that cannot be measured directly exist in assessment and analysis. Chen et al. (2010) designed a food safety risk assessment indicator system by drawing on the food safety alert indicator system and the comprehensive, sensitive, practical, exercisable, and dynamic principles of indicator design during food safety risk assessment. According to the method of Chen et al. (2011), our system included 11 assessment indicators reflecting food safety risks: sampling qualified rates of veterinary drug residues; pesticide residues in vegetables, aquatic products, and pigs (clenbuterol) in production; pass rates in national food quality checks; pass rates in regular health monitoring of drinking water; sampling qualified rates of food in circulation; number of food complaints received by the Consumers Association in circulation; and of consumers poisoned by food, food poisoning deaths, and poisoning incidents in consumption (Table 5.3).

5.2 Food Safety Risk Assessment during 2006–2012

Based on national-level macroeconomic indicators related to food safety risks and catastrophe models, risk assessment can be conducted on Chinese food safety in three subsystems, production, circulation, and consumption, during 2006–2012, and the overall food safety risk level in the corresponding year can be assessed on this basis.

5.2.1 Data Sources and Processing

Assessment of food safety risks with the above-mentioned indicators required collection and processing of data as outlined in Sections 5.2.1.1 and 5.2.1.2.

TABLE 5.3
Food Safety Risk Assessment Indicator Values

No.	Indicator	2006	2007	2008	2009	2010	2011	2012
1	Sampling qualified rates of veterinary drug residues (%)	75.00	79.20	81.7	99.5	99.6	99.6	99.7
2	Sampling qualified rates of pesticide residues in vegetables (%)	93.00	95.30	96.3	96.4	96.8	97.4	97.9
3	Sampling qualified rates of aquatic products (%)	98.80	99.80	94.7	96.7	96.7	96.8	96.9
4	Sampling qualified rates of pigs (clenbuterol) (%)	98.50	98.40	98.6	99.1	99.3	99.5	99.7
5	Pass rate in national food quality checks (%)	80.80	83.10	87.3	91.3	94.6	95.1	95.4
6	Pass rate in regular health monitoring of drinking water (%)	87.70	88.60	88.6	87.4	88.1	92.07	92.07[a]
7	Sampling qualified rate of food in circulation (%)	80.19[b]	80.19	93	93	93[b]	93[b]	93.06
8	Number of food complaints received by the National Consumers Association (n)	42,000	37,000	46,000	37,000	35,000	39,000	29,200
9	Number of consumers poisoned by food (persons)	18,063	13,280	13,095	11,007	7,383	8,324	6,685
10	Food poisoning deaths (persons)	196	258	154	181	184	137	146
11	Number of poisoning incidents (n)	596	506	431	271	220	189	174

Source. *China Health Statistics Yearbook, 2007–2013; China Statistical Yearbook; China Food Industry Yearbook; China Development Report on Food Safety, 2012.*

Data for 2012 were mainly adapted from data published in news reports or on government websites; however, since the pass rates in regular health monitoring of drinking water in 2012 could not be verified, data from 2011 were used instead.

[a] Data of sampling qualified rates of food in circulation in 2007–2009 and 2012 could be found in relevant literature and statistical material; however, since those in 2006, 2010, and 2011 were not available, an approximation method was used; specifically, data from 2007 were used for 2006, and data from 2009 were used for both 2010 and 2011.

5.2.1.1 Data Sources   Table 5.3 shows the food safety risk assessment indicator values during 2006–2012. Data were adapted from the *China Health Statistics Yearbook* (2007–2010), *China Statistical Yearbook*, *China Food Industry Yearbook*, and *China Development Report on Food Safety* (2012). Since relevant authorities had not officially published the relevant Statistical Yearbook at the time this study was conducted, data of 2012 were mainly collected from data published on government websites or in news reports. It should be noted that the final data in official publications of relevant authorities shall prevail.

5.2.1.2 Data Processing   Production, circulation, and consumption are represented as butterfly, swallowtail, and butterfly catastrophes, respectively, and all the three subsystems are in a complementary relationship. Therefore, the indicators must first be normalized to values between 0 and 1 and corrected by linear transformation. If an indicator is a positive indicator, the correction formula is as follows:

$$y_{ij} = \frac{\max P_{ij} - P_{ij}}{(\max P_{ij} - \min P_{ij})} \tag{5.4}$$

If an indicator is a negative indicator, the correction formula is as follows:

$$y_{ij} = \frac{P_{ij} - \min P_{ij}}{(\max P_{ij} - \min P_{ij})} \tag{5.5}$$

After the corresponding corrected data were obtained, a quantitative recursive algorithm was applied to each of the normalized indicator values using the normalization equation given earlier for the corresponding type of catastrophe. The processed data are shown in Table 5.4.

The total membership function value of catastrophe in each subsystem, that is, overall risk level in each food safety subsystem, was calculated according to the complementary and noncomplementary principles. As the subindicators of production, circulation, and consumption are complementary, the average of the corresponding catastrophe progression of control variables was taken as the total catastrophe membership function value. Details are shown in Table 5.5.

The above-mentioned steps were repeated and the state variable values from bottom to top level were integrated using the normalization equation until the "overall food safety risk values of the highest level" were obtained. These values could then be used to assess the food safety risks in each year between 2006 and 2012. The final overall food safety risk values during 2006–2012 are shown in Table 5.6.

TABLE 5.4 Corrected and Normalized Data								
Sector	*Indicator*	*2006*	*2007*	*2008*	*2009*	*2010*	*2011*	*2012*
Production	Sampling qualified rates of veterinary drug residues (%)	1.000	0.837	0.729	0.008	0.004	0.004	0.001
	Sampling qualified rates of pesticide residues in vegetables (%)	1.000	0.531	0.327	0.306	0.224	0.102	0.001
	Sampling qualified rates of aquatic products (%)	0.196	0.000	1.000	0.608	0.608	0.588	0.569
	Sampling qualified rates of pigs (clenbuterol) (%)	0.923	1.000	0.846	0.462	0.308	0.154	0.003
Circulation	Pass rate in national food quality checks (%)	1.000	0.842	0.555	0.281	0.055	0.021	0.001
	Pass rate in regular health monitoring of drinking water (%)	0.936	0.743	0.743	1.000	0.850	0.000	0.004
	Sampling qualified rate of food in circulation (%)	1.000	1.000	0.004	0.004	0.004	0.004	0.006
	Number of food complaints received by the National Consumers Association (*n*)	0.762	0.464	1.000	0.464	0.345	0.583	0.003
Consumption	Number of consumers poisoned by food (persons)	1.000	0.580	0.563	0.380	0.061	0.144	0.011
	Food poisoning deaths (persons)	0.488	1.000	0.140	0.364	0.388	0.002	0.074
	Number of poisoning incidents (*n*)	1.000	0.787	0.609	0.230	0.109	0.036	0.002

5.2.2 Risk Measurement Standard

Measurements of overall food safety risks in different years were obtained after performing complicated calculations. Food safety risks in China can be judged based on the risk measurement results. However, as the catastrophe progression method does not provide a method to determine threshold, the generally accepted standard methods, the extreme-mean method and the expert advice-based judgment method, were used to provide threshold evaluation criteria. A normalization standard for food safety risk levels was obtained using the above-mentioned methods (Tables 5.7 and 5.8).

TABLE 5.5
Data Processed by Catastrophe Progression

Sector	Indicator	Years						
		2006	2007	2008	2009	2010	2011	2012
Production	x_a	1.000	0.837	0.729	0.008	0.004	0.004	0.001
	x_b	1.000	0.531	0.327	0.306	0.224	0.102	0.001
	x_c	0.196	0.000	1.000	0.608	0.608	0.588	0.569
	x_d	0.923	1.000	0.846	0.462	0.308	0.154	0.003
	Total	0.780	0.592	0.726	0.346	0,286	0.212	0.142
Circulation	x_a	1.000	0.842	0.555	0.281	0.055	0.021	0.001
	x_b	0.936	0.743	0.743	1.000	0.850	0.000	0.004
	x_c	1.000	1.000	0.004	0.004	0.004	0.004	0.006
	x_d	0.762	0.464	1.000	0.464	0.345	0.583	0.003
	Total	0.925	0.762	0.576	0.437	0.314	0.152	0.004
Consumption	x_a	1.000	0.580	0.563	0.380	0.061	0.144	0.003
	x_b	0.488	1.000	0.140	0.364	0.388	0.000	0.011
	x_c	1.000	0.787	0.609	0.230	0.109	0.036	0.074
	Total	0.829	0.789	0.437	0.325	0.186	0.060	0.025

TABLE 5.6
Overall Food Safety Risk Values during 2006–2012

Indicator	2006	2007	2008	2009	2010	2011	2012
x_a	0.780	0.592	0.726	0.346	0.286	0.212	0.142
x_b	0.925	0.762	0.576	0.437	0.314	0.152	0.004
x_c	0.829	0.789	0.437	0.325	0.186	0.060	0.025
Overall risk value	0.937	0.875	0.832	0.700	0.624	0.496	0.258

TABLE 5.7
Classification Standards of Food Safety Risks in Subsystems

Objective Layer	Criterion Layer	Normalization Standards for Food Safety Risk Levels				
		Potential Risk	Mild Risk	Moderate Risk	Severe Risk	Crisis Range
Food safety risk level	Production	0–0.227	0.227–0.471	0.471–0.673	0.673–0.877	0.877–1.000
	Circulation	0–0.471	0.471–0.615	0.615–0.758	0.758–0.902	0.902–1.000
	Consumption	0–0.382	0.382–0.544	0.544–0.706	0.706–0.868	0.868–1.000

TABLE 5.8
Classification Standards of Overall Food Safety Risks

Risk Value	Safety Level	Food Safety Level Judgment
0.918–1.000	Level I	In crisis range; at the highest risk
0.848–0.918	Level II	In very severe risk range; at very high risk
0.778–0.848	Level III	In severe risk range; at high risk
0.708–0.778	Level IV	In moderate risk range; at moderate risk
0.500–0.708	Level V	In mild risk range; at low risk
0.000–0.500	Level VI	In safe range; with a potential risk

5.2.3 Risk Assessment Results

Thresholds of food safety risk level were distributed between 0 and 1. The closer the value was to 1, the higher the risk, the higher the fragility, and the lower the safety. The closer the value was to 0, the lower the risk, the lower the fragility, and the higher the safety. A food safety risk level of 0 meant that food safety conditions were optimal and the food safety system was in an ideal and optimal state. A food safety risk level of 1 meant that the food safety system was very fragile and was almost always in crisis. Food safety risk values and levels in production, circulation, and consumption during 2006–2012 were determined based on the above-mentioned assessment method (Table 5.9).

According to the classification standard of food safety risks in Table 5.8, general assessment results of food safety risks during 2006–2012 in China were obtained (Table 5.10).

TABLE 5.9
Risk Assessment of Food Safety in Subsystems during 2006–2012 in China

	Production		Circulation		Consumption	
Year	Risk Value	Risk Level	Risk Value	Risk Level	Risk Value	Risk Level
2006	0.780	Severe risk	0.925	Crisis range	0.829	Severe risk
2007	0.592	Mild risk	0.762	Moderate risk	0.789	Severe risk
2008	0.726	Moderate risk	0.576	Mild risk	0.437	Safe range
2009	0.346	Safe range	0.437	Safe range	0.325	Safe range
2010	0.286	Safe range	0.314	Safe range	0.186	Safe range
2011	0.212	Safe range	0.152	Safe range	0.060	Safe range
2012	0.142	Safe range	0.004	Safe range	0.025	Safe range

TABLE 5.10 **General Risk Assessment of Food Safety during** **2006–2012 in China**							
Year	*2006*	*2007*	*2008*	*2009*	*2010*	*2011*	*2012*
Overall risk value	0.937	0.875	0.832	0.700	0.624	0.496	0.258
Risk level	Level I	Level II	Level II	Level V	Level V	Level VI	Level VI
Range of risk	Crisis range	Very severe risk range	Very severe risk range	Mild risk range	Mild risk range	Safe range	Safe range

5.3 Analysis of Changes in Food Safety Risk during 2006–2012

Based on data in Tables 5.9 and 5.10, the food safety risk level and its change in trajectory in the overall Chinese food safety system and in the three subsystems, that is, food production, circulation, and consumption, can be presented, and the changes in food safety risks during 2006–2012 can be further analyzed on this basis.

5.3.1 General Characteristics of Food Safety Risks

As shown in Tables 5.9 and 5.10, the empirical results of this chapter revealed a clear decreasing trend in the overall risk in China's food safety system. The overall risk value in the food safety system was 0.496 in 2011, which was in a relatively safe range, despite various potential risks; it decreased to 0.258 in 2012, which was its lowest value and was relatively safe. However, no 100% safe foods exist, as inevitable potential risks exist even in the safe range. Therefore, it can be concluded from the study of this chapter that the level of food safety in China has been generally stable and gradually improving in recent years.

5.3.2 Risk Characteristics in Major Subsystems

The food safety risks in the three subsystems, production, circulation, and consumption, were in a relatively safe range in 2012, and changes in risk levels in these subsystems during 2006–2012 are discussed in Sections 5.3.2.1 through 5.3.2.3.

5.3.2.1 Production Risk value in production peaked at 0.780 in 2006, which was within the severe risk range. Analysis of the overall risk in subsequent years revealed an overall decreasing trend, despite a fluctuation in 2008, and the lowest level, 0.142, was achieved in 2012.

5.3.2.2 Circulation Risk value in circulation decreased from 0.925 in 2006 to 0.004 in 2012 and has remained in a relatively safe range since 2009.

5.3.2.3 Consumption Similar to that in circulation, risk value in consumption decreased from 0.829, within the severe risk range, in 2006 to 0.025, within the safe range, in 2012. It has remained in a relatively safe range since 2008.

Although the risk values in food production, distribution, and consumption all generally decreased during 2006–2012, the three subsystems, in descending order of risk value in 2012, are ranged as follows: production > consumption > distribution. Therefore, food safety in production should be the focus of government regulators. Of course, this is a conclusion drawn at the macrolevel. As the situation varies with regions and food varieties, food supervision should be carried out based on reality.

5.3.3 Future Trends of Food Safety Risks

Based on the macroeconomic environment of food safety in China, with the existing production technology and government regulations (especially with the implementation of structural reform in food safety supervision by the State Council in March 2013 and the downward movement of the focus of food safety regulatory power), it can be assumed, according to the study in this chapter, that food safety risks in China will remain generally stable and continue to gradually improve in the future. Although the possibility of rebound or even great fluctuations cannot be excluded in some food industry sectors or local production areas, this overall basic trend is unlikely to change, provided no irresistible, sudden catastrophic event occurs at a large scale.

5.3.4 Limitations of the Study

It should be noted that, limited by the availability of data and other objective factors, the accuracy of the Chinese food safety risk assessment during 2006–2012 has the following limitations: (1) In addition to the *China Statistical Yearbooks*, data for 2012 used in this chapter came from government websites and internet literature. The accuracy of the conclusions of this chapter depends, to some extent, on the authenticity of data cited. However, it is clear that authenticity of some data cannot be determined. (2) The missing data affect the scientificity of results. Specifically, in Table 5.3, because the pass rate of drinking water in regular health monitoring in 2012 could not be verified, it was approximated by data from 2011. Similarly, since data of sampling qualified rates of food in circulation in 2006, 2010, and 2011 were not available, data for 2006 were approximated by data from 2007, and data for both 2010 and 2011 were approximated by data from 2009. Pass rates of drinking water in regular health monitoring and sampling qualified rates of food in circulation are important indicators of food safety risks in circulation. The lack of data not only made it difficult to accurately calculate the trend of safety risks in circulation, it also affected the scientific calculation of the value of overall food safety risk. (3) Data used in the final calculation of overall food safety risk values and food safety risk values in the three subsystems are closely related to the Chinese food safety inspection standards, strictness of law enforcement, and consumer awareness of food safety. If the standard is increased, the risk value will increase correspondingly. Therefore, the Chinese food safety risk levels and range of risk determined in this chapter can only be used as a reference. The study described in this chapter provides only a relative trend of food safety risks and has obvious limitations. Nevertheless, the conclusion that "the level of food safety in China has been generally stable and gradually improving," drawn by using catastrophe models from the perspective of the entire food supply chain, is reliable.

5.4 Major Characteristics of Food Safety Risks at This Stage: Human Factors

Absolutely risk-free foods do not and cannot exist in the world. Despite many uncertain potential risks, food safety has been generally stable and gradually improving in recent years in China, which reflects the remarkable achievements made by China in food safety risk prevention and management. Only with a comprehensive and objective understanding of the main characteristics of food safety risks in China, can more effective prevention and control measures be implemented to further improve the overall level of food safety in China. This section aims to summarize the main characteristics of food safety risks in China based on the previous literature research.

5.4.1 Classification of Causes of Food Safety Incidents

Analyzing the main cause of food safety incidents in China based on comparison between China and developed countries is important in the investigation of the major characteristics of food safety risks in China at this stage.

Because of the freshness, perishability, specificity of processing, and high demand for circulation of food, hazardous food safety factors may exist in all aspects of the food supply chain that can lead to food safety incidents.

The possible causes of food safety risks in China are biological, chemical, physical, and human factors. Biological, chemical, and physical factors are natural factors that can directly cause food safety risks. These factors, in a sense, cannot be completely eliminated. In addition to biological, chemical, and physical factors, human misconduct and institutional factors, such as producer characteristics, information asymmetry, interest among stakeholders, and government regulation, may also lead to food safety risks. Human misconduct and institutional factors are henceforth referred to as human factors. Table 5.11 compares the characteristics of food safety risks.

TABLE 5.11
Comparison of Food Safety Risks with Different Characteristics

Risk Type	Food Safety Risks Caused by Natural Factors	Food Safety Risks Caused by Human Factors
Hazard type	Chemical pollution by veterinary drug residues, heavy metals, and other chemical substances introduced during farming, feeding, and processing Microbial contamination Physical pollution by foreign bodies Cannot be completely avoided Predictable and preventable or can be reduced to an acceptable level	Abuse of food additives Illegal addition of chemical substances Misconduct due to low scientific literacy Neither predictable nor preventable in traditional management
Characteristics of occurrence	Often occur in individual sectors, individual commodities, and individual batches	Often occur in a systematic, continuous, and industry-wide manner
Scope of occurrence	Present in all countries	Specific to developing stage
Nature of cause	Negligence	Lack of integrity and ethics

5.4.2 Causes of Chinese Food Safety Incidents

The sources of typical food safety cases occurring in China during 2002–2011 are summarized in Table 5.12. Wen and Liu (2012) demonstrated that 68.2% of food safety incidents were the result of food quality and safety problems knowingly caused by stakeholders in the supply chain that were motivated by a desire for personal gain or profit. This fully demonstrates that "knowing violations" by food producers and traders are currently the main cause of food safety problems; that is to say, most Chinese food safety incidents are caused by moral failure of food producers and traders in the face of interests.

Zhou (2013) suggested that, in general, food safety incidents could be roughly divided into two types. The first type of food safety problem results from the lack of integrity of producers motivated by interests in selecting and using inputs during cultivation, production, and manufacturing of food, for example, the illegal addition of food additives. The second type of food safety problem is caused by negligence in management or ignorance of producers and traders as well as the limitations of existing technologies, for example, crops produced by farmers have potential safety problems related to soil contamination that the farmers are not aware of. Ji (2012) summarized typical food safety incidents occurring during 2000–2012 and concluded that, in general, they were caused by man-made pollution.

Wu and Qian (2012) analyzed the characteristics of 24 influential food safety incidents occurring in China in 2011. Although some of these food safety incidents were caused by environmental pollution, a lack of government regulation, and imperfect legislation, most of them resulted from

TABLE 5.12
Sources of Typical Food Safety Cases Occurring in China during 2002–2011

Source of Case	Number
National Food Safety Information Center	712
National Food Safety Resource Database	75
Safety letters	130
Other media	84
Total	1001

Source. Wen, X.W. and Liu, M.L. Cause, dilemma and supervision of food safety from the year 2002 to 2011, *Reform*, 9, 37–42, 2012.

improper use of agrochemicals, use of shoddy, false certification, and use of nonfood raw materials during production. In other words, the majority of food safety incidents, in essence, were caused by violation of laws and regulations by producers and traders who were motivated by pursuit of economic profit to distribute foods with serious quality problems or unqualified foods. This not only endangered the health of consumers, but also seriously disrupted the food market order. Therefore, the fundamental problem lies in the fact that illegal activities by producers and traders motivated by pursuit of economic profit cannot be eliminated despite repeated prohibitions.

5.4.3 Brief Comparison and Main Conclusions

Food safety is a worldwide problem, and food safety incidents occur in almost all countries around the world, not only in developing countries, but also in developed countries. Food safety cannot be completely guaranteed in any country, but the cause, nature, manifestation mode, and number of food safety incidents vary among countries. Along with differences in economic development, food safety hazard factors also vary significantly between countries. In developed countries, food safety incidents are mainly caused by biological factors as a result of environmental pollution and food chain contamination, instead of intentional contamination, that is, human factors. In China, although some food safety incidents are the result of technical deficiencies and environmental pollution, most of them are caused by human factors, that is, misconduct, noncompliance or partial compliance to existing food technical specifications and standard systems, and other illegal acts by producers and traders. To some extent, this is related to the fact that China is at the developing stage. Thus, to prevent food safety risks and safety incidents in China, it is imperative to restrain the negative behaviors of food producers and traders.

In summary, physical, chemical, biological, and human factors result in different food safety risks. New technologies can also lead to new food safety risks. Absence of technologies, specifications, and standards breeds food safety risks. Trade internationalization has also accelerated the spread of food safety risks. However, it has been generally accepted that human factors are the major cause of food safety risks in China at present. Moreover, the fact that most food safety risks are caused by human factors will remain for a very long period in the future and cannot be fundamentally changed in the short term. Thus, prevention of food safety risks will be a long-term arduous process in China.

5.5 Major Characteristics of Human-Induced Food Safety Risks

The characteristics of Chinese food safety risks primarily caused by human factors are worthy of attention because analysis of these characteristics contributes to more effective management.

5.5.1 All Sectors of the Food Supply Chain Are Involved

The food industry has a long industry chain and involves many people, businesses, and areas. Chinese food safety risks primarily caused by human factors, to varying degrees, are present in all sectors of the food supply chain. Table 5.13 shows the distribution of typical Chinese food safety incidents that occurred in the supply chain during 2002–2011. As shown in Table 5.13, human-induced incidents occurred in primary agricultural production, primary processing of agricultural products, deep processing, food circulation, sales and catering, and consumption. Unfortunately, such instances are numerous and frequent.[*]

TABLE 5.13
Distribution of Chinese Food Safety Incidents in the Supply Chain during 2002–2011

Sector	Incident Number[a]	Frequency (%)
Primary agricultural production	130	10.87
Primary processing of agricultural products	240	20.07
Deep processing	452	37.79
Food distribution	82	6.86
Sales and catering	205	17.14
Consumption	32	2.68
Undistinguishable	55	4.60
Total	1196	100.00
Total after removing repetitions	1001	

Source. Wen, X.W. and Liu, M.L., Cause, dilemma and supervision of food safety from the year 2002 to 2011, *Reform*, 9, 37–42, 2012.
[a] As the same food safety incidents can occur in two or more sectors, the total incident number of 1196 is correct.

---

[*] See Wen and Liu (2012). All 1001 food safety incidents were not caused by human factors. However, Wen and Liu (2012) demonstrated that 68.2% of food safety incidents were the results of food quality and safety problems knowingly caused by stakeholders in the supply chain motivated by a desire for personal gain or profit. Table 5.17 shows exactly the same situation.

As human-induced food safety risks are present in all sectors in China, the sources and characteristics of Chinese food safety risks are different from those in other countries, especially developed countries. This particularity is caused by the basic national conditions of China. In terms of agricultural production, small, decentralized private farmers are still the backbone of agricultural production in China. It is difficult to supervise agricultural production by these small, decentralized private farmers. In terms of food production and processing, logistics and distribution, more than 90% of the over 0.4 million of food producers are small businesses, and there are too many small workshops to count. In terms of food circulation, individual businesses hold an absolutely dominating position and continue to spread in urban and rural areas throughout the country, especially in rural areas and remote mountainous areas; this sector is characterized by a wide distribution of many small decentralized businesses. A large number of food producers and traders are small-scale, decentralized, and have a low level of intensive management and poor ability to ensure quality and safety management. In the backdrop of the market failure caused by the absence of government regulation and information asymmetry, the cause, time, place, manner, and damage of the improper or illegal behaviors of food producers and traders motivated by relentless pursuit of economic interests are complex. Therefore, it is difficult to predict and prevent food safety risks resulting from such behaviors.

5.5.2 The Biggest Hazard Is Present in Deep Processing

The China *Food Safety Incident Response Plan* classifies food safety incidents into four levels, special major event (I), major event (II), large event (III), and ordinary food safety incident (IV). The degree of hazard of food safety incidents occurring during 2002–2011 in different sectors of the supply chain, represented by the level of food safety incidents, is shown in Figure 5.1. The different sectors of the food supply chain, ranked in descending order of incident number of the four levels, are as follows:

1. *Special major food safety incidents (I)*. Deep processing, agricultural production, food distribution, primary processing of agricultural products, sales and catering, and consumption
2. *Major food safety incidents (II)*. Deep processing, sales and catering, food distribution, agricultural production, primary processing of agricultural products, and consumption

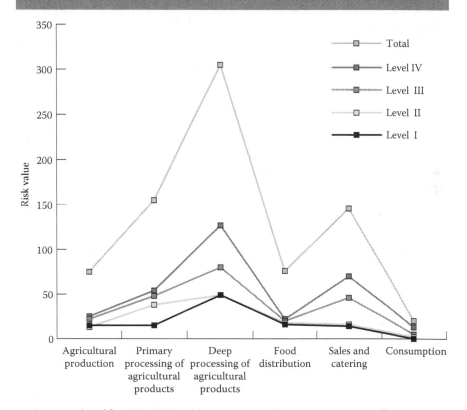

FIGURE 5.1

Degree of Hazard of Food Safety Incidents Occurring during 2002–2011 in Different Sectors of the Supply Chain

Source. Adapted from Wen, X.W. and Liu, M.L. Cause, dilemma and supervision of food safety from the year 2002 to 2011, *Reform*, 9, 37–42, 2012.

3. *Large food safety incidents (III).* Deep processing, primary processing of agricultural products, sales and catering, agricultural production, food distribution, and consumption

4. *Ordinary food safety incidents (IV).* Deep processing, primary processing of agricultural products, sales and catering, agricultural production, food distribution, and consumption

The degree of hazard is significantly different in different sectors of the food supply chain and the biggest hazard is present in deep processing. Food safety incidents occurring in the sector of deep processing not only involve a wide range of areas, but also do harm to a large number of consumers. Deep processing is the critical control point in the food supply chain. Monitoring of food safety in food processing can play an important role in reducing the hazards of Chinese food safety incidents. At the same time, food safety hazards in primary processing of agricultural products and sales and catering cannot be underestimated.

5.5.3 Food Safety Incidents Occur More Frequently in Developed Regions

According to the *China Statistical Yearbook*, the top 16 GDP provinces and cities from 2004 to 2011 were Guangdong, Shandong, Jiangsu, Zhejiang, Henan, Hebei, Shanghai, Liaoning, Sichuan, Beijing, Fujian, Hunan, Hubei, Heilongjiang, and Anhui. According to the website Defenestration (http://www.zccw.info/), the top 15 provinces and cities with the largest number of news items regarding food safety incidents during 2004–2011 were Beijing, Guangdong, Shandong, Shanghai, Zhejiang, Jiangsu, Chongqing, Hubei, Sichuan, Liaoning, Shannxi, Fujian, Hunan, Henan, and Hebei. It can be concluded that the number of food safety scandals is associated with the severity of the food safety crisis. The two ranking lists are highly consistent. With the exception of Heilongjiang and Anhui, all the top 16 GDP provinces and cities are in the top 15 rankings of food safety scandals, and except for Chongqing and Shannxi, all the top 15 provinces and cities with the largest number of food safety scandals are in the top 16 GDP rankings. More economically developed provinces and cities have more serious food safety risks. Geographically speaking, the coastal region has the highest food safety risk, followed by the central region; the western region has the lowest risk. Relatively advanced economic development, a large mobile population, high population density, intensive food processing industry, and a wide distribution range of food are the main causes of the higher frequency of food safety incidents in the developed regions in China.

It can be seen that China had a distinct geographical distribution of food safety incidents during 2002–2011. The possible reasons for the relatively high frequency of food safety incidents in developed provinces are quite complex. One possible reason is that consumers have a high awareness of food safety and therefore have more complaints about food problems, resulting in an increased exposure rate. Another possible reason is that government regulators have made stronger efforts in food safety law enforcement and have actively exposed food safety incidents. However, as shown in Table 5.14,

TABLE 5.14
Number of Food Safety Incidents Occurring in Chinese Provinces/Regions during 2002–2011

Province/Region	Incident Number[a]	Proportion (%)
Beijing	184	10.37
Guangdong	150	8.45
Zhejiang	82	4.62
Jiangsu	80	4.51
Shanghai	79	4.45
Shandong	76	4.28
Hubei	66	3.72
Sichuan	60	3.38
Chongqing	60	3.38
Liaoning	56	3.15
Hunan	52	2.93
Fujian	52	2.93
Shaanxi	50	2.82
Henan	49	2.76
Hebei	46	2.59
Jiangxi	44	2.48
Qinghai	43	2.42
Heilongjiang	42	2.37
Gansu	41	2.31
Anhui	40	2.25
Hainan	40	2.25
Jilin	40	2.25
Guangxi	39	2.20
Guizhou	35	1.97
Ningxia	35	1.97
Tianjin	35	1.97
Yunnan	34	1.92
Xi'an	33	1.86
Xinjiang	31	1.75
Tibet	30	1.69
Inner Mongolia	29	1.63
Hong Kong	22	1.24
Taiwan	20	1.13
Total	1775	

Source. Wen, X.W. and Liu, M.L., Cause, dilemma and supervision of food safety from the year 2002 to 2011, *Reform*, 9, 37–42, 2012.

[a] As the same food safety incidents can occur in two or more provinces or regions, the total incident number of 1775 is correct.

food safety incidents occurred at a low frequency in Hong Kong and Taiwan during the same period. The above-mentioned reasons cannot explain the low frequencies of food safety incidents in Hong Kong and Taiwan, two regions that are economically developed, strict in enforcement of laws, and effectively supervised by public opinion.

5.6 Main Contradiction in Preventing Food Safety Risks

Food safety risk is a global issue, and no country can stand aloof. Food safety risks must be comanaged by all countries because of their global characteristics. Some of the Chinese food safety risks are global problems, but most are problems unique to China. As Chinese food safety risks are of international concern and attract a significant amount of global attention, China plays an important role in defending against the global challenge of food safety. To this end, China must focus on the principal contradiction in preventing food safety risks.

Prevention of Chinese food safety risks is an eternal proposition and a very complex, highly systemic, and important issue, especially today. Different researchers investigated the main contradictions in preventing Chinese food safety risks from different points of view, with varied and even significantly different research emphasis and content. Nevertheless, according to the previous analyses of the main characteristics of food safety risks, the most fundamental issue should be the contradiction between decentralized, small-scale food production and risk management, no matter what point of view is taken. Optimizing and restraining the behaviors of food producers and traders and changing the food production mode are the only ways to prevent food safety risks in China at present and in the far future.

A long process is required to solve the principal contradictions in preventing food safety risks. This is because of the large base, long industrial chain, and involvement of many people and businesses of the Chinese food industry. Furthermore, unspoken rules are more likely to be developed in China's food industry, due to the misconduct of food producers and traders, as well as delayed and insufficient penalties and legal sanctions. With food safety risks rising day by day, according to the broken windows theory,[*]

---

[*] The broken windows theory was first introduced by the American political scientist James Q. Wilson and the criminologist George L. Kelling. If a window is broken by someone and not repaired, the tendency is for vandals to break a few more windows. Broken windows left unrepaired are a clear sign that the society has accepted this disorder and displays a lack of defense against the situation, which will ultimately lead to escalation into a more serious crime.

the inevitable result is a high level of revealed food safety risks and a high probability of food safety accidents in China. Therefore, the limited regulatory power and the major social resources in food safety risk monitoring should be centralized to solve the principal contradiction. The current focus is to strive to solve the lack of food safety supervision, alert system, and ineffective sanctions.

6

Food Safety Evaluation and Concerns of Urban and Rural Residents

To better understand food safety evaluation, awareness of the main causes of food safety risks, top concerns about food safety risks of urban and rural residents, as well as their evaluation of food safety regulation and law enforcement by the government, and to investigate solutions of food safety risks, we conducted a special survey in 96 sites in 12 provinces, autonomous regions, and municipalities (hereinafter referred to as provinces and regions). This chapter summarizes the main findings of the survey and provides a preliminary analysis.

6.1 Survey Description and Respondent Demographics

With the exception of Taiwan, Hong Kong, and Macau, the 31 provinces and regions in the Chinese mainland have different levels of economic and social development, food production, circulation and consumption environments, and food safety situations. Considering this, it is difficult to set up survey sites covering all provinces in the Chinese mainland and establish a scientific, reasonable, balanced, dynamic, and well-organized survey network in a short period of time. Therefore, it was extremely important to select representative survey sites to investigate the objective food safety evaluation of urban and rural residents, to provide an approximate description of the national situation.

6.1.1 Selection of Survey Areas

A wide distribution of survey sites and a large sample size at each survey point are required for a special questionnaire survey of food safety evaluation of urban and rural residents. As a result, a heavy workload and uncontrollability of accuracy exist from the use of simple random sampling. The main features of the survey design are described in Sections 6.1.1.1 and 6.1.1.2.

6.1.1.1 Principles of Sampling Design   The sample survey followed the scientific, efficient, and convenient principles, which are as given below.

1. *Science.* The overall design of survey plan was in strict accordance with the probability sampling method, which requires the sample to be representative of specified cities or regions of the country.
2. *Efficiency.* The sampling plan was designed to maximize the accuracy of the survey with the same sample size and minimize the sampling error of target variable estimation.
3. *Convenience.* Emphasis was also placed on operability. The survey design aimed not only to facilitate the specific implementation of sample survey, but also to facilitate subsequent data processing.

6.1.1.2 Stratified Multistage Random Sampling   In view of the regional differences in awareness of food safety risks, as well as the food consumption habits of Chinese urban and rural residents, a stratified multistage random sampling method was used in the survey to obtain ideal, objective, and true survey results. In general, a comparative sample survey of urban and rural residents can be conducted in two ways as follows: (1) urban and rural residents are independently sampled as two research fields, and (2) sampling units (such as districts and counties) are first selected without distinction between urban and rural areas, and a distinction is made between urban and rural areas in the subsequent sampling. Both methods have their advantages and disadvantages. The first method is more convenient but requires a larger number of sample points, and survey regions will be dispersed; thus, the requirements cannot be met at this stage. The advantage of the second method is relative concentration of the sample points, but the subsequent data processing is more complex. After comprehensive consideration, the second method was used for this survey. After stratified sampling, 96 survey sites were finally selected in 12 provinces and regions, including Fujian, Guizhou, Henan, Hubei, Jilin, Jiangsu, Jiangxi, Shandong, Shaanxi, Shanghai, Sichuan, and Xinjiang (Table 6.1).

TABLE 6.1
Stratification of 31 Provinces in the Chinese Mainland

Layer	Provinces and Regions Included	Selected Province or Region
First layer: Eastern China region	Sublayer 1: Shanghai, Beijing, Tianjin	Shanghai
	Sublayer 2: Liaoning, Shandong	Shandong
	Sublayer 3: Jiangsu, Zhejiang	Jiangsu
	Sublayer 4: Fujian, Guangdong, Hainan	Fujian
Second layer: Central China region	Sublayer 5: Heilongjiang, Jilin	Jilin
	Sublayer 6: Hebei, Henan, Shanxi	Henan
	Sublayer 7: Anhui, Jiangxi	Jiangxi
	Sublayer 8: Hubei, Hunan	Hubei
Third layer: Western China region	Sublayer 9: Neimenggu, Xinjiang, Ningxia	Xinjiang
	Sublayer 10: Shaanxi, Gansu, Qinghai	Shaanxi
	Sublayer 11: Sichuan, Chongqing	Sichuan
	Sublayer 12: Guangxi, Yunnan, Guizhou, Tibet	Guizhou

Respondents aged 18 years or older were selected in a completely random manner at each survey site. The survey is not further detailed here due to space limitations.

6.1.2 Respondent Demographics

A preliminary survey was conducted in the urban areas of Wuxi, Jiangsu Province, and Anyang County, Henan Province, in December 2011. Based on experience from the preliminary survey, examination of questionnaires, and training of investigators, a formal survey was performed from January to April in 2012. At each survey site, 50 urban or rural residents were selected randomly by trained investigators. It was planned to select approximately 2400 urban and 2400 rural residents. Invalid questionnaires were excluded. This produced an effective sample of 4289, with an 89.35% response rate. Of the valid questionnaires, 2143 and 2146 were from urban and rural respondents, respectively.

The descriptive statistics are shown in Table 6.2.* Basic demographics of the urban and rural residents surveyed (hereinafter referred to as "respondents") are discussed in Sections 6.1.2.1 through 6.2.1.5.

---

* Urban and rural respondents were residents surveyed in urban and rural areas, respectively, but not defined by registered permanent residence or place of employment. It should be noted that all demographics of respondents and their families are not provided in Table 6.2.

TABLE 6.2
Descriptive Statistics of Respondents' Relevant Demographics

Description	Total Sample (n = 4289) (%)	Urban Sample (n = 2143) (%)	Rural Sample (n = 2146) (%)
Gender			
Male	52.6	53.4	51.7
Female	47.4	46.6	48.3
Age			
18–25 years	28.1	27.2	28.9
26–40 years	39.4	43.1	35.7
41–55 years	25.1	23.5	26.8
≥56 years	7.4	6.2	8.6
Marital status			
Unmarried	31.5	32.2	30.7
Married	68.5	67.8	69.3
Education			
Graduate degree	4.3	4.2	4.4
Bachelor's degree	29.0	35.4	22.6
Junior college	16.0	20.7	11.2
Senior high school (including secondary vocational)	24.3	23.4	25.3
Junior high school or lower	26.4	16.3	36.5
Individual annual income			
≤10,000 yuan	32.5	29.02	35.88
10,000–20,000 yuan	20.9	17.17	24.65
20,000–30,000 yuan	21.8	24.59	18.92
30,000–50,000 yuan	13.9	16.47	11.42
>50,000 yuan	10.9	12.75	9.13
Number of household members			
1	1.59	2.19	0.98
2	5.62	7.14	4.10
3	40.80	46.80	34.81
4	29.26	25.85	32.67
≥5	22.73	18.02	27.44
Total	100.00	100.00	100.00

6.1.2.1 A Male-Biased Sex Ratio   A male-biased sex ratio was observed in the 4289 respondents. The proportions of males and females were 52.6% and 47.4%, respectively. The proportion of males was higher in urban respondents (53.4%) than in rural respondents (51.7%).

6.1.2.2 A Much Higher Proportion of Married Respondents   The proportion of married respondents was 68.5%, much higher than that of unmarried respondents (31.5%). The proportions of married respondents were 67.8% and 69.3%, respectively, in urban and rural respondents.

6.1.2.3 A Higher Proportion of the Respondents Aged between 26 and 40 Years   In total, 92.6% of the respondents were aged 55 years or younger. Respondents aged between 26 and 40 years accounted for the largest proportion (39.4%). The proportions of the respondents aged 18–25 years (28.1%) and 41–55 years (25.1%) were similar. The proportion of the respondents aged 26–40 years was higher in urban respondents than in rural respondents (43.1% vs. 35.7%).

6.1.2.4 A Higher Education Level in Urban Respondents   The proportions of the respondents with junior college or higher degrees (49.3%) and with senior high school or lower degrees (50.7%) were similar. The proportion of the respondents with junior college or higher degrees was higher in urban respondents (60.3%) than in rural respondents (38.2%).

6.1.2.5 A Large Urban–Rural Income Gap   The national per capita disposable income of urban residents was 21,810 yuan in 2011. Nearly one-half (46.6%) of the respondents had a per capita income of more than 20,000 yuan (i.e., reaching or nearly reaching the per capita disposable income of urban residents). The proportion of the respondents with a per capita income of more than 20,000 yuan was higher in urban respondents (53.81%) than in rural respondents (39.47%).

6.2 Food Safety Evaluation and Confidence in Its Future Trend

The respondents' evaluation of food safety and confidence in its future trend were analyzed in the following five aspects based on survey data.

6.2.1 Concern about Food Safety

As shown in Figure 6.1, of the 4289 respondents, 60.97% were concerned about food safety, including 18.70% who were "very concerned" and 42.27% who were "concerned." The total proportion of the respondents that were "not concerned" and "absolutely not concerned" was lower than 10%. It was very apparent that the total proportion of "very concerned" and "concerned" respondents was far higher than that of "not concerned" and "absolutely not

FIGURE 6.1
Urban and Rural Respondents' Concerns about Food Safety

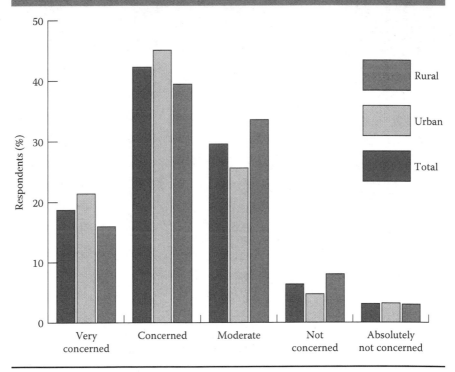

concerned" respondents. In terms of the differences between urban and rural residents, the proportion distribution of different levels of concerns about food safety was similar in urban and rural residents. The total proportion of "very concerned" and "concerned" respondents and of "not concerned" and "absolutely not concerned" respondents was 66.50% and 7.88%, respectively, in urban respondents, and 55.45% and 10.99%, respectively, in rural respondents. Thus, it is clear that both urban and rural respondents were generally concerned about food safety problems despite the differences.

It is worth noting that more than a quarter of the respondents had a "moderate" attitude toward food safety. The proportion of "moderate" respondents was 29.58%, 25.62%, and 33.56% in the total, urban, and rural samples, respectively. This may be related to the fact that a considerable number of the respondents still had ambiguous concerns about food safety due to the accelerated pace of work and life and asymmetric information. With the spread of food safety incidents and the consequent impacts, this group is most likely to become "concerned" or "very concerned" about food safety.

FIGURE 6.2
Respondents' Evaluation of Local Food Safety

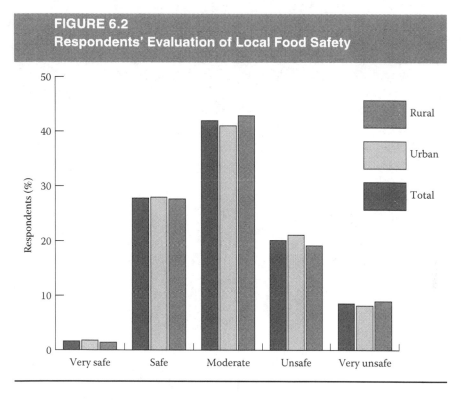

6.2.2 Evaluation of Local Food Safety

As shown in Figure 6.2, the respondents' evaluation of local food safety was almost exactly the same in the total, urban, and rural samples. The total proportion of "safe" and "very safe" was 29.38% versus 29.72% versus 29.03%, and the total proportion of "unsafe" and "very unsafe" was 28.68% versus 29.26% versus 28.10%, in the total, urban, and rural samples, respectively. Notably, less than 30% of the respondents believed that local food was safe or unsafe, regardless of whether they were in the total, urban, or rural samples. By contrast, more than 40% of the respondents had a "moderate" attitude toward local food safety, with the proportions being 41.94%, 41.02%, and 42.87%, respectively, in the total, urban, and rural samples. These results indicate that a considerable proportion of the respondents believed that local food safety "did not meet their expectations." Overall, urban and rural respondents did not have an optimistic evaluation of local food safety. Nevertheless, the overall evaluation of food safety has begun to improve as compared to the findings of similar existing surveys.[*]

---

[*] According to the *2010–2011 Report on Consumer Confidence in Food Safety* issued by China Overall Well-off Research Center, almost 70% of consumers felt insecure about food: 52.3% felt "quite insecure" and 15.6% felt "especially insecure."

6.2.3 Evaluation of Improvement in Food Safety

Figure 6.3 shows the respondents' evaluation of improvement in local food safety. In response to whether local food safety has improved compared with the past (e.g., 2011), 31.02% of all respondents believed that it had improved, while 24.88% of the respondents believed that it had not improved, but had become worse. Urban and rural respondents also had a very similar evaluation of improvement in food safety. In total, 30.15% of urban respondents believed that it had improved, and 26.37% believed that it had become worse. Similarly, 31.83% of the rural respondents believed that it had improved, and 23.99% believed that it become worse. In addition, 44.00%, 43.48%, and 44.18% of the respondents believed that there was "virtually no change" in local food safety in the total, urban, and rural samples, respectively. Therefore, it is clear that only a small proportion of the respondents believed in a definite improvement in food safety, while the majority of the respondents had a neutral attitude toward improvement in food safety.

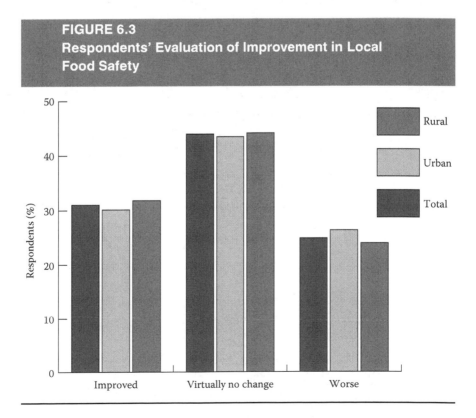

FIGURE 6.3
Respondents' Evaluation of Improvement in Local Food Safety

6.2.4 The Effect of Frequent Food Safety Incidents on Respondents' Confidence in Food Safety

In recent years, major food safety incidents have occurred frequently in China. As shown in Figure 6.4, when asked about "whether their confidence in food safety has been affected by recent food safety incidents, such as the swill-cooked dirty oil scandal," 47.05% and 23.15% of the respondents stated that their confidence had been affected and seriously affected, respectively. Only 18.91% of the respondents believed that food safety incidents were unavoidable and that their confidence was not affected. More than 70% of the respondents' confidence in food safety has been affected by frequent food safety incidents. In terms of differences between urban and rural areas, 73.03% and 69.02%, respectively, of urban and rural respondents' confidence in food safety was affected by frequent food safety incidents. Therefore, minimizing the rate of food safety incidents and providing proper guidance for public opinion on food safety are the most efficient ways to enhance respondents' confidence in food safety.

FIGURE 6.4
Proportion Distribution of Respondents with Confidence in Food Safety Differently Affected by the Frequent Incidents

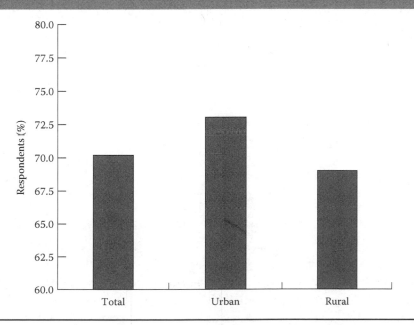

TABLE 6.3
Respondents' Confidence in Improvement of Food Safety in the Future

Description	Total Number of Respondents	Total Effective Proportion (%)	Urban		Rural	
			Number of Respondents	Effective Proportion (%)	Number of Respondents	Effective Proportion (%)
Definitely not improved	285	6.64	133	6.21	152	7.08
Probably not improved	732	17.07	400	18.67	332	15.47
Moderate	1323	30.85	635	29.63	688	32.06
Probably improved	1681	39.19	847	39.52	834	38.87
Definitely improved	268	6.25	128	5.97	140	6.52

6.2.5 Future Trends of Food Safety

As shown in Table 6.3, when asked about the "future trend of food safety," 45.44% of the respondents had confidence in the improvement of food safety, believing that food safety will "probably improve" or "definitely improve." However, approximately 23.71% of the respondents had no confidence in the improvement of food safety, believing that food safety will "definitely not improve" or "probably not improve." In addition, 30.85% of the respondents had an ambiguous attitude toward the future trend of food safety, considering the trend to be "moderate." A small difference between urban and rural respondents existed, with 45.49% and 45.39%, respectively, of urban and rural respondents having confidence in the improvement of food safety. These results indicate that respondents were not optimistic about the prospects for improvement.

6.3 The Most Important Food Safety Risk Factor and Level of Worry

Whether they can distinguish or identify the possible risk factors of food and their level of worry about health hazards due to these risk factors are more direct and accurate reflections of urban and rural residents' awareness of food safety. To this end, the most important food safety risk factor and the levels of worry about four major food safety risks, that is, abuse of food additives and intentional addition of nonfood substances, excessive pesticide and veterinary drug residues, bacterial and microbial contamination, and heavy metal contamination, were assessed in the survey.

6.3.1 The Most Important Food Safety Risk Factor

As shown in Figure 6.5, of the 4289 respondents, 36.33% considered the most important food safety risk factor to be "abuse of food additives and intentional addition of nonfood substances," while 23.51%, 16.02%, 14.64%, and 9.50%, respectively, considered it to be "excessive pesticide and veterinary drug residues," "bacterial and microbial contamination," "contamination by inherent toxic and hazardous substances contained in food," and "heavy metal contamination." Two conclusions can be drawn from these findings, which are as follows:

1. Respondents had a high awareness of "abuse of food additives and intentional addition of nonfood substances," which was the only option with a proportion higher than 30%. This finding may be related to major food safety events caused by food additives or other nonfood substances, such as the melamine milk powder incident.
2. Respondents had a wide selection of the most important food safety risk factors. In addition to "abuse of food additives and intentional addition of nonfood substances," the respondents were aware of

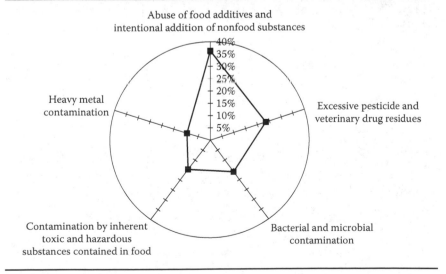

FIGURE 6.5
Respondents' Judgment about the Most Important Food Safety Risk Factor

"excessive pesticide and veterinary drug residues," "bacterial and microbial contamination," and other risk factors. This indicates that respondents were concerned about a wide range of food safety risk factors.

The ranking of the most important food safety risk factor was exactly the same between urban and rural respondents. However, a higher proportion of rural (24.72%) than urban (22.47%) respondents were worried about "excessive pesticide and veterinary drug residues," because of a greater knowledge of the use of pesticides and veterinary drugs.

6.3.2 Level of Worry about Major Food Safety Risks

External contamination of food by toxic and hazardous substances during production, transportation, storage, and sales, that is, food contamination, is the most common source of hazards in food. Therefore, further investigation was not carried out on worry regarding contamination by inherent toxic and hazardous substances contained in food. Emphasis of the investigation was placed on the external safety risks. The levels of worry about "abuse of food additives and intentional addition of nonfood substances," "excessive pesticide and veterinary drug residues," "bacterial and microbial contamination," and "heavy metal contamination" were further investigated.*

6.3.2.1 Abuse of Food Additives and Intentional Addition of Nonfood Substances   As shown in Figure 6.6, of the 4289 respondents, 76.94% of the respondents were worried about abuse of food additives and intentional addition of nonfood substances. Overall, 38.10% and 38.84% of the respondents, respectively, were "very worried" and "worried," and only 4.78% and 3.38% were "not worried" and "absolutely not worried," respectively. Clearly, respondents had a high level of worry about the abuse of food additives and the intentional addition of nonfood substances.

The proportions of "very worried" and "worried" were higher in urban respondents than in rural respondents (38.36% > 37.84% and 39.57% > 38.12%), but the difference was not significant. Therefore, urban and rural residents had universal and consistent levels of worry about the abuse of food additives and the intentional addition of nonfood substances.

---

*The total proportion of "very worried" and "worried" was referred to as the level of worry in this chapter.

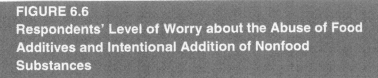

FIGURE 6.6
Respondents' Level of Worry about the Abuse of Food Additives and Intentional Addition of Nonfood Substances

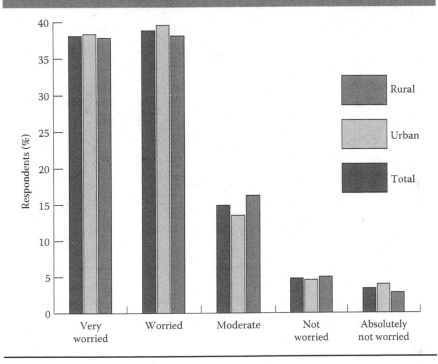

6.3.2.2 Excessive Pesticide and Veterinary Drug Residues As shown in Figure 6.7, 34.41%, 38.52%, 5.20%, and 2.10% of the respondents, respectively, were "very worried," "worried," "not worried," and "absolutely not worried" about excessive pesticide and veterinary drug residues; 19.77% of the respondents had a "moderate" attitude. In total, 72.93% of the respondents were worried about excessive pesticide and veterinary drug residues. The total proportion of "very worried" and "worried" was approximately 3% higher in urban respondents (74.62%) than in rural respondents (71.25%). Only 7.36% of rural respondents were "not worried" or "absolutely not worried." This reflects an urgent need to reduce pesticide and veterinary drug residues.

6.3.2.3 Bacterial and Microbial Contamination As shown in Figure 6.8, 29.91% and 35.02% of the respondents, respectively, were "very worried" and "worried" about bacterial and microbial contamination, with

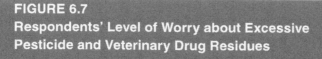

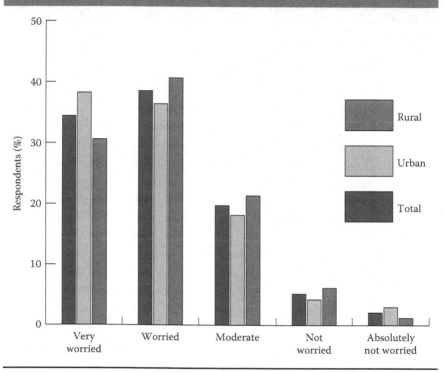

FIGURE 6.7
Respondents' Level of Worry about Excessive Pesticide and Veterinary Drug Residues

a total proportion of 64.93%. Only 6.92% and 2.56% of the respondents, respectively, were "not worried" or "absolutely not worried." Moreover, 25.58% of the respondents had a "moderate" attitude, and more urban than rural respondents were "very worried" (30.89% vs. 28.94%) and "worried" (36.86% vs. 33.18%) about bacterial and microbial contamination.

6.3.2.4 Heavy Metal Contamination As shown in Figure 6.9, 67.78% of the respondents recognized the dangers of heavy metal contamination and were "very worried" and "worried," while only 8.42% were "not worried" or "absolutely not worried." In addition, 23.81% of the respondents had a "moderate" attitude, and more urban than rural respondents were worried about heavy metal contamination. The total proportion of "very worried" and "worried" was approximately 7.6% higher in urban respondents (71.58%) than in rural respondents (63.98%).

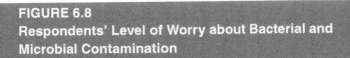

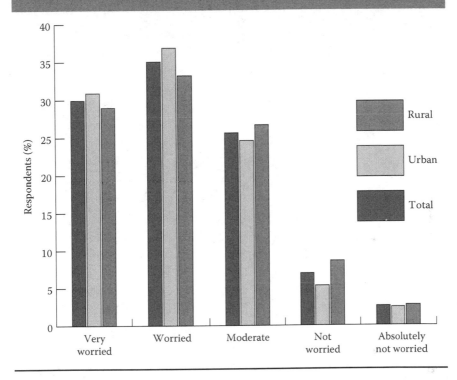

FIGURE 6.8
Respondents' Level of Worry about Bacterial and
Microbial Contamination

6.3.3 Comparison of Levels of Worry about Food Safety Risks between Urban and Rural Respondents

As shown in Figure 6.10, the rankings of levels of worry about the abuse of the four major food safety risks were exactly the same in the total, urban, and rural samples, that is, abuse of food additives and intentional addition of nonfood substances > excessive pesticide and veterinary drug residues > heavy metal contamination > bacterial and microbial contamination. Moreover, more than 64% of the respondents were worried about the four external food safety risks, whether in the total, urban, or rural samples.

Because of the differences in risk perception of food safety, more urban than rural respondents were worried about the four major food safety risks. The four major food safety risks, in descending order of difference between urban and rural respondents, were heavy metal contamination, bacterial and microbial contamination, excessive pesticide and veterinary drug residues, and abuse of food additives and intentional addition of nonfood substances.

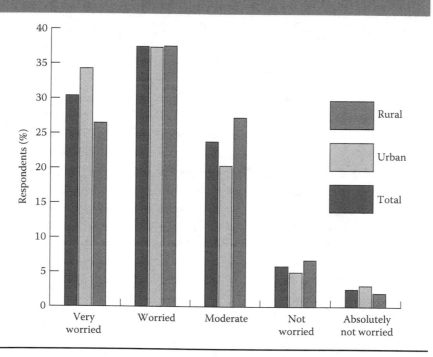

FIGURE 6.9
Respondents' Level of Worry about Heavy Metal
Contamination

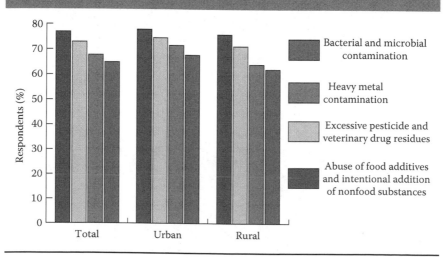

FIGURE 6.10
Comparison of Levels of Worry about Food Safety
Risks between Urban and Rural Respondents

The largest difference of 7.6% was observed in heavy metal contamination, and the smallest difference of 1.97% was observed in the abuse of food additives and the intentional addition of nonfood substances.

6.4 Judgment Regarding the Main Cause of Risks and Evaluation of Governmental Regulation and Law Enforcement Capacity

Based on the investigation of the respondents' judgment about the main cause of food safety risks, food safety regulation and enforcement capacity of the government were intensively investigated.

6.4.1 Judgment Regarding the Main Causes of Food Safety Risks

As shown in Figure 6.11, 27.12% of the respondents believed that the main cause of food safety risks was that consumers and government regulators could not gain complete relevant information of food production process and food contents, which was exploited by unscrupulous food producers, processers, and traders. Overall, 24.56% of the respondents considered the main cause to be

**FIGURE 6.11
Respondents' Judgment about the Main Cause of
Food Safety Risks**

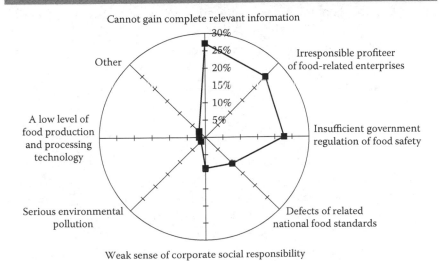

irresponsible profiteering of food-related enterprises, 22.71% considered it to be insufficient government regulation of food safety, and 10.80% considered it to be defects of related national food standards. In addition, 8.97%, 1.68%, 1.71%, and 2.45%, respectively, of the respondents considered the main cause to be a weak sense of corporate social responsibility, serious environmental pollution, a low level of food production and processing technology, and other causes.

6.4.2 Evaluation of the Government's Ability to Supervise and Manage Food Safety Risks

Five questions regarding policy and legal system construction effects, the government's efforts in food safety supervision and law enforcement, ability of the government and social organizations to guide consumer awareness of food safety, the government's ability to manage food safety incidents, and supervisory function of government-run news media were set to investigate the respondents' evaluation of the government's ability to supervise and manage food safety risks.

6.4.2.1 Policy and Legal System Construction Effects
As shown in Figure 6.12, 40.34% of the respondents had a "moderate" attitude toward the policy and legal system construction effects in safeguarding food safety. Furthermore, 32.62% were "satisfied" or "very satisfied," which was approximately 5.6% higher than those were "very dissatisfied" or "dissatisfied" (27.04%). This finding may be related to the respondents' level of understanding. Although the satisfactory evaluation varied, respondents generally recognized the policy and legal system construction effects in safeguarding food safety. In addition, urban and rural respondents generally had the same satisfactory evaluation in this aspect. Although limited by income and education levels, rural respondents did not fall behind urban respondents in terms of satisfaction with policy and legal system construction effects in safeguarding food safety. These results indicate that rural residents also had high expectations for the improvement of food safety by the government.

6.4.2.2 Governmental Efforts in Food Safety Supervision and Law Enforcement
As shown in Figure 6.13, 19.47% and 4.24%, respectively, of the respondents were "satisfied" and "very satisfied" with the government's efforts in food safety supervision and law enforcement, while 27.00% and 13.13%, respectively, were "dissatisfied" and "very dissatisfied." The total proportion of "dissatisfied" and "very dissatisfied" respondents was 16.42% higher than that of "satisfied" and "very satisfied" respondents. Apparently, the respondents' satisfaction with this aspect was much lower than that with

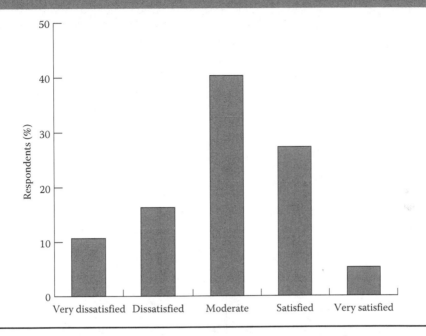

FIGURE 6.12
Respondents' Satisfaction with Policy and Legal System Construction Effects in Safeguarding Food Safety

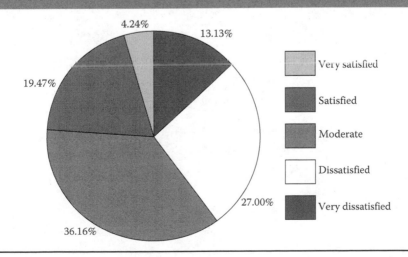

FIGURE 6.13
Respondents' Satisfaction with the Government's Efforts in Food Safety Supervision and Law Enforcement

policy and legal system construction effects. Therefore, it is important for the government to increase efforts in food safety supervision and law enforcement to enhance urban and rural residents' confidence in food safety.

Moreover, urban and rural respondents had highly consistent satisfactory evaluation of the government's efforts in food safety supervision and law enforcement. However, the proportion of rural respondents that was dissatisfied (41.33%) was 2.41% higher than that of urban respondents (38.92%). This finding indicates that rural respondents' food safety requirement and consumer safety awareness has been gradually improving. The government's food supervision and law enforcement forces should be further extended to rural areas to strengthen rural food market supervision and law enforcement.

6.4.2.3 Ability of Government and Social Organizations to Guide Consumer Awareness of Food Safety Improvement of consumer awareness of food safety requires the guidance of government and social organizations. Increasing publicity efforts to form a safe environment for consumption and guide urban and rural consumer awareness of food safety is a basic step to enhance food safety. As shown in Figure 6.14, the respondents' satisfaction with the ability

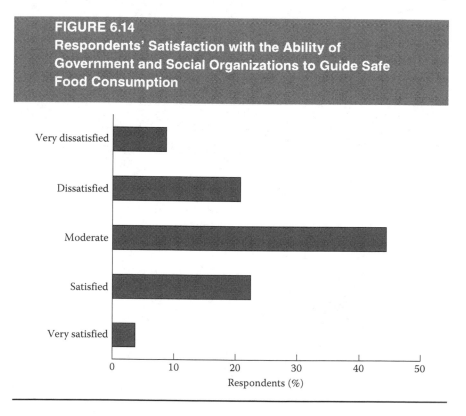

FIGURE 6.14
Respondents' Satisfaction with the Ability of Government and Social Organizations to Guide Safe Food Consumption

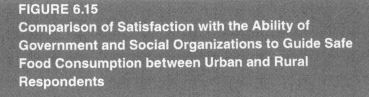

FIGURE 6.15

Comparison of Satisfaction with the Ability of Government and Social Organizations to Guide Safe Food Consumption between Urban and Rural Respondents

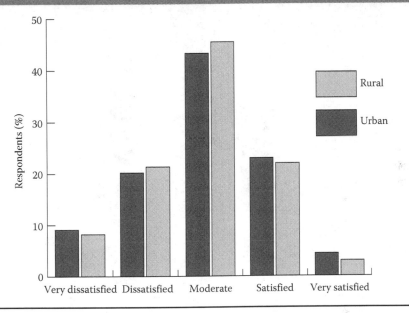

of government and social organizations to guide consumer awareness of food safety was normally distributed. The total proportion of "dissatisfied" and "very dissatisfied" was 3.24% higher than that of "satisfied" and "very satisfied" (29.42% vs. 26.18%).

As shown in Figure 6.15, although the total proportion of "dissatisfied" and "very dissatisfied" urban respondents (27.4%) was higher than that of rural respondents (24.97%), it was approximately 2% lower than the total proportion of "satisfied" and "very satisfied" urban respondents (29.31%). It was clear that, urban respondents had a slightly higher satisfaction with the ability of government and social organizations than rural respondents, but the satisfaction was not high. Therefore, as Chinese urban and rural residents generally have low awareness and scientific literacy of food safety, the government and social organizations must make great efforts to increase consumer awareness of food safety.

6.4.2.4 The Government's Ability to Manage Food Safety Incidents A question was set to investigate the respondents' satisfaction with the government's

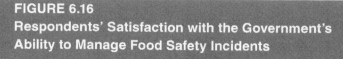

FIGURE 6.16
Respondents' Satisfaction with the Government's Ability to Manage Food Safety Incidents

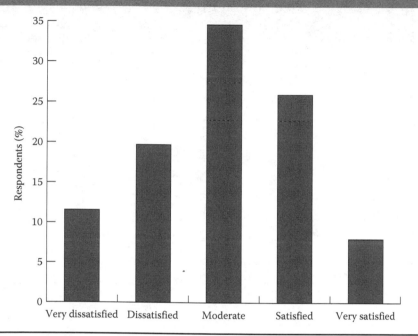

ability to manage food safety incidents. Respondents were asked to consider a scenario in which the government promptly published news of the occurrence of a food safety incident. As shown in Figure 6.16, most of the respondents (34.62%) had a "moderate" attitude in this respect, but the total proportion of "satisfied" and "very satisfied" respondents was 2.66% higher than that of "dissatisfied" and "very dissatisfied" respondents (33.93% vs. 31.27%).

As shown in Figure 6.17, the total proportion of "satisfied" and "very satisfied" urban respondents was very close to that of "dissatisfied" and "very dissatisfied" urban respondents (32.48% vs. 32.53%), while it was 35.37% versus 30.38% in rural respondents. Rural respondents had a relatively higher satisfaction with the government's ability to manage food safety incidents than urban respondents. This finding is closely related to the low food safety requirements and relatively limited horizons of rural residents.

6.4.2.5 Supervisory Function of Government-Run News Media   As shown in Figure 6.18, 5.29%, 27.00%, 19.58%, and 11.17% of the respondents were "very satisfied," "satisfied," "dissatisfied," and "very dissatisfied" with

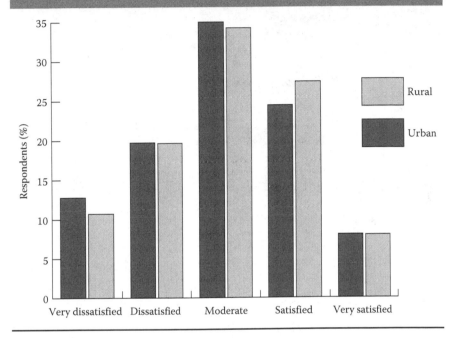

FIGURE 6.17
Comparison of Satisfaction with the Government's Ability to Manage Food Safety Incidents between Urban and Rural Respondents

the supervisory function of government-run news media, respectively, and 36.96% had a "moderate" attitude. The total proportion of "satisfied" and "very satisfied" was higher than that of "dissatisfied" and "very dissatisfied" (32.29% vs. 30.75%).

In general, urban and rural respondents had the same pattern of satisfaction in this respect. In urban respondents, the total proportion of "satisfied" and "very satisfied" was approximately 3.4% higher than that of "dissatisfied" and "very dissatisfied" (32.99% vs. 29.64%; Figure 6.19). In contrast, the total proportion of "satisfied" and "very satisfied" was very close to that of "dissatisfied" and "very dissatisfied" (31.52% vs. 31.87%) in rural respondents (Figure 6.20).

6.4.3 Summary of Related Evaluations

Scores of 1 through 5, respectively, were assigned to the five options, "very dissatisfied," "dissatisfied," "moderate," "satisfied," and "very satisfied," of the five questions regarding "evaluation of the government's ability to supervise

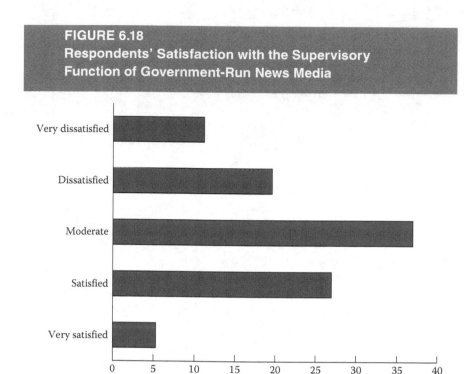

FIGURE 6.18
Respondents' Satisfaction with the Supervisory Function of Government-Run News Media

and manage food safety risks." The scores were then averaged over the sample size of each option. The calculated average score was the satisfaction score of the certain aspect of "the government's ability to supervise and manage food safety risks."

Results obtained by this method are shown in Table 6.4. As the satisfaction score of the intermediate level of the five options, "moderate," was 3, a score of 3 was taken as the standard of acceptability. As can be seen in Table 6.4, the satisfaction scores are as follows: (1) In rural respondents, the satisfaction scores of the government's ability to manage food safety incidents and policy and legal system construction effects were 3.02 and 3.01, respectively, which were just over the standard; the satisfaction scores of the government' efforts in food safety supervision and law enforcement, the ability of government and social organizations to guide consumer awareness of food safety, and the supervisory function of government-run news media were all substandard. (2) In urban respondents, the satisfaction scores of all the five options were lower than 3. Therefore, it can be concluded that urban respondents generally considered the government's ability to supervise and manage food safety risks to be unacceptable. Furthermore, although the

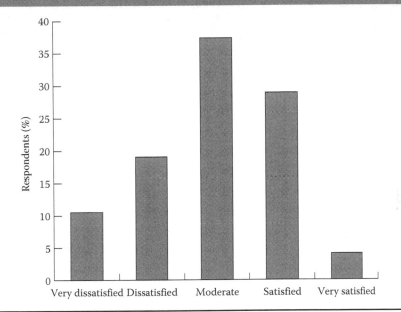

FIGURE 6.19
Urban Respondents' Satisfaction with the Supervisory Function of Government-Run News Media

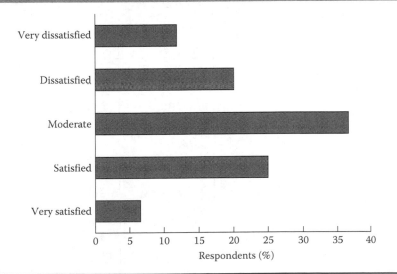

FIGURE 6.20
Rural Respondents' Satisfaction with the Supervisory Function of Government-Run News Media

TABLE 6.4

Evaluation of the Government's Ability to Supervise and Manage Food Safety Risks

Items	Options	Total Number of Respondents		Satisfaction Score	
		Rural	Urban	Rural	Urban
Legal system construction effects	1	214	245	3.01	2.99
	2	365	336		
	3	865	865		
	4	581	592		
	5	121	105		
Government efforts in food safety supervision and law enforcement	1	273	290	2.70	2.79
	2	614	544		
	3	808	743		
	4	385	450		
	5	66	116		
Ability of government and social organizations to guide consumer awareness of food safety	1	176	195	2.90	2.93
	2	458	433		
	3	976	928		
	4	472	493		
	5	64	94		
Government's ability to manage food safety incidents and policy	1	230	274	3.02	2.95
	2	422	423		
	3	735	750		
	4	588	524		
	5	171	172		
Supervisory function of government-run news media	1	253	226	2.94	2.97
	2	431	409		
	3	784	801		
	4	538	620		
	5	140	87		

satisfactory evaluations of urban and rural respondents were not exactly the same, they both had the lowest satisfaction with the government' efforts in food safety supervision and law enforcement.

6.5 Suggestions to the Government on Improving Food Safety Risk Management

Reduction of food safety risks is associated with systems, technology, standards, social responsibility, and other aspects, with many complex factors blending together. This determines the enormity and complexity of food safety risk management. To better amass the people's wisdom, a question was specifically set in the questionnaire to allow respondents to provide a suggestion to the government to improve food safety risk management.

6.5.1 Collection of Suggestions

A question "What do you think is the most important measure of food safety risk management?" was specifically set in the questionnaire, with nine options for respondents to select from, including "disclosing information of food production and processing and other relevant aspects," "increasing punishment for enterprises contributing to the occurrence of food safety incidents," "strengthening education of corporate social responsibility for producers," "completing food-related national standards," "severely punishing government officials and responsible persons for dereliction of food safety supervision," "strengthening environmental protection," "strengthening government supervision and inspection of food safety," "improving group reporting mechanism," and "other."

As shown in Table 6.5, as respondents had different points of view, none of the nine options was selected by more than 30% of the respondents. This suggests that respondents had a relatively wide range of different views regarding measures for improving food safety risk management in the future. Most of

TABLE 6.5
Respondents' Choices of Measure to Improve Food Safety

Options	Total Respondents (%)	Urban (%)	Rural (%)
Disclosing information of food production and processing and other relevant aspects	15.06	15.40	14.73
Increasing punishment for enterprises contributing to the occurrence of food safety incidents	21.22	23.05	19.38
Strengthening education of corporate social responsibility for producers	9.72	9.71	9.74
Completing food-related national standards	7.37	6.86	7.88
Severely punishing government officials and responsible persons for dereliction of food safety supervision	24.04	24.83	23.25
Strengthening environmental protection	0.98	1.03	0.93
Strengthening government supervision and inspection of food safety	17.28	15.35	19.20
Improving group reporting mechanism	2.77	2.28	3.26
Other	1.56	1.49	1.63
Total	100.00	100.00	100.00

the respondents (24.04%) selected "severely punishing government officials and responsible persons for dereliction of food safety supervision." Other options, in descending order, were "increasing punishment for enterprises contributing to the occurrence of food safety incidents" (21.22%), "strengthening government supervision and inspection of food safety" (17.28%), "disclosing information of food production and processing and other relevant aspects" (15.06%), "strengthening education of corporate social responsibility for producers" (9.72%), "completing food-related national standards" (7.37%), "improving group reporting mechanism" (2.77%), "other" (1.56%), and "strengthening environmental protection" (0.98%).

6.5.2 Main Suggestions of the Respondents and the Basic Characteristics

Further analysis revealed that the respondents' suggestions to the government on improving food safety risk management had two main features.

First, urban and rural respondents generally had the same understanding of government measures for improving food safety risk management. Although urban and rural respondents had different perceptions regarding the food safety situation and different demands for food safety, as well as different individual and family characteristics, the priority ranking of the government measures for improving food safety risk management was the same in urban and rural respondents.

Second, although respondents held a wide range of different views on management measures, emphasis was placed on three measures. The first, second, and third options were "severely punishing government officials and responsible persons for dereliction of food safety supervision," "increasing punishment for enterprises contributing to the occurrence of food safety incidents," and "strengthening government supervision and inspection of food safety," respectively, in the total, urban, and rural respondents. Respondents selecting one of these three measures accounted for 62.54%, 63.23%, and 61.83%, respectively, of the total, urban, and rural samples. From this perspective, strengthening supervision, strict law enforcement, and increasing penalties should be the basic measures to fully implement the *Food Safety Law* and its supporting laws and regulations and to improve the overall level of food quality and safety in the future.

7

Evolutionary Development of the Chinese Food Safety Legal System

It is a common practice in numerous countries, especially developed countries, to minimize food safety risks and guarantee food safety as much as possible by making every effort to improve the food legal system* and maintain strict enforcement of food safety laws. The implementation of the *Food Safety Law of the People's Republic of China* (hereinafter referred to as *Food Safety Law*) on June 1, 2009, marked the entry into a new stage of the Chinese food safety legal system construction. This chapter attempts to analyze the effects of the current food safety legal system based on a brief review of the developmental history of the Chinese food safety legal system from the point of view of severe punishment of criminal acts that endanger food safety. Furthermore, suggestions are provided on the Chinese food safety legal system based on the authors' views.

7.1 Developmental Course of the Chinese Food Safety Legal System

Before the reform and opening up in 1978, relevant departments of the Chinese government had successively formulated, enacted, and implemented appropriate food hygiene standards, food hygiene inspection methods, and

---

*Legal system is also called "system of law" or "law system" in China. It is a systematic, unified whole combining and grouping of all the current legal norms by legal branches. See Zhan, W.X. 2007. *Nomology* (3rd edition). Beijing, People's Republic of China: Law Press (in Chinese). According to the Legislation Law of the People's Republic of China, the Chinese legal system includes laws, administrative and local regulations, and rules.

other rules and regulations based on the then national conditions. After the reform and opening up, the Chinese food industry developed rapidly with the deepening of economic reform. Since the beginning of the new century, the Chinese food safety legal system has been gradually established and improved along with the development and changes in agriculture, food processing industry, and international food trade, and in particular, a tremendous change in governing ideas has occurred. China initially established a relatively complete legal system for food safety as indicated by the implementation of the *Food Safety Law* on June 1, 2009.

7.1.1 Before Implementation of the Food Safety Law

7.1.1.1 Formulation and Implementation of the Food Hygiene Law (For Trial Implementation)   After the founding of New China, the concept of food safety was limited to quantity and the primary goal of the government was to supply adequate food and clothing for an extended period of time. In the 1950s and 1960s, most Chinese food safety incidents were, in fact, poisonings that occurred in the stage of food consumption. Therefore, in a sense, food quality and safety was mainly focused on food hygiene issues at that time in China. The *Administrative Regulations on Food Hygiene for Trial Implementation* jointly formulated and implemented in 1965 by the then Ministry of Health, Ministry of Commerce, First Ministry of Light Industry, Central Administration for Industry and Commerce, and All China Federation of Supply and Marketing Cooperatives (ACFSMC) was the first comprehensive regulation for food hygiene management at the central level after the founding of New China.

As the situation developed, the then Ministry of Health revised the *Regulations for Food Hygiene Management for Trial Implementation* and enacted the *Administrative Regulations of the People's Republic of China on Food Hygiene* in 1979. During 1978–1992, the Chinese government began to implement major reforms to the economic system. Food production and marketing channels became increasingly diverse and complex. However, during this period, acute food poisoning incidents occurred more and more frequently, hazardous substances caused serious contamination of food in some regions, and some serious violations of food hygiene were not punished properly by the law. An increasingly urgent need existed for improving the food hygiene environment, which created a new requirement for the legal construction on food hygiene.* Based on the new situation, the *Food*

---

* See "An explanation for the Food Hygiene Law of the People's Republic of China (Draft)" made by the then deputy minister of health Wei Wang.

Hygiene Law of the People's Republic of China (For Trial Implementation) was promulgated by the Standing Committee of the National People's Congress (NPC) in November 1982 and put into trial implementation on July 1, 1983.

7.1.1.2 Formulation and Implementation of the Food Hygiene Law   In 1992, the Chinese government proposed to establish a market economy. It was decided to implement an institutional reform of the State Council at the First Meeting of the Standing Committee of the Eighth NPC in March 1993. Food industry enterprises gradually got into the market and were no longer directly managed by the government. During the 10 years from the implementation of the *Food Hygiene Law of the People's Republic of China (For Trial Implementation)* on July 1, 1983, to the institutional reform of the State Council in March 1993, the Chinese food industry developed rapidly, with the number of food industry enterprises increasing from 51,734 to 75,362 and the number of employees from 2.132 million to 4.846 million. Moreover, some new foods, health foods, and foods produced by development and utilization of new resources emerged in large quantities. The *Food Hygiene Law for Trial Implementation* was unable to adapt to the new situation and was, therefore, reformulated to a formal law, the *Food Hygiene Law of the People's Republic of China* at the 16th Meeting of the Standing Committee of the Eighth NPC in October 1995. The formulation and implementation of the *Food Hygiene Law* had a positive effect on the then food hygiene supervision. The number of food poisoning incidents decreased from 1,861 in 1991 to 522 in 1997, the number of people poisoned decreased from 47,367 in 1990 to 13,567 in 1997, and the number of food poisoning deaths decreased from 338 in 1990 to 132 in 1997 (Liu 2010a).

7.1.1.3 Establishment of a Segmented Legal System for Food Safety   In the new institutional reform of the State Council implemented in March 1998, the responsibilities of the State Bureau of Quality and Technical Supervision (SBQTS), Ministry of Health, Food Bureau, State Administration for Industry and Commerce (SAIC), and Ministry of Agriculture were adjusted. In the other institutional reform of the State Council in 2003, the original State Drug Administration was changed to the State Food and Drug Administration (SFDA) and was charged with comprehensive supervision and coordination of food safety and investigation and management of major food safety incidents. The substandard milk powder incident in Fuyang, Anhui, during 2003–2004 spawned the segmented food safety regulatory system. In September 2004, the State Council issued the *Decision to Further Strengthen Food Safety Management ([2004] No. 23)* to explicitly state, for the first time, the implementation of a segmented regulatory model supplemented by supervision of food categories according to the principle that one single stage was managed by one single department. The division of regulatory functions was as follows. The agriculture department was responsible for the supervision of primary agricultural production, the quality inspection department for the supervision of food production and processing, the industrial and commercial department for the supervision of food circulation, the health department for the supervision of consumption stage, such as catering and canteens, and the food and drug administrative department for the comprehensive food safety supervision and coordination, and investigation and management of major food safety incidents.

Therefore, the *Food Hygiene Law* implemented as of October 1995 appeared to partially lag behind the multidepartmental segmented management system. Moreover, the *Food Hygiene Law* only regulated food hygiene and safety-related activities in production, processing, transportation, circulation, and consumption, but not primary agricultural production stages, such as planting, breeding, fishing, gathering, and hunting. Furthermore, no norms were available for food safety risk analysis and assessment, food recall system, and supervision of food additives. Also, modern regulation patterns with high technological and legal contents in a market economy, such as food advertising regulation, were not defined. To this end, the 21st Meeting of the Standing Committee of the 10th NPC adopted the *Law of the People's Republic of China on Agricultural Product Quality and Safety*, which came into force on November 1, 2006. The *Law on Agricultural Product Quality and Safety* is considered to be the first food safety law concerning the health and safety of Chinese people. The promulgation and implementation of this law marked the official establishment of a complete legal system for the segmented food safety management model in China.

7.1.2 Implementation of the Food Safety Law

7.1.2.1 Change in the Legislative Idea   Along with its overall rapid development, the Chinese food industry has been extended to agriculture, agricultural product processing, food manufacturing and processing, food circulation, and catering, essentially involving a complete industry chain. A series of new changes also arose in the planting and catering industries. The traditional concept of food hygiene, mainly confined to catering and consumption, could no longer adapt to the extension of the food industry and was far from satisfying the requirements of the Chinese people for food safety and quality. An integrated food safety concept emphasizing food safety in all four major stages, planting and breeding, production and processing, circulation and sales, and catering and consumption, would be more in line with the social and public food safety standards and requirements (Liu 2010a). Moreover, during the implementation of the *Food Hygiene Law*, the legislators were mainly concerned about food-borne diseases, food poisoning, street vendors, and small workshops and took unlicensed vendors, self-employed entrepreneurs, and private entrepreneurs as the main regulatory targets. However, the Sanlu milk powder incident in 2008 changed the legislators' overall view on the legal regulation of food safety problems. More importantly, the incident exposed the drawbacks of the segmented food safety management and a lack of a unified and coordinated system for food safety standards, food safety information disclosure, and food risk monitoring and assessment. These factors jointly changed the idea of legislators from modifying the *Food Hygiene Law* to formulating a *Food Safety Law*.

The conversion from the *Food Hygiene Law* to the *Food Safety Law* is more of a change in legislative idea than a simple change in concept. In *Guidelines for Strengthening a National Food Safety Programme (1996)* published by the World Health Organization (WHO), food safety is interpreted as an assurance that food will not cause harm to the consumer when it is prepared and/ or eaten according to its intended use, and food hygiene is defined as all conditions and measures necessary to ensure the safety and suitability of food at all stages of the food chain. Therefore, food safety and food hygiene are two different concepts. In short, the replacement of the *Food Hygiene Law* with the *Food Safety Law* aims to lay down comprehensive provisions for food safety issues throughout the entire process "from farm to fork" and to avoid intersection and overlap between the food hygiene, quality, and nutritional standards by unifying all food standards with food quality and safety standards in a more scientific system. Based on this change in understanding, the Office of Legislative Affairs of the State Council changed the name of the *Food Hygiene Law (Revised Draft)* to *Food Safety Law (Draft)* during drafting.*

7.1.2.2 Implementation and Main Content of the Food Safety Law The change in legislative idea has a direct impact on the content of the law. The *Food Safety Law* was overwhelmingly passed by the Seventh Meeting of the Standing Committee of the 11th NPC by 158 votes in favor, with three against and one abstention on February 28, 2009; the law has been in effect since June 1, 2009. The *Food Safety Law* has more extensive content than the *Food Hygiene Law* and involves eight systems (Table 7.1). Its main highlights are as follows:

---

* See "An explanation for the food Safety Law of the People's Republic of China (Draft)" made by Kangtai Cao at the 31st Meeting of the Standing Committee of the 10th NPC on December 26, 2007 (in Chinese).

TABLE 7.1
Eight Systems Included in the *Food Safety Law* and Their Main Contents

No.	System / Framework	Existing System and Drawbacks	Future Development
1	Management	Segmented management. Each segment of the production and circulation managed by the agriculture administrative, quality inspection, industry and commerce administrative, health administrative, food and drug supervision and administration departments, respectively; poor coordination	First, the specific responsibilities of various departments in segmented management shall be further clarified. The health department is committed to comprehensive coordination of food safety; the quality inspection, industry and commerce administrative, and food and drug administrative departments manage food production, circulation, and catering, respectively; and the agriculture department manages agricultural product quality according to the *Law on Agricultural Product Quality*, but establishes edible agricultural product quality and safety standards and publishes information on edible agricultural product safety in accordance with the relevant provisions of the *Food Safety Law*. Second, the State Council shall establish a Food Safety Committee as a high-level agency for discussing and coordinating food safety issues and for coordinating and guiding food safety supervision based on segmented management. Third, the supervision responsibilities of local governments and relevant departments shall be further strengthened.
2	Risk assessment and monitoring	Not available	The health administrative department of the State Council shall be responsible for organizing the food safety risk assessment work. It shall form a food safety risk assessment expert committee composed of experts in medical science, agriculture, food, nutrition, etc., to assess food safety risks. The health administrative department shall collect and analyze the information, and give a timely warning of food safety risk and make an announcement when necessary.
3	Safety standards	Inconsistent and incomplete	The national food safety standards shall be formulated uniformly. Except for the food safety standards, no other mandatory food standards shall be set down.
4	Responsibility of producers and traders	Incomplete	Primary responsibility for food safety of food producers and traders is ensured from four aspects: food production and marketing licensing, certificate and invoice claiming, food safety management system, and recall of unsafe food and discontinued operation of business involved.
5	Safety of food additives	Nonstandard use/abuse of food additives; 1812 food additives falling into 22 categories allowed for use	No food additive may be listed in the scope of allowed use unless it is absolutely technically necessary and has been proven as safe and reliable upon risk assessment. A food producer shall use food additives under food safety standards on the varieties, extent of use, and dosages of food additives and shall not, during the process of food production, use any nonfood-additive chemical substance or any other substance which is potentially hazardous to human health.

6	Management of health food	Lack of supervision resulting in a disordered market of health food	The state shall stringently supervise foods claimed to have particular effects on human health. The relevant supervision and administration departments shall perform their functions according to law and undertake the responsibilities. No food claimed to have particular effects on human health shall cause any acute, subacute, or chronic harm to the human health. The labels and instructions of such food shall not include the effect of prevention or treatment of any disease, and the contents thereof shall be true and indicate applicable groups of people, inapplicable groups of people, effective ingredients or symbolic ingredients and contents thereof. The effects and ingredients of a product shall be consistent with the indications in the labels and instructions.
7	Incident	Loopholes in the reporting system	First, the reporting system. Daily supervision departments shall promptly notify the health administrative department after getting the information about the relevant food safety risks. No entity or individual may conceal, make false report, or delay the report of any food safety accident, or destroy relevant evidence. Second, incident handling. As soon as a health administrative department receives a report of food safety accident, it shall, jointly with the relevant supervision departments, investigate and deal with it and take measures to prevent or mitigate its hazards. In the case of a major food safety accident, the people's government at or above the county level shall promptly form a command body for handling the food safety accident, initiate the emergency plan and promptly deal with the accident. Third, liability investigation. In the case of a major food safety accident, the health administrative department of the people's government at or above the level of a districted city shall, jointly with the relevant departments, investigate the liabilities for the accident, urge the relevant departments to perform their functions, and propose to the people's government at the same level a report about the investigation and handling of accident liabilities.
8	Punishment	Not severe enough	A violator shall be subject to relevant criminal, administrative, and civil liabilities for any serious violation. Civil liability has surpassed the previous concept of civil damages in China, and a punitive compensation system has been established that besides claiming damages, a consumer may require the producer, who produces food which does not conform to the food safety standards, or the seller who knowingly sells food that does not conform to the food safety standards, to pay 10 times the money paid. Civil compensation takes precedence, meaning that a violator of this Law shall bear the civil compensation liability and pay the fine or pecuniary penalty. If its (his) property is insufficient to cover all the payment at the same time, it (he) shall first bear the civil compensation liability.

Source. Relevant content of the *Food Safety Law.*

1. A Food Safety Commission of the State Council was established to coordinate and guide food safety supervision.
2. The systems for safety issues, incident reporting and handling, and accountability have been improved.
3. Supervision emphasis has been placed on hazardous substances and production and use of food additives.
4. The scope of supervision has been extended to health food and not just limited to ordinary food.
5. The supervision process has been traced back to the source of food production, the quality and safety management of primary agricultural products is regulated by the agricultural product quality and safety laws, and the quality and safety standards of edible agricultural products and their announcement are regulated by the *Food Safety Law*.
6. Special restrictions have been imposed on food advertising to prohibit exaggeration of food functions.
7. A punishment system combining punitive compensation with fines and civil with criminal penalties was established.

7.2 Main Characteristics of the Chinese Food Safety Legal System

At present, China has basically established an integrated food safety legal system. It centers on the *Food Safety Law*, is supported by relevant special laws, and connects with the laws on environmental protection, product quality, import and export commodity inspection, and quarantine of animals and plants, as well as other relevant aspects. Its three main characteristics are discussed in Sections 7.2.1 through 7.2.3.

7.2.1 Diversity in Sources of Law and a Rich Hierarchy of Legal Force

The Chinese food safety legal system features a large number of normative documents in various forms and a rich hierarchy of legal force. It covers various legislative forms defined by the *Legislation Law*, including laws, regulations, and rules. This indicates the diversity of legislatures and the different power of different sources of food safety laws. The major laws, administrative regulations, department rules, and local legislation on food safety are briefly summarized in Sections 7.2.1.1 through 7.2.1.4 to provide a basic understanding of the food safety legal system.

7.2.1.1 Laws   In China, the *Food Safety Law* and the *Law on Agricultural Product Quality and Safety* are the most direct and basic laws for ensuring food safety. In addition, some laws regulate specific contents of food safety. They are as follows:

1. The production and business operations of food-related products, including food packaging materials, containers, detergents, and disinfectants, as well as tools and equipment, shall be subject to the *Law of the People's Republic of China on Product Quality.*
2. The *Standardization Law of the People's Republic of China* shall apply to the supervision over the implementation of the food safety standards.
3. The *Criminal Law of the People's Republic of China* and the relevant provisions of its amendment shall apply to the violations of food safety provisions that should incur criminal responsibility.
4. The contents related to administrative penalty, administrative licensing, and administrative enforcement in the *Food Safety Law* and the *Law on Agricultural Product Quality and Safety* shall be regulated by three administrative procedure laws, that is, the *Law of the People's Republic of China on Administrative Penalty*, the *Law of the People's Republic of China on Administrative Permission*, and the *Law of the People's Republic of China on Administrative Enforcement.*
5. The *Law of the People's Republic of China on Import and Export Commodity Inspection*, the *Law of the People's Republic of China on Animal Epidemic Prevention*, and the *Law of the People's Republic of China on the Entry and Exit Animal and Plant Quarantine* shall apply to import and export food inspection and plant and animal epidemic prevention and quarantine.

7.2.1.2 Administrative Regulations   Administrative regulations include both executive legislation and creative legislation involving specific areas of food safety. Executive legislation includes the *Implementing Regulations of the Food Safety Law, Implementing Regulations of the Standardization Law, Implementing Regulations of the Law on Import and Export Commodity Inspection*, and *Implementing Regulations of the Law on Entry and Exit Animal and Plant Quarantine*. To regulate and control some specific foods, the State Council also enacted the *Regulations on Supervision and Administration of Dairy Products Quality and Safety, Regulations on Administration of Agricultural Genetically Modified Organisms Safety, Pig Slaughter Regulations*, and *Regulations on Administration of Salt Industry*. In addition, the *Regulations on Open Government Information* provides an important legal basis for regulating food safety information disclosure. In 2007, the State Council also enacted the *Special Provisions on Strengthening the Regulation of the Safety of Products Such as Foodstuffs*. It is also an important legal document on food safety regulation and part of it has become legally provisioned.

7.2.1.3 Department Rules Competent authorities have promulgated several rules on the food safety areas under their administration or on special food safety issues. For example, the Administration of Quality Supervision, Inspection and Quarantine (AQSIQ) promulgated the *Administrative Regulations on Risk Warning and Rapid Response for Entry–Exit Inspection and Quarantine* in 2001; the Ministry of Commerce promulgated the *Operation Specifications on Food Safety in Supermarket (For Trial Implementation)* in 2006; the Ministry of Health promulgated the *Measures for the Archival Filing of the Enterprise Food Safety Standards* in 2009; the SAIC promulgated the *Measures for Administration of Food Safety in Circulating Field* in 2009; and the SFDA promulgated the *Measures for Cognizance of Rapid Examination Methods of Food Safety in Catering Services* in 2011.

7.2.1.4 Local Legislation In China, the Standing Committees of the provinces, autonomous regions, and municipalities directly under the central government can develop and promulgate local regulations, which constitute an important component of the Chinese legal system. After the implementation of the *Food Safety Law* on June 1, 2009, the Standing Committees of the relevant provinces, autonomous regions, and municipalities directly under the central government have promulgated and implemented their respective regulations. For example, the 28th Meeting of the Standing Committee of the 13th Shanghai People's Congress adopted the *Measures for Implementation of "Food Safety Law of the People's Republic of China" in Shanghai* on July 29, 2011; the 26th Meeting of the Standing Committee of the 11th People's Congress of Zhejiang Province adopted the *Measures for Implementation of "Food Safety Law of the People's Republic of China" in Zhejiang Province* on July 29, 2011. At the level of local rules, executive legislation for the *Food Safety Law* in 2011 includes the *Measures for Administration of Food Safety in Chongqing* and *Administrative Regulations on Food Safety in Tangshan*. Local rules focusing on special areas of food safety include the *Measures for Administration of Food Safety in Rural Group Dinner in Shijiazhuang, Measures for Administrative Accountability for Food Safety in Ningxia, Measures for Administration of Fresh Food Safety in Xiamen, Measures for Administration of Food Safety in Guangzhou*, and *Administrative Regulations on Food Safety in Beijing*.

In addition to the legislation and regulation on relevant stages of food safety, the Chinese government at all levels and its functional departments have also formulated a large number of normative documents in diverse forms. These normative documents play a large role in the practice of food safety supervision.

7.2.2 Clear Priorities and Reasonable Structure

After the enactment of the *Food Safety Law*, China has basically developed a large group of food safety laws and regulations. This group of laws and regulations centers on the *Food Safety Law* and the *Law on Agricultural Product Quality and Safety* and is supported by a large number of laws and regulations on the management of special areas. It strives to provide a legal basis for all areas of food safety, while reducing conflicts between the legal norms and promoting harmony within the legal system.

7.2.2.1 The Food Safety Law *and the* Law on Agricultural Product Quality and Safety *Are the Core* The *Food Safety Law* and the *Law on Agricultural Product Quality and Safety* are two core laws for the regulation of comprehensive food safety management. Before the enactment of the *Food Safety Law*, the food production and business operations stages, food safety authorities, and laws for regulation of food safety management basically corresponded one-to-one. Table 7.2 shows this corresponding relationship.

At present, the Chinese food safety legal framework is basically constructed according to the segmented management model. After the enactment of the *Food Safety Law*, the Chinese legal framework for the regulation of food safety

TABLE 7.2
Relationships between Management Stages, Competent Authorities, and Applicable Laws before Enactment of the *Food Safety Law*

Food Production and Marketing Stage	Competent Authority	Applicable Law
Primary agricultural production	Agriculture department	*Law on Agricultural Product Quality*
Food production, processing, and circulation	Quality inspection and industry and commerce department	*Law on Product Quality*
Food consumption	Health department or state food and drug administrative department	*Food Hygiene Law*

Source.   Relevant laws and regulations.

TABLE 7.3

Relationships between Management Stages, Competent Authorities, and Applicable Laws after Enactment of the *Food Safety Law*

Food Production and Marketing Stage		Competent Authority	Applicable Law
Edible agricultural products	Management of edible agricultural product quality and safety	Agriculture department	*Law on Agricultural Product Quality*
	Agricultural product quality and safety standards and disclosure of edible agricultural product safety information	Health department	
Food production, processing, and circulation		Quality inspection, industry and commerce, and health departments	*Food Safety Law*
Food consumption		State food and drug administrative and health departments	

Source.   Relevant laws and regulations.

management retained the basic structure of the segmented management system. The provisions of Article 2 of the *Food Safety Law* clearly state this structural division. The relationships between management stages, competent authorities, and applicable laws after the enactment of the *Food Safety Law* are shown in Table 7.3. Of course, the amendment of the legal system is still underway after the reform of the food safety management system in March 2013.

7.2.2.2 The Product Quality Law *and the* Standardization Law *Are the General Laws in Particular Areas of Food Safety*   In the management of food quality and relevant standards, the *Product Quality Law* and the *Standardization Law* are general laws as to the special law, the *Food Safety Law*.*

Article 2 of the *Product Quality Law* provides that "anyone who manufactures or sells any product within the territory of the People's Republic of China

---

*Article 83 of the *Legislation Law* provides that "With regard to laws, administrative regulations, local regulations, autonomous regulations, separate regulations or rules, if they are formulated by one and same organ and if there is inconsistency between special provisions and general provisions, the special provisions shall prevail; if there is inconsistency between the new provisions and the old provisions, the new provisions shall prevail." According to this provision, the relationship between special laws and general laws can basically be summarized as follows: (1) Special laws prevail; when both special laws and general laws apply, special laws shall prevail. (2) General laws supplement special laws, and when special laws are not applicable, general laws shall apply.

shall abide by this Law." Therefore, in the supervision of quality of food, food additives, and food related products, the *Food Safety Law* is a special law as to the general law, the *Product Quality Law.** Because edible agricultural products do not meet the definition of "product" in the *Product Quality Law*, this law does not apply to the management of edible agricultural product quality. Similarly, in the matter of food safety standards, the *Food Safety Law* is also a special law as to the general law, the *Standardization Law*.

7.2.2.3 Relevant Laws Play a Complementary Role   In China, many laws play a role in guaranteeing food safety and protecting public health and safety in their specific regulating areas. These laws are all food safety-related laws. Some laws have already included the protection of food safety in their legislative intent. For example, in Article 1, the *Animal Husbandry Law of the People's Republic of China* provides that it is enacted for the purpose of "ensuring the quality and safety of livestock and poultry products." Some laws provide rules and measures to ensure food safety, which also play a role in ensuring food safety. For example, the *Law of the People's Republic of China on Import and Export Commodity Inspection* specifies the competent authorities and procedures for inspection of import commodities (including food, food additives, and food-related products). Some laws supplement the food safety legal system from a specific perspective. For example, in the *Law of the People's Republic of China on Protection of the Rights and Interests of Consumers*, the provision on labeling of commodity information includes the management of the food labeling, and its provision on double compensation for frauds complements the content of civil compensation in the *Food Safety Law*. Also, the *Criminal Law* has certain provisions on food safety offenses.

7.2.2.4 Administrative Regulations Made under the Delegation of a Law Regulate Administration of Specific Areas   Some administrative regulations were enacted under the delegation of the *Food Safety Law* and provide requirements for food safety administration in specific areas. According to Article 101 of the *Food Safety Law*, "the food safety administration of dairy products, genetically modified food, slaughtering of live pigs, spirits and common salt shall be governed by

---

*Before the enactment of the *Food Safety Law*, Chinese scholars considered that both the *Product Quality Law* and *Food Hygiene Law* were about product quality requirements, but comparatively speaking, the *Product Quality Law* was a general law and the *Food Hygiene Law* was a special law, and that when it comes to food quality, the *Food Hygiene Law* shall apply. See Civil Law Office of the Commission of Legislative Affairs of the Standing Committee of the National People's Congress. Wang, S.M. 2010. *Interpretation of the Tort Liability Law of the PRC*. Beijing, People's Republic of China: China Legal Publishing House (in Chinese).

this Law. If it is otherwise provided for by any other law or administrative regulation, that law or administrative regulation shall prevail." Currently, administrative regulations that provide requirements for the above special foods include the *Regulations on Supervision and Administration of Dairy Products Quality and Safety*, *Regulations on Administration of Agricultural Genetically Modified Organisms Safety*, *Pig Slaughter Regulations*, and *Regulations on Administration of Salt Industry*.* The provisions of these administrative regulations are to those of the *Food Safety Law* what special provisions are to general provisions.

7.2.2.5 A Large Number of Executive Legislations Are an Important Part of the Food Safety Legal System   The basic principle of legislation in China is the unity of principle and flexibility. Laws usually require a large number of supporting regulations and rules to be operational, and the field of food safety is no exception. A large number of administrative and local regulations and rules have been made. Most of them are executive legislations of the *Food Safety Law*, the *Law on Agricultural Product Quality and Safety*, and other food safety-related laws. These laws play a large and direct role in regulating the field of food safety. The *Food Safety Law* makes explicit provisions for delegated legislation (Table 7.4) and requires the State Council, the Central Military Commission, relevant ministries, and commissions and local legislatures to formulate appropriate rules and regulations on specific administration issues.

7.2.3 An Aggregate of Legal Norms of Various Natures

In terms of legal nature, the current Chinese food safety legal system is an aggregate of a variety of legal norms, including administrative law norms, economic law norms, civil law norms, and criminal law norms. The norms of administrative law and economic law constitute the main part of the aggregate of Chinese food safety laws, regulations, and norms.

7.2.3.1 Administrative Law Norms Are a Main Content of the Legal System   It is a common thread in most countries to ensure food safety by clearly defining the responsibilities of food safety supervision and administration departments. Similarly, in the *Food Safety Law* and the *Law on Agricultural Product Quality Safety*, articles defining the power and responsibilities of food safety supervision

---

*Commercial administrative departments are responsible for the administration of spirits circulation. The *Measures for Administration of Alcohol Circulation* (come into force as of January 1, 2006) enacted by the Ministry of Commerce defines an archival filing and registration system and a spirits circulation traceability system. However, no administrative regulations have been enacted to provide special requirements on spirits safety.

TABLE 7.4
Provisions on Delegated Legislation in the *Food Safety Law*

Article	Delegated Department	Content
Article 21	Health administrative and agriculture administrative departments of the State Council	Provisions on limits of pesticide residues and veterinary medicine residues, and the inspection methods and procedures thereof
Article 21	Relevant competent department and health administrative department of the State Council	Inspection procedures for slaughtered livestock and poultry
Article 29	Standing committee of the people's congress of the province, autonomous region, or municipality directly under the central government	Specific measures for administration of small food production or processing workshops and food vendors in food production or business operation
Article 51	State Council	Concrete administrative measures for food claimed to have particular effects on human health
Article 102	Relevant competent department and health administrative department of the State Council	Administrative measures for the food safety in the railway business operations
Article 102	Central Military Commission	Administrative measures for the food safety of the exclusive food and self-supplied food of the army

Source. Relevant content of the *Food Safety Law.*

and administration departments account for the greatest proportion. The current Chinese food safety legal system is constituted mainly by administrative law norms, including the administrative organization law norms, the administrative act law norms, and administrative responsibility law norms that are extremely rich in content.

7.2.3.2 Economic Law Norms Occupy an Important Position in the System   Economic law norms are another major part of the Chinese food safety legal system that imposes various legal obligations and responsibilities on food producers and traders. Food is a special product, which is directly related to public health and human safety. However, a large information asymmetry exists between the producers and traders, as well as the consumers. Therefore, it is one of the compulsory social responsibilities of food producers and traders to ensure food safety. However, food producers and traders will not voluntarily and actively assume this social responsibility due to their pursuit of profit. Civil and commercial law norms are focused on the protection and maintenance of market mechanisms and cannot effectively solve this problem. Therefore, a special system of food production and marketing must be constructed by the economic law norms,* and special duties and responsibilities must be imposed on food producers and traders. In view of this, Article 3 of the *Food Safety Law* stipulates clearly that food producers and traders are primarily responsible for food safety (Administrative Law Office of the Commission of Legislative Affairs of the Standing Committee of the National People's Congress 2009). The law provides that food producers and traders shall assume social responsibilities for relevant stakeholders (including consumers, employees, society, and the environment) when creating profits and being responsible for shareholders.

7.2.3.3 The System Also Includes a Number of Civil and Criminal Law Norms   Article 96 of the *Food Safety Law*, for example, is exactly a typical civil law norm. It provides that a consumer may require the violator to "pay 10 times the money paid," which is a special form of punitive compensation. A violation of the *Food Safety Law* and the *Law on Agricultural Product Quality and Safety* that constitutes a crime shall be governed by the relevant provisions of the *Criminal Law*.

7.3 Effects of the Existing Food Safety Legal System

During the three years from the implementation of the *Food Safety Law* on June 1, 2009, to December 2012, the overall level of food quality and safety in China improved. The food safety legal system has played a very large role in safeguarding food safety. As a whole, effects of the Chinese food safety legal system have been gradually reflected. Considering the large amount of details regarding this aspect, this section only discusses the punishment of criminal behaviors endangering food safety in recent years.

In recent years, the Chinese judicial system has carried out special law enforcement inspection of key food safety problems, such as illegal addition

---

*Some scholars pointed out, based on the social responsibilities stipulated in the *Food Safety Law,* that "It is an inevitable requirement for the economic laws in modern society to define the role of producers, traders, and related public organizations in social and economic life, declare the social nature of their behaviors, and make them assume active, forward-looking, 'predetermined social responsibilities.'" This is obviously different from the relief and punishment afterward and negative "backward-looking responsibility" pattern in the traditional civil, administrative, and criminal laws. See Liu, S.L. 2010. From individual rights to social responsibility: A holistic consideration of the Food Safety Law of the PRC. *Modern Law Science* 3:32–47 (in Chinese).

of chemical substances, abuse of food additives, illegal use of clenbuterol, and production and selling of swill-cooked dirty oil as edible oil, and has investigated a number of typical cases. Furthermore, relevant law enforcement departments have further enhanced the relevant legal system and strived to strengthen the connection between administrative law enforcement and criminal justice. The relevant law enforcement departments have also focused on further improving the transfer procedures of suspected criminal cases, providing interconnection between law enforcement and judicial information interoperability, preventing nontransfer or difficulty in transfer of cases and replacement of criminal punishment with administrative penalty, and ensuring adequate investigation and punishment of food safety crimes in pursuance of the *Food Safety Law*. Through these efforts, new achievements have been made in punishing food safety crimes.

7.3.1 Public Security Organizations

In recent years, with a focus on edible oil, meat, wines and spirits, spices, local specialty food products, and other best-selling food products, the public security organizations at all levels have cooperated actively with the food safety supervision, agriculture, industry and commerce, quality inspection, and state food and drug administrative departments to address outstanding problems that seriously affect the sense of safety of the masses, endanger their health, and cause serious damage to their vital interests. They have cracked down on food safety crimes that involve a wide range of areas, have a large social impact, have evoked a strong public response, and involve "unspoken industry rules." They firmly demolished illegal factories, workshops, dens, and markets producing and selling toxic and hazardous foods and have broken down the chain of illegal profits. Moreover, they have strengthened the cooperation and coordination between different security forces and have taken the initiative to strengthen the cooperation with the procuratorate and court. Through these efforts, good results have been achieved in safeguarding food safety (Ministry of Public Security of the PRC 2012). In 2011, the public security organizations at all levels solved more than 5200 cases of food safety crimes and arrested more than 7000 people. On February 3, 2013, the Ministry of Public Security announced 10 typical food safety crimes among a number of cases that occurred in 2012.

7.3.1.1 Liaoning Shengtai Meat Processing Plant Sealed for Producing and Selling Toxic and Hazardous Sliced Mutton On January 29, 2013,

the Ministry of Public Security deployed the public security organization of Liaoning Province and nine other provinces, autonomous regions, and municipalities and solved the extremely serious case of the production and selling of toxic and hazardous sliced mutton by the Shengtai Meat Processing Plant in Liaoyang County. Thirty-four suspects were arrested, two production dens were destroyed, three production lines and more than 10 sets of production equipment were closed down, and more than 50 tons of toxic sliced mutton and raw materials were detained. The total value involved in this case was more than 30 million yuan. Through investigation, it was found that the plant produced toxic and hazardous sliced mutton by using duck meat and mutton fat as raw materials and illegally purchasing and adding water-retaining agents and other substances. The hazardous mutton was then sold to Liaoning, Jilin, and other places. Test results revealed that the water-retaining agents contained excessive nitrites and preservatives.

7.3.1.2 Fake Sliced Mutton Factory Unearthed in Dalian, Liaoning On January 25, 2013, the public security organization of Dalian successfully destroyed a large illegal factory that was found to be producing and selling fake mutton rolls. Four storage and processing dens were taken out, eight suspects were arrested, and four tons of fake mutton and two sets of processing equipment were detained. The total value involved in this case was more than 4 million yuan. The suspects, Xu et al., produced fake mutton with duck meat and sold it to Lushun and other places on a long-term basis.

7.3.1.3 Beijing Sunshine 100 Biotechnology Development Co., Ltd. Seized for Manufacturing and Selling Harmful Health Products On January 29, 2013, under the unified command of the Ministry of Public Security, the public security organizations of Jiangsu, Beijing, and Hebei jointly seized the criminal gang of Beijing Sunshine 100 Biotechnology Development Co., Ltd. This gang was suspected of adding harmful substances to health food. Six suspects were arrested and five production, sales, and storage dens were destroyed. The company added banned substances to its product, wild ginseng capsules, and sold the product to 23 provinces (autonomous regions and municipalities). The total value involved in this case was more than 30 million yuan.

7.3.1.4 Suspects Selling Fake Imported Liquor Arrested in Wenzhou, Zhejiang The public security organization of Zhejiang solved a serious case of selling fake imported liquor and seized more than 2000 bottles of Chateau De Lafite, Carruades de Lafite, and other brands of fake liquor, with a total value involved of more than 10 million yuan. Further investigation revealed that the suspects, Li et al., set up shell companies and sold fake liquor after purchasing through illegal channels.

7.3.1.5 Suspects Producing and Selling Water Injected Beef Arrested in Shijiazhuang, Hebei On January 28, 2013, the Ministry of Public Security sent officers to supervise the public security organization of Hebei and promptly solved the case of the production and selling of water injected beef which had been exposed by China Central Television (CCTV). As a result, six suspects were arrested. The criminal suspects, Di et al., produced and sold water injected beef since June 2012. On average, they slaughtered two head of cattle per day and injected 50 kg of water into each slaughtered animal. Until the arrest, they had slaughtered and sold more than 200 head of water-injected cattle.

7.3.1.6 Fake Beef Jerky Workshop Seized in Hohhot, Inner Mongolia On January 27, 2013, the public security organization of Hohhot destroyed an illegal workshop producing fake beef jerky and arrested four suspects. Since August 2012, the workshop had produced fake beef jerky by smoking other frozen meats over coal fires. The total value involved in this case was more than 1 million yuan.

7.3.1.7 Two Fake Beverage Factories Seized in Xiangyang, Hubei The public security organization of Xiangyang, Hubei, seized two fake beverage factories, ferreted out more than 2,000 carton boxes of fake beverages, including fake Sprite and Coca-Cola, along with branded dairy drinks, more than 50,000 pieces of counterfeit trademarks, and two sets of production equipment. In total, 26 suspects were arrested in this case. Since August 2012, the suspects had purchased counterfeiting equipment, counterfeit trademarks, cheap cola flavors, sodium citrate, and other inferior materials to produce fake beverages. After purification, mixing, and filling, the fake beverages were sold as famous brand names. More than 2000 boxes were produced per day and wholesaled to Shaanxi, Henan, Hubei, and other places. The total value involved in this case was more than 10 million yuan.

7.3.1.8 Counterfeit Chinese Liquor Den Unearthed in Nanning, Guangxi The public security organization of Nanning, Guangxi, unearthed a counterfeit Chinese liquor den, arrested 10 suspects, and ferreted out more than 2000 bottles of counterfeit Chinese liquor, such as counterfeit Red Star Erguotou, more than 0.2 million pieces of counterfeit trademarks, and four sets of counterfeiting equipment. Since August 2011, the suspects had purchased counterfeiting equipment and materials to produce more than

10,000 boxes of counterfeit Chinese liquor. The total value involved in this case was more than 2 million yuan.

7.3.1.9 Two Tainted Bean Sprout Workshops Seized in Yinchuan, Ningxia   The public security organization of Yinchuan successfully seized two tainted bean sprout workshops and detained 6.3 tons of tainted bean sprouts, illegal additives, such as growth-promoting agents, freshness improving powder, injectable drugs (veterinary medicine), and injection syringes, in conjunction with the agriculture and animal husbandry department and industry and commerce department.

7.3.1.10 Case of Selling Pork from Ill and Dead Pigs Unearthed in Weifang, Shandong   On January 26, 2013, the public security organization of Weifang unearthed a case of selling pork from ill and dead pigs and arrested nine suspects. Since June 2012, the suspects purchased a large number of dead pigs from the pig farm of Shouguang Fenggu Animal Husbandry Co., Ltd. as well as other farms. The company sold to local cold meat shops, meat pie shops, and pork stalls after slaughtering. The total value involved in this case was more than 2.6 million yuan.

7.3.2 Procuratorate

Statistics show that the People's procuratorates at all levels prosecuted 11,251 criminal suspects manufacturing and selling fake or substandard drugs, or toxic and hazardous food, and investigated 465 public officials suspected of malfeasance in the incidents of tainted milk powder, clenbuterol, swill-cooked dirty oil, and toxic capsules in the last five years to severely punish crimes endangering people's life and health (China.org.cn 2013). The Supreme People's Procuratorate also requires procuratorates at all levels to stress both punishment and prevention when investigating cases of malfeasance. In other words, when handling cases, the procuratorates should also make efforts to prevent malfeasance, make suggestions on rectification to relevant departments against problems discovered in the regulatory systems and mechanisms, and then jointly develop management measures to prevent and reduce criminal activities at the source.

7.3.3 Court

Statistics show that the national court system concluded 14,000 criminal cases of production and marketing of tainted milk powder, clenbuterol, swill-cooked dirty oil, and other toxic and hazardous foods, counterfeit products, or products that did not meet safety (hygiene) standards and sentenced more than 20,000 criminals involved in the last five years to severely punish crimes endangering people's life and health (Xinhuanet 2013c). The Supreme People's Court held a press conference on May 3, 2013, to circulate information on the trials of cases of food safety crimes in recent years and announce relevant typical cases. For the first time, the press conference was broadcasted live to the public by all media outlets, including CCTV, China National Radio, People's Daily Online, and the website of the Supreme People's Court, as well as microblogs of the People's Daily, Xinhua Viewpoint, CCTV News, Voice of China, People's Court Daily, Henan Provincial Higher People's Court, Higher People's Court of Shanghai, and Guangdong Provincial Higher People's Court. This live broadcast marked a new step of the Supreme People's Court in promoting judicial publicity (People's Daily Online 2013).

7.3.3.1 Overall Situation of Trials of Food Safety Crimes In recent years, the people's courts at all levels have tried a number of crimes endangering food safety and have punished numerous criminals involved. From 2010 to 2012, the people's courts nationwide concluded 1533 criminal cases of production and marketing of toxic and harmful foods or foods that did not meet safety/hygiene standards and effectively sentenced 2088 criminals. Of them, 39, 55, and 220 cases of production and marketing of foods that did not meet safety standards were concluded and 52, 101, and 446 people were effectively sentenced in 2010, 2011, and 2012, respectively; 80, 278, and 861 cases of production and marketing of toxic and harmful foods were concluded and 110, 320, and 1,059 people were effectively sentenced in 2010, 2011, and 2012, respectively.

7.3.3.2 Basic Characteristics of Food Safety Crimes Based on the trial practice, the current food safety situation remains very serious. It is mainly manifested in the following six aspects:

1. The number of criminal cases involving food safety has increased substantially. The number of criminal cases of production and marketing of foods that do not meet safety/hygiene standards or toxic and harmful food concluded by the people's courts nationwide registered a year-on-year growth of 179.83% and 224.62% in 2011 and 2012, respectively; correspondingly, the number of people effectively sentenced registered a year-on-year growth of 159.88% and 257.48%, respectively.
2. Vicious and serious food safety crimes, such as clenbuterol, substandard milk powder, tainted bean sprouts, waste oil, swill-cooked dirty

oil, toxic capsules, and pork from ill and dead pigs, occurred frequently and have triggered strong responses from the public.

3. The crimes involved not only ordinary consumers, but also infants and young children.

4. The way of committing food safety crimes has been changing and has become more difficult to detect. In essence, it is becoming more and more difficult to identify the nature of such criminal cases. In the past, criminals simply mixed toxic and hazardous substances with the food. However, a variety of techniques are in use today to circumvent the law; for example, the products are sold through the Internet and couriers to evade the supervision of administrative departments.

5. The abuse of additives not only occurs in the stage of food processing, but also occurs during production, circulation, and storage of food and agricultural products.

6. More crimes are organized by gangs and operated in the form of a chain, and criminals have had a growing awareness of evading criminal prosecution. Compared with other crimes, a higher proportion of food safety crimes have been organized by gangs. The members of criminal gangs have a division of labor responsibility. The stages of production, transportation, storage, and marketing coordinate with each other, while at the same time, they are relatively independent. The number of transregional crimes has increased significantly. Moreover, evidence is seldom left during the criminal activity. All these features set up obstacles for tracing the source of crimes.

7.3.3.3 Improvement of Relevant Laws   To fully implement the *Food Safety Law,* China's judicial and administrative authorities issued a range of provisions to punish activities endangering food safety and malfeasance relating to food safety in recent years, including *A Notice on Severely Punishing Food Safety Crimes* (document number: No. 38 [2010] of the Supreme People's Court of the People's Republic of China; the Supreme People's Court, the Supreme People's Procuratorate, the Ministry of Public Security, and the Ministry of Justice, September 15, 2010), *A Notice on Strengthening the Investigation and Transfer of Cases of Processing Foods with Non-food Substances* (Ministry of Health, Ministry of Public Security, Ministry of Agriculture, SAIC, AQSIQ, and SFDA, January 28, 2011), *A Notice on Severely Cracking Down on Illegal Addition to Food and Strengthening Regulation of Food Additives* (document number: No. 20 [2011] of the General Office of the State Council of the People's Republic of China; the General Office of the State Council of the People's Republic of China, April 20, 2011), and

A Notice on Severely Punishing Food Safety Crimes and Relevant Malfeasances (the Supreme People's Procuratorate, March 28, 2011). These provisions serve essentially the same purpose, that is, to strengthen coordination, cooperation, and communication between each department and implement the principle of "punishing in a heavy and quick way" to severely punish criminal cases involving food safety. *A Decision to Strengthen Food Safety Supervision* issued by the State Council on June 23, 2012, also emphasized the principle of "stressing management of disorder and heavy punishment."

The Supreme People's Court published two typical cases of food safety crimes in 2012 to guide the understanding and application of relevant criminal law provisions. The Supreme People's Court published four typical cases of food safety crimes in March 2012 and issued the *Typical Cases of Food and Drug Safety Crimes*, six of which were food safety crimes, on July 31, 2012. Many of the typical cases of food safety crimes include crimes of endangering the public security with a dangerous method, crimes of producing and selling toxic and harmful food, food that do not meet hygiene standards, and fake products, crimes of illegal business operations, and crimes of dereliction of duty. The Supreme People's Court has also provided guidance on conviction and sentencing for selling out-of-date milk powder; producing pork from ill and dead pigs; adding lemon yellow, pericarpium papaveris, formaldehyde, or other illegal substances; adding clenbuterol to feed that is to be produced; excessive use of food additives; failure of regulators to inspect the products as required; and other criminal behaviors that occur in practice by publishing the typical cases.

The Ministry of Public Security, the Supreme People's Procuratorate, and the Supreme People's Court jointly issued *A Notice on Severely Punishing Crimes Relating to Swill-cooked Dirty Oil* and made specific provisions on conviction and sentencing for production, processing, and marketing of swill-cooked dirty oil and other related activities on January 9, 2012. In terms of sentencing, the notice requires accurately controlling the application of the criminal policy of combining punishment with leniency in the field of food safety and clearly provides the following:

1. A criminal punishable by death shall resolutely be sentenced to death.
2. The source of a food safety crime shall be severely punished.
3. Traders of illegal food and individuals who are responsible shall be prosecuted and severely punished.
4. Conditions where suspension of sentence or exemption from criminal punishment applies shall be strictly controlled.
5. When suspension of sentence applies, injunction shall be declared at the same time in principle.

The improvement of the relevant legal system has provided conviction and sentencing standards for food safety crimes, unified the advice on the application of law to new difficult cases, and created a tight net of criminal justice for punishing food safety crimes. Furthermore, it will play an important role in further increasing the number of crackdowns on food safety crimes and creating a good atmosphere of preventing and punishing food safety crimes in the whole society.

7.4 Thoughts and Suggestions on Improving the Chinese Food Safety Legal System

Since the enactment of the *Food Safety Law*, the central and local departments have conducted a number of special rectifications on food safety and have made great achievements. However, many hidden dangers and problems remained to be solved for dairy products, pork from ill and dead pigs, and swill-cooked dirty oil. Moreover, new problems are constantly emerging. For example, the recurrence of the plasticizer incident in 2012 took a heavy toll on the Chinese liquor industry. At present, the Chinese food safety legal system and regulation model essentially maintains a top-down approach (Qi 2011). The *Food Safety Law* has strengthened the administration pattern led by government regulation. In the system, emphasis has been placed on the allocation and implementation of regulatory responsibilities of administrative organs in the field of food safety. In practice, over-reliance has been placed on the regulation approach of special rectification. Although the special rectification approach can manage disorders using heavy penalties, it lacks universality and continuity, cannot fundamentally solve food safety problems, and can provide only stop-gap measures in most cases. Therefore, to complement government regulation with the social power to address food safety problems should become the general trend and orientation of the construction of the Chinese food safety legal system. In addition, it is imperative to achieve a seamless connection of management and supervision throughout the food chain "from farm to fork" and to fill the gap of regulatory responsibilities in order to improve governmental management. This should be the general idea and direction in improving the legal system. The Chinese food safety management system was reformed in March 2013 and the amendment of the *Food Safety Law* and relevant laws is still underway. However, efforts should be made to address the three outstanding issues discussed in Sections 7.4.1 through 7.4.3 based on the current situation and more than three years of practice of the *Food Safety Law*, in addition to modifying the food safety management system and addressing the relevant problems.

7.4.1 Improve the System of Punitive Compensation

Improving tort liability mechanisms and increasing civil liability of food producers and traders for compensation for violations can effectively mobilize and utilize social power to supervise food safety violations and alleviate the burden and pressure of administrative organizations in food safety management. The *Food Safety Law* set up a tenfold punitive compensation mechanism and provides, in Article 96, that "Besides claiming damages, a consumer may require the producer, who produces food which does not conform to the food safety standards, or the seller who knowingly sells food which does not conform to the food safety standards, to pay 10 times the money paid." However, this punitive compensation system has barely functioned in practice. The court generally does not support the plaintiff's claim for tenfold compensation. The current punitive compensation system prescribed in the *Food Safety Law* can hardly achieve substantial results.

To give full play to the function of the punitive compensation system, the current *Food Safety Law* can be further amended by the following measures:

1. *Give a broad interpretation of "consumer."* It means that anyone who purchased a particular food can be identified as a consumer. This allows professional fake fighters to impose tort liability upon producers and traders under the punitive compensation provision, which will give full play to the role of the punitive compensation system in supervision and prevention.
2. *Reverse the burden of proof.* The producer shall bear the burden of proof for the conformity of the food with the food safety standards; the seller shall bear the burden of proof for his/her state of being uninformed of the safety defect of the food, and the consumer only needs to prove the existence of any apparent safety defect of the food he/she purchased to start proceedings.
3. *Give due consideration to public interest litigation.* Cases relating to food consumption are usually small in the value of an individual case, but large in the quantity of total cases. The litigation cost is too high for individual consumers. However, if a large number of cases for the same food are gathered into one case to share the cost of

litigation, it will stimulate the enthusiasm of consumers to safeguard their legal rights. All along, many scholars in the field of economic law have advocated the establishment of economic public interest litigation (Meng 2002; Lv and Yan 2003; Qi 2011). In the public interest litigation system, the range of plaintiffs will be expanded and the right to sue will be transferred from the victim to a third party, thus enhancing litigation force. The United States has had developed a relatively complete system of public interest group litigation, of which the "Declaration of Withdrawal" system and the system of rewarding lawyers from the funds recovered are a good example for China to learn from.*

7.4.2 Implement a Compulsory Liability Insurance System for Food Producers

In the area of food safety, liability for negligence occurring in the raw material supply, production, processing, storage, transportation, and sales of food and other stages shall be allocated to the perpetrator and victim according to the *Tort Liability Law*. However, due to the complexity of the food production and marketing chain, food safety accidents are often caused by a combined effect of various factors. It is difficult to clarify the liability allocation under the *Tort Liability Law* because of the complex causality. As a result, on the one hand, the legitimate rights and interests of consumers cannot obtain timely and effective protection, and on the other hand, the erring producer and trader are not subject to tort liability. Liability insurance is a commercial insurance purchased from a qualified insurer based on the insured's own free will. The insurer will compensate the victim in place of the insured when an accident incurring any legal liability occurs. Liability insurance can compensate for the disadvantage of the *Tort Liability Law* in addressing food safety incidents (Tu and Xu 2012). Therefore, liability insurance should be an inevitable choice for timely protection of the interests of victims in food safety incidents. To give full play to the role of liability insurance, the current *Food Safety Law* must be amended to legalize

---

*In 1986, the US Congress amended the *False Claims Act* to increase rewards for whistleblowers up to 30%, increase the power of whistleblowers to sue, and impose treble damages and civil fines on the defendant. Because of the large number of consumers, the rewards are sometimes quite lucrative, which is very attractive to lawyers. Lawsuits brought by whistleblowers increased rapidly thereafter. See Xu, B. 2008. Public interest law and public interest litigation. http://www.haongo.cn/news.asp?unid=30 (accessed April 20, 2013) (in Chinese).

the implementation of a compulsory liability insurance system in the field of food safety.*

7.4.3 Improve the Legal Foundations for Public Participation

Public participation is very important for the development of the Chinese food safety legal system. However, the current food safety information disclosure lags far behind the public demand for information on food safety in China. A large amount of information on food safety that should be disclosed is not publicly available (Li 2012). This section will provide a brief analysis of the case of application for information disclosure on national dairy standards in 2012 as an example.

Article 23 of the applicable *Food Safety Law* provides that "The national food safety standards shall be examined and adopted by the National Food Safety Standard Review Committee." On January 20, 2010, the Ministry of Health established the first National Food Safety Standard Review Committee, issued the *Constitution of the National Food Safety Standard Review Committee*, provided the basic procedures for establishing national food safety standards, and stated that the establishment of national food safety standards must be open for comments, according to the *Food Safety Law*. However, no specific procedure has been provided on the public consultation. On December 2, 2011, citizen Zhengjun Zhao applied to the Government Affairs Disclosure Office of the Ministry of Health for information disclosure. Zhao requested the Ministry of Health to disclose drafting organizations and drafters of the 66 national food safety standards regarding raw milk and other foods, the agreement on authorization of formulation/revision of national food safety standards entered into between the drafting organization of raw milk standards and the Ministry of Health, data and information on feedback unaccepted by the drafting organization, and minutes of the meeting prepared by the professional subcommittee for review, along with discussion of the raw milk standards and other government information. Emphasis was placed on the minutes of meeting for establishment

---

*Compulsory liability insurance for food safety is a new measure for the regulation of food safety issues and has been implemented in some countries in a limited manner. For example, the *Food Hygiene Law of Taiwan* provides that "Producers and traders of certain foods and a certain scale specified by competent authorities shall purchase product liability insurance." As another example, Germany provides that the producer of genetically modified food shall be liable for damages incurred by the genetically modified food, regardless of whether or not the producer is at fault, and that purchase of liability insurance shall be a precondition for the production and marketing of genetically modified food. See Guo, F. et al. 2009. *Research on Legislation of Compulsory Insurance*. Beijing, People's Republic of China: People's Court Press (in Chinese).

of the new national raw milk standards. However, Zhao's application was not approved (Xinhuanet 2012), which made the people reflect on this issue.

Therefore, the current *Food Safety Law* must be further amended and improved to clarify examination and approval procedures of the national food safety standards, increase transparency, expand the scope of public participation, and protect the public's right to know and participate (Qi 2011). In short, the ongoing amendment of the *Food Safety Law* should further clarify the scope of public information on food safety, identify institutions and news spokesmen for releasing authorized food safety information, and provide what information should be reported to and handled by the supervision department and that the supervision department shall bear legal liability for not handling the information.

8

Evolution and Reform of the Chinese Food Safety Management System

Prior to March 2013, the Chinese food safety management system had always been very controversial. This chapter briefly describes the historical changes of the Chinese food safety management system and analyzes the significance of its reform in March 2013, as well as the potential problems it is currently faced with. In view of the fact that the reformed food safety management system continues to emphasize the local management model of "local governments being fully responsible for food safety," research was carried out on the features of food safety management system reform in Shenzhen to assess whether local management systems were compatible with local reality.

8.1 Evolutions of the Food Safety Management System

In this section, the evolution of the Chinese food safety management system according to the history of China's development and reform with the founding of New China in 1949 as the starting point has been described. In this way, a panoramic view of the development process and the actual state of the Chinese food safety management system can be observed.

8.1.1 Directive Management System during the Planned Economy Period (1949–1978)

During the 29-year period from the founding of New China in 1949 to the reform and opening up in 1978, a centrally planned economic system was implemented in China. Correspondingly, the Chinese food safety management in this period was mainly controlled by the competent departments with the support from the health administrative department, which integrated food hygiene management into administrative management. Since this management system was adopted under the directive planned economy, it can be referred to as a "directive management system."

8.1.1.1 Formation of the Directive Management System During 1949–1978, China's primary goal of food safety was to address the problem of food and clothing. Food safety incidents also occurred during this period, but generally consisted of food poisonings during consumption. China's food quality and safety was, in fact, mainly affected by food hygiene issues at that time. The Chinese economic system implemented at that time determined the high dependence of food business management on competent authorities. Enterprises did not and could not generate relatively independent pursuits of commercial interests. The operational goals of enterprises were almost completely replaced by administrative goals. Therefore, food safety risks during this period of time were mainly caused by nonmarket competition factors, which in essence were a premarket risk. In addition, government authorities mainly regulated food producers, processors, and traders through means of plan targets and internal control. Economic incentives and punishments, judicial adjudication, information disclosure, technical standards, and other modern food regulatory policy tools were seldom used. In addition, China was affected by the epidemic prevention system implemented in the Soviet Union at that time. Thus, the health department naturally took charge of food hygiene management.

8.1.1.2 Main Characteristics of Food Safety Incidents in the Directive Management System Food safety incidents also occurred in China during this period. However, most were caused by objective constraints such as production technology and technical standards rather than unethical behavior regarding materials and workmanship, as well as counterfeiting and adulteration by food enterprises motivated by interests. For example,

107 food poisoning incidents affecting 4237 people occurred in Xuhui District, Shanghai, in the 1960s. The major causes of these incidents were cross-contamination (48.60%), prolonged storage period (23.36%), and food spoilage (14.95%) (Chen 1990). Along with environmental improvements, the number of food poisoning incidents and people poisoned decreased substantially to 71 and 2058, respectively, in the 1970s. During the three years from 1974 to 1976, there were 177, 133, and 96 food poisoning incidents, respectively, in Jiangsu Province, with a death rate of 0.17%; of these deaths, 89% were peasants who were killed by eating poisonous plants and animals by mistake (Guangxi Institute of Medical Scientific Information 1980).

8.1.1.3 Main Characteristics of the Directive Management System   In the directive management system, the food safety management authority was divided and held by the competent authorities of food enterprises (China Food News 1983). The competent authorities and food enterprises were in a superior–subordinate administrative relationship. Food management and supervision was achieved by administrative appointment, dismissal, education, persuasion, quality competitions, and other internal control measures and mass movements, rather than legal, economic, and professional standards. The main characteristics of food safety management and supervision in the directive management system were clearly reflected in the New China's first comprehensive central food hygiene regulations, *Administrative Regulations on Food Hygiene for Trial Implementation*, issued by the then Ministry of Health, Ministry of Commerce, First Department of Light Industry, Industry and Commerce Administration, and National Supply and Marketing Cooperative in 1965.[*][†]

---

[*]Prior to this, the Chinese central and local governments had issued relevant regulations and standards for the sanitary control of a specific variety of food. For example, the Ministry of Health issued the "Notice on the standard of sodium glutamate content in seasoning powder" and "Interim measures of management of cooling drink and food" in March 1953 and "A provision on the dosage of saccharin in food" in 1954. In 1957, the Tianjin health departments detected high levels of arsenic in soy sauce and proposed that the arsenic content should not be more than 1 mg/kg of soy sauce, and this standard was implemented nationwide by the Ministry of Health. In 1958, the then Ministry of Light Industry, Ministry of Health, and the second Ministry of Commerce issued ministerial standards and test methods of milk and dairy products, which have been in place since August 1, 1958. See Chen, Y.J. 1999. Standardization of food hygiene during 50 years in China. *Chinese Journal of Food Hygiene* 11(6): 17–19 (in Chinese).

[†]Citations in the section "Main characteristics of the directive management system" were from Tianjin People's Committee. 1965. A notice on implementation of the Provisional Regulations of Food Sanitation Management approved by the State Council. *People's Government Gazette of Tianjin Municipality* 17: 2–7 (in Chinese).

8.1.2 Hybrid Management System during the Economic Transition Period (1979–1992)

During 1979–1992, China's food safety management was still focused on food hygiene management, but the management system existed as a hybrid between a planned and market economy and between the traditional control model and the modern supervision model. The Chinese food hygiene management system during this period can be referred to as a "hybrid management system."

8.1.2.1 Coexistence of Diverse Forms of Ownership Along with China's reform and opening up in 1978, the Chinese food industry experienced rapid development, had an output of 113.4 billion yuan in 1987, and became the third biggest industry in the national economy (Yang and Xu 1988). At this time, the number of food producers, processors, and traders, as well as catering enterprises, increased sharply. The rapid development of the food industry was mainly attributed to the reform measures allowing common development of food enterprises under diverse forms of ownership. The development of the food industry broke down not only the barriers between sectors and regions, but also the barriers among forms of ownership. The exclusive dominance of state-owned enterprises over the years gradually resolved. The coexistence of diverse forms of ownership eclipsed the food hygiene management system that was based on control by competent authorities with support from the health department, which integrated food hygiene management into industrial administration. The emergence of the coexistence of diverse forms of ownership in the food industry had directly shaken the directive food hygiene management system established in the planned economy.

8.1.2.2 Main Characteristics of the Hybrid Management System The main characteristics of the hybrid management system are as follows: (1) The *Food Hygiene Law (For Trial Implementation)*, implemented on July 1, 1983, clearly stated that health departments at all levels took charge of food hygiene supervision, indicating that the health department became the primary law enforcer for food hygiene supervision. (2) The *Food Hygiene Law (For Trial Implementation)* did not completely abolish the right of food hygiene management of other administrative authorities. The competent authorities of food producers and traders were still in charge of food hygiene supervision in the respective systems. Therefore, the health department only achieved nominal dominance in food hygiene supervision during this period. In fact, the health department was in an awkward position as their dominant regulatory power was fragmented, because the combination of governmental function and enterprise management was not completely separated in the food industry and the management right of other competent authorities was partially retained. (3) The *Food Hygiene Law (For Trial Implementation)* empowered some nonhealth departments to supervise food hygiene in select areas. For example, the industry and commerce administrative department was in charge of the food hygiene management of urban and rural markets; the animal husbandry and fishery department was in charge of livestock and veterinary health inspections; the national import and export commodity inspection department was in charge of supervision and inspection of food exports; food hygiene supervision of railways, transportation, factories, and mines was the responsibility of the corresponding health and epidemic prevention department; the Ministry of Commerce was in charge of the production, marketing, and hygiene of grains and oils, subsidiary foods, local specialty food products, and food services; and the newly created State Bureau of Technical Supervision was responsible for the setting and implementation of food quality standards. The industry and commerce, standard measurement, environmental protection, sanitation, animal husbandry and veterinary, and food hygiene supervision departments were involved in food quality supervision (Xu 1992). Thus, food hygiene management was decentralized.

8.1.3 Regulatory System in the Market Economy (1993–2012)

In October 1992, China determined the goal of economic reform to be the development of a socialist market economic system. Along with the establishment and continuous improvement of the market economy system, the Chinese food safety management system has been in a constant state of change and adjustment. A market economy-based regulatory system has been gradually created. The evolution of the Chinese food safety management system during 1993 2012 will be discussed in Sections 8.1.3.1 through 8.1.3.3.

8.1.3.1 Management System Led by the Health Department   In March 1993, the *Reform Plan of the State Council* was passed on China's Eighth National People's Congress to abolish seven ministries and commissions, such as the Ministry of Light Industry and the Ministry of Textile Industry, and reorganize them as industry associations, such as the Association of Light Industry and the Association of Textile Industry (Luo 1993). Government departments no longer had direct management over food enterprises, and these food enterprises gradually became independent entities for production and

operation. This marked the transformation of the food safety management system to an external and third-party regulatory system. In October 1995, the Sixteenth Session of the Eighth National People's Congress formally adopted the revised *Food Hygiene Law* and stipulated that the health department of the State Council was in charge of nationwide health food hygiene supervision and management (Chen 1995). In terms of the management system, although a full right of food hygiene supervision and management was not granted to the health department under this law, the dominance of the health department was established and the relevant management authorities of competent departments were abolished. A relatively centralized and uniform food safety management system was thus developed. The number of food poisoning incidents and people poisoned in China decreased substantially during this period.*

8.1.3.2 Multidepartmental Segmented Management System (2003–2008) Since 1992, the deepening of China's economic reform has not only greatly promoted the development of the food industry, but also effectively improved the food industry chain system. Accordingly, a multidepartmental comprehensive food safety concept is more in line with the development of a modern food industry. Subtle changes in regulatory concepts were gradually presented in the reform of the management system. In the institutional reform of the State Council in 1998, the newly established State Bureau of Quality and Technical Supervision (SBQTS) was put in charge of the approval and release of national food hygiene standards, and the development of grain and oil quality standards and detection systems and methods, which were originally the responsibilities of the Ministry of Health and State Administration of Grain, respectively; the industry and commerce department replaced the quality and technical supervision department, taking charge of food quality supervision and management in circulation. The department of agriculture was still responsible for the quality and safety supervision and management of primary agricultural products. In the institutional reform of the State Council in 2003, the State Council decided to change the original State Drug Administration to the State Food and Drug Administration (SFDA) and put it in charge of comprehensive supervision and coordination of food safety, and investigation and management of major food safety incidents.

The inferior milk powder incident in Fuyang, Anhui, spawned the segmented food safety management system. The milk powder incident resulted in mild to moderate malnutrition of 189 infants and the death of

---

*Data from the *China Health Statistics Yearbooks*, 1991 and 1998.

12 infants from severe malnutrition and brought the central government's unprecedented attention to food safety management. As a result, the Chinese State Council officially established a new management system in September 2004 to hold the department of agriculture responsible for the supervision of primary agricultural production; the quality control department for the supervision of food processing; the industry and commerce department for the supervision of food circulation; the health department for the supervision of catering and consumption; and the food and drug supervision and administration department for the comprehensive food safety supervision and coordination, and the investigation and management of major food safety incidents. At this point, the Chinese food safety management system was officially transformed from a management system led by the health department to a segmented management system cogoverned by five departments.

8.1.3.3 Multidepartmental Segmented Management System under Comprehensive Coordination (2009–February 2013)   Since 2007, food safety incidents have occurred frequently in China, people have paid unprecedented attention to food safety problems, and the government has implemented a series of reform measures. In 2008, to address the overlapping responsibilities between departments, the State Council decided that the Ministry of Health supervise the SFDA, which took charge of the comprehensive coordination of food safety, and investigation and management of major food safety incidents. Also, the SFDA was responsible for food safety management in catering and consumption, and quality control of health products and cosmetics. In February 2009, the *Food Safety Law* was passed on the Seventh Session of the 11th National People's Congress. In February 2010, the Food Safety Commission was set up under the State Council as a high-level agency for the central government to discuss and coordinate food safety issues. In 2011, the State Council approved the move of three of the responsibilities of the Ministry of Health, including comprehensive coordination of food safety, investigation of major food safety incidents, and disclosure of important food safety information, under the Office of Food Safety Commission of the State Council. At this point, a food safety management system, with the Food Safety Commission as its coordinating body (with responsibilities divided among government departments) that held local governments fully responsible, was formally established.

The historical changes and features of the Chinese food safety management system, since the founding of New China in 1949, are summarized in Table 8.1.

TABLE 8.1

Changes in the Chinese Food Safety Management System and Differences between the Directive, Hybrid, and Regulatory Systems

	Directive System (1949–1977)	Hybrid System (1978–1992)	Regulatory System (1993–February 2013)
Attitude toward market mechanism	To eliminate the market	To expand the market	To supervise the market
Management subjects	Based on competent authority control; supplemented by health department supervision; combination of government and enterprises	Led by the health department in a fragmented pattern; partial combination of government and enterprises	Segmented supervision by multiple departments; complete separation between government and enterprises
Management objects	Public–private partnership; combination of government and enterprises; soft budget constraints; premarket risks	Denationalization of ownership; government and enterprises began to separate; hard budget constraints; mixed risks (premarket and market risks)	Diversification of ownership; complete separation between government and enterprises; hard budget constraints; transition from food hygiene in a single stage to food safety in the entire process; market risks being the primary risks
Major policy instruments	Persuasion and education; political campaigns; direct administrative intervention	Prohibited by law; judicial adjudication; economic sanctions	Product and technical standards; licensing system; providing information

8.2 General Framework and Key Responsibilities of the New Food Safety Management System

Before the reform in February 2013, a food safety management system was implemented in China. Despite its gradual development after several reforms, this system failed to fundamentally address the overlap of and gap in responsibilities caused by the multidepartmental segmented regulation. In March 2013, the Chinese food safety management system was reformed again to restructure regulatory bodies, in order to address the "buck passing" and unclear division of powers, as well as responsibilities among food safety management departments.

8.2.1 Major Drawbacks of the Existing Food Safety Management System

Food safety is directly related to people's health and safety, and the Chinese government has always attached great importance to food safety. With the continuous improvement in economic and cultural levels, increasing concerns exist regarding food safety. However, food safety in China has not been satisfactory and remains a difficult and complex issue.

Before the reform in February 2013, the Chinese food safety management system adopted a segmented regulation approach and a number of regulatory authorities were involved. For example, the General Administration of Quality Supervision, Inspection and Quarantine (AQSIQ) was responsible for production, the State Administration for Industry and Commerce (SAIC) for circulation, and the SFDA for catering (Figure 8.1). Practice has revealed that the more the regulatory authorities are involved, the more ambiguous the regulatory boundaries are. This results in an overlap of and gap in responsibilities, making it difficult for multiple departments to work seamlessly together and to effectively fulfill regulatory responsibilities (Dong and Chu 2013). Moreover, multidepartmental management also caused dispersion of regulatory resources, resulting in the weak power of each department, low utilization of resources, and low efficiency of law enforcement. Therefore, it is necessary to promote the effective integration of appropriate authorities and their responsibilities to achieve unified food safety management, to further improve food safety management.

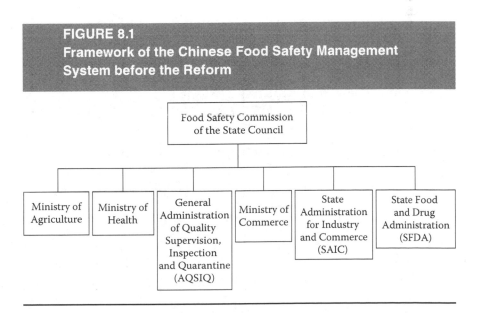

FIGURE 8.1
Framework of the Chinese Food Safety Management System before the Reform

8.2.2 General Framework of the New Food Safety Management System

According to the *Institutional Reform and Functional Transformation Plan of State Council* approved by the First Session of the 12th National People's Congress and *Notice on Organizational Structure* issued by the State Council, this reform of the food safety management system integrated the responsibilities of the Food Safety Commission Office of the State Council, SFDA, AQSIQ, and SAIC for food safety supervision in circulation and set up the China Food and Drug Administration (CFDA) directly under the State Council (General Office of the Chinese State Council 2013a). The key responsibilities of the newly created CFDA are to perform unified supervision and management of food and drug safety and to improve the efficacy in production, circulation, and consumption. The food safety supervision and management teams and inspection agencies of the industry and commerce department and of the quality and technical supervision department were transferred to the food and drug supervision and administration department accordingly. The Food Safety Commission of the State Council is retained with its specific responsibilities being taken over by the CFDA. In addition, to clarify responsibilities and to achieve seamless integration of food safety supervision and management among different departments, the newly created National Health and Family Planning Commission (NHFPC) was put in charge of food safety risk assessment and food safety standards setting; the Ministry of Agriculture was put in charge of agricultural product quality safety supervision; and the responsibilities of the Ministry of Commerce for pig slaughter supervision and management have been assigned to the Ministry of Agriculture.

The CFDA was formally established on March 22, 2013, and has taken on the function of the Food Safety Commission Office of the State Council. The State Council issued the *Key Responsibilities, Internal Structure, and Staffing of the China Food and Drug Administration* on March 31, 2013, which marked the official establishment of the new Chinese food safety management system. The new food safety management system has fundamentally changed compared with the previous system. This system has realized the effective integration of various regulatory resources and has unified the supervision and management of food production, distribution, and consumption. The segmented regulatory model supplemented by supervision of individual foods was converted to a centralized regulatory model. The general framework of China's new food safety management system is shown in Figure 8.2.

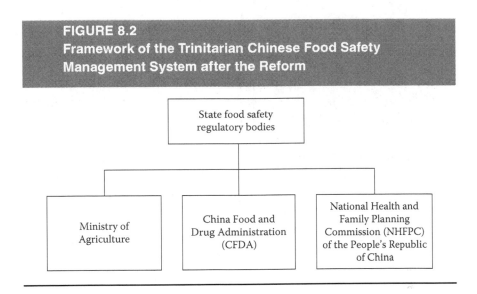

FIGURE 8.2
Framework of the Trinitarian Chinese Food Safety Management System after the Reform

8.2.3 Key Responsibilities of Relevant Authorities after the Reform

As shown in Figure 8.2, the newly reformed food safety management system implements food safety management mainly through three departments: the Ministry of Agriculture is in charge of the nationwide supervision of primary agricultural production; the NHFPC is responsible for food safety risk assessment and formulation of national standards; and the CFDA performs unified supervision and management of food production, circulation, and consumption (Feng 2013). In addition, the AQSIQ, SAIC, Ministry of Commerce, Ministry of Public Security, and other departments still need to provide support and coordination in their respective sectors.

8.2.3.1 China Food and Drug Administration   The CFDA integrates the relevant responsibilities of the Food Safety Commission Office of the State Council, SFDA, AQSIQ, and SAIC to perform unified supervision and management of food and drug safety and efficacy in production, circulation, and consumption (General Office of the Chinese State Council 2013a).

8.2.3.2 National Health and Family Planning Commission   The NHFPC is primarily responsible for carrying out food safety risk monitoring and assessment; developing and publishing food safety standards; performing safety reviews of new raw materials and varieties of food, food additives, and related products; and developing and implementing food safety risk monitoring plans in conjunction with the CFDA and other departments.

8.2.3.3 Ministry of Agriculture The key responsibilities of the Ministry of Agriculture are quality and safety supervision and management of agricultural products, specifically including quality and safety supervision and management of edible agricultural products from planting and growing to the entry into wholesale and retail markets or production and processing enterprises. The Ministry of Agriculture is also involved in supervision and management of the use and quality of veterinary drugs, feed, feed additives, pesticides, fertilizers, and other agricultural inputs, and quality and safety supervision and management in livestock, poultry slaughter, and raw and fresh milk procurement.

8.2.3.4 Administration of Quality Supervision, Inspection and Quarantine The AQSIQ is primarily responsible for the supervision and management of the production and processing of food-related products, such as packaging materials, containers, and production and trade tools. It is also responsible for the safety and quality supervision and inspection of imported and exported food.

8.2.3.5 State Administration for Industry and Commerce, Ministry of Commerce, and Ministry of Public Security The SAIC is responsible for the supervision and inspection of health food advertising, while the review of health food advertising content is performed by the CFDA. The SAIC and CFDA now work jointly to establish and improve the supervision of food safety.

The Ministry of Commerce is responsible for developing plans and policies to promote catering and liquor circulation, while the CFDA is responsible for the supervision and management of food safety in catering and liquor safety.

The Ministry of Public Security is responsible for the investigation of food cases. In addition, the CFDA and the Ministry of Public Security have established a bridging mechanism between administrative law enforcement and criminal justice. The food and drug supervision and administration department must report illegal conduct related to food to the public security department in a timely manner once detected in accordance with the relevant provisions. The public security department shall then promptly review the case and decide whether to file the case according to the law. The public security department may request the food and drug supervision and administration department to provide testing, verification, identification, or other assistance according to the law, and it shall provide the assistance requested.

8.3 Significance of the Food Safety Management System Reform

The previous Chinese food safety management system was developed during the transition from a planned economy to a market economy. It evolved from the directive management system in the planned economy period, through the hybrid management system in the economic transition period, to the regulatory system in the market economy. This system has played a positive role in improving food safety. On the contrary, the newly reformed management system has made a new step toward the exploration and final solution of overlapped and segmented food safety management, and "buck passing," and unclear division of powers and responsibilities among different departments. It has also had a positive effect on developing an integrated, extensive, specialized, efficient food safety management system, and comanagement of food safety by the government and civil society.

8.3.1 A New Step toward Seamless Management

Compared with the previous multidepartmental, segmented management system, the new management system, developed through the functional transformation-based "super-ministry reform," is able to better integrate various regulatory resources to effectively address the overlap of and gap in responsibilities. Theoretically, the new system can achieve the centralized management of food safety; unified supervision and management of food safety and drug safety and efficacy in production, circulation, and consumption; and seamless integration and management of each stage in the complete food industry chain "from farm to fork." The integration of food production and circulation into the food and drug administration system will contribute to clear division and unified management of responsibilities and avoid "buck passing" between regulators.

8.3.2 Integration of Regulatory Power

According to the general requirements of the State Council on institutional reform and functional transformation, the food safety supervision and management teams and inspection bodies of the industry and commerce department and quality and technical supervision department from central to local governments were transferred to the food and drug supervision and administration department. This was done to ensure that the CFDA had sufficient power and resources to perform its duties effectively. Specifically, the industry and commerce departments at the provincial, municipal, and county levels, and their agencies at the grassroots level, should assign the appropriate regulation and law enforcement personnel, headcount, and related funds. In addition, the quality and technical supervision departments at the provincial, municipal, and county levels should assign the appropriate regulation and law enforcement personnel, inspection agencies, number of qualified personnel, equipment involved in food safety, and related funds. Moreover, regional inspection and testing centers have been established by integrating food safety inspection and testing resources at the county level.

Integration of the food safety inspection and monitoring capabilities of food safety supervision departments helps resolve the inconsistencies in inspection and monitoring information and in relevant standards. A state food safety laboratory system affiliated with the food safety supervision departments has been developed by integrating the food safety inspection and monitoring agencies of each department through institutional adjustment. Based on the integration of food safety inspection and monitoring capabilities, great efforts have been made to develop a robust top-level design scheme to guide the complementary and characteristic development of food safety inspection and monitoring capabilities, avoid redundancy, and build a unified, shared food safety inspection and monitoring information database (Jiao 2013).

8.3.3 Shifting of the Focus of Regulatory Power to Lower Levels

The new management system encourages the county-level food regulatory authorities to establish food regulatory agencies in villages and towns and to increase grassroots regulatory power and allocate necessary technical equipment to fill the gaps in grassroots regulation and law enforcement. This ensures the enhancement of grassroots food supervision capacities through regulatory resources integration. Food regulatory coordinators have been appointed in administrative villages and urban communities to assist law enforcement and work on troubleshooting, information reporting, advocacy, and guidance, moving inspection points forward along the chain, shifting the focus of food safety supervision to lower levels, and accelerating the development of a complete food management system that has a full coverage (Xinhuanet 2013b).

In the previous management system, the food safety regulatory agencies were only set up at district and county levels in large- and medium-sized cities. Administration and daily supervision was absent at the grassroots level, where supervision is most needed. It was difficult for the food safety regulatory agencies to take down ubiquitous, long-standing, illegal factories, dens, and workshops in a timely and effective manner, not to mention implementing preventive measures (Xinhuanet 2013a). As food safety regulatory power had long been absent at the grassroots level, the food safety regulatory agencies could only deal with problems by unscheduled inspections. The food safety regulatory agencies would hastily arrange special inspection and rectification only in the event of large regional food safety incidents that were usually exposed by the media.

8.4 Potential Problems the Food Safety Management System Is Faced with after the Reform

The new management system has integrated relevant resources and, to some extent, addressed the unclear division and overlap of and gap in responsibilities caused by the multidepartmental segmented regulation. However, a number of problems regarding the division of labor and responsibilities remain unsolved. Substantial efforts are still required to achieve seamless multidepartmental collaboration and comanagement by the government and civil society. The food safety management system still needs to be further reformed and improved.

8.4.1 *Difficulties in Source Supervision*

As mentioned earlier, the new management system is comprised of three main bodies: the CFDA, the Ministry of Agriculture, and the NHFPC. The SAIC, AQSIQ, and the Ministry of Commerce are now excluded. However, the hardest nut, the supervision of edible agricultural products during production, which is the source of current food safety risks (and will be more risky in the future), is excluded from the responsibilities of the CFDA and is left to the Ministry of Agriculture. From the perspective of institutional arrangements, the supervision of both quantity and quality will lead to severe problems (Zheng 2013).

There are more than 200 million small households of peasants in China. Thus, it is the most difficult production unit to be managed in food safety supervision. In the new food management system, the Ministry of Agriculture has to supervise not only the agricultural production, but also the safety of agricultural products. Management of both by the same department is highly difficult. Abuse of antibiotics in breeding, plant hormones in fruits, pesticide residues in vegetables, and excessive use of fertilizers in cereals not only lead to quality problems in the raw materials of food, but also lead to leaching of a large number of toxic and hazardous substances into the soil and ground-water, which generates long-term danger and is more difficult to manage. It is clearly impossible to manage the source contamination simply by relying on self-management of the agricultural sector. The main task of the Ministry of Agriculture is to protect the supply of agricultural products, that is, the quantity. However, quantity and quality are contradictory to some extent, especially in the short term.

Several major food safety incidents that occurred recently in China are associated with the source of food production. For instance, in the "melamine milk powder incident," dairy farmers colluded with milk stations to add water and melamine to the milk. In the "Shineway clenbuterol scandal," farmers added clenbuterol to the feed to meet consumer preferences for lean meat due to irresponsible profiteering. In the "KFC incident," farmers used excessive antibiotics and hormones to obtain higher yields and prevent poultry dis-eases. In short, difficulties in food safety supervision will gradually move up along the chain and will be reflected in the source in the future. The source will be the most difficult link for management in food safety supervision and will bring about many problems. Currently, the new food safety manage-ment system still assigns the supervision of source of food production to the Ministry of Agriculture, which may be building up problems for the future.

8.4.2 Seamless Management Is Yet to Be Tested

A clear division of responsibilities is a prerequisite in food safety manage-ment. Moreover, to avoid regulatory gaps, regulatory bodies should have regular and irregular interdepartmental meetings, as well as excellent com-munication and coordination. It is critical to achieve effective coordination among the three coregulators, that is, the CFDA, Ministry of Agriculture, and NHFPC. A good coordination mechanism can improve the effectiveness of comanagement and avoid fragmentation. However, organic coordination among regulators has been the weakness of the Chinese food safety manage-ment system for quite some time. Under the new management system, the CFDA is ultimately responsible for food safety issues. However, food safety encompasses numerous areas. The CFDA has to coordinate not only with the Ministry of Agriculture, which is responsible for the source supervision, and the NHFPC, which is responsible for food risk assessment and standard management, but also with the AQSIQ, SAIC, Ministry of Commerce, Ministry of Public Security, and other departments involved in food safety supervision. Therefore, a long process will be required to establish a good coordination mechanism among the multiple food safety regulators, and seamless management is yet to be tested in practice.

8.4.3 The Local Management Model Is Facing Challenges

To consolidate the established pattern of local governments being fully responsible for local food safety, the new reform continues to emphasize the full responsibility of local governments for local food safety and implement the local management model. The biggest problem of local management is the possibility of local protectionism. In other words, it will be very challenging for the CFDA to supervise the food safety law enforcement by local governments to effectively prevent local protectionism. The frequent occurrence of serious food safety problems in recent years in China has fully revealed that the fulfillment of food safety supervision responsibilities is far beyond the regulatory capacity of local governments. The local management model of the new management system will inevitably result in a high dependence of the central government on local governments. Efforts must be devoted to effectively enhancing the direct supervision of food safety by the central government.

8.4.4 Difficulties in Comanagement by the Government and Civil Society

Fundamentally, the final solution to Chinese food safety problems requires the participation of social subjects to give full play to the market mechanism, industry self-regulation, and social supervision to achieve comanagement of food safety by the government and civil society. However, the new food safety management system fails to develop an effective incentive mechanism to give full play to the role of civil supervisors.

8.4.4.1 Lack of Knowledge of the Positive Effects of Other Social Subjects   In the Chinese food safety management system, the government has always played the role of an active mighty supervisor and treated other social subjects as passive vulnerable managed objects or government-affiliated agencies. They have not developed a partnership of equality and cooperation.

The government often undertakes all supervision affairs in an almighty way and even crosses the line to directly interfere with the behaviors of social subjects. In fact, Chinese social subjects have had the ability to participate in food safety supervision and come to the forefront. For example, many Chinese food safety incidents were reported by citizens, exposed by the media, and solved by corporate self-regulation. The Chinese government needs to change its position in the food safety management system, essentially, moving away from trying to run the whole show and setting the stage for the participation of multiple subjects in social management.

8.4.4.2 Insufficient Participation in Food Safety Supervision of Social Subjects   Food safety supervision is an onerous and complex task. Only by giving full play to the positive role and advantages of social subjects can seamless management be truly achieved. For a long time, the participation of social subjects in food safety supervision has been in a fragmented and unordered state in China. Social subjects have a limited right to speak and play an extremely limited role in improving food safety supervision. The primary cause of this is the failure to establish effective policy mechanisms to motivate social subjects to participate in food safety supervision. Thus, the participation of industry associations and social groups is obviously inadequate. Moreover, some governmental departments have not reduced the control of and even interference with the media, and effective channels and incentives to motivate consumer participation in food safety supervision are absent. Although some social organizations have displayed their supervision capacity, social supervision has not been guaranteed by effective incentive mechanisms and systems, which has severely restricted the development of a comanagement system.

8.5 Exploration of the Innovation of Food Safety Management System by the Shenzhen Municipal Government

As mentioned earlier, the local management model of food safety management is yet to be tested after the reform of the Chinese food safety management system. However, regardless of the reform, the emphasis of management should be placed at the local level, the focus should be directed at the grassroots level, and the primary responsibility should be taken on by the local governments. During the transition between the old and new systems, it is important to explore and develop local government food safety systems suitable to local realities. This section is devoted to a discussion of this issue with the exploration of the innovation of a food safety management system by the Shenzhen municipal government as an example.

8.5.1 New Background

Optimizing food safety management system to improve food safety management efficiency is the key strategic issue of concern for the Chinese government with respect to food safety management in recent years. Since the Sanlu milk powder incident in 2008 and the promulgation and implementation of the *Food Safety Law* in 2009, the local governments have had an increasingly important role and responsibility in food safety management. The issuance of the official document *A Decision to Strengthen Food Safety Management* by the Chinese State Council in July 2012 marked the official establishment of a local government accountability system for food safety issues. The *Institutional Reform and Functional Transformation Plan of the State Council* approved by the First Session of the 12th National People's Congress in March 2013 has made significant reforms in the food safety management system. The *Instruction on the Reform and Improvement of Local Food and Drug Administration Systems* issued by the State Council in April 2013 continues to emphasize that the local governments should take total responsibility for local food safety and be responsible for the reform of the local food safety management system (General Office of the Chinese State Council 2013b). Therefore, local governments will be given greater powers in the reform and innovation of food safety management system and mechanism. Hence, research of the reform and innovation of local government food safety management systems is of great practical significance. As a pioneer in food safety management reform, Shenzhen has actively explored the new model of food safety management by the local government.

8.5.2 Evolution of the Food Management System Reform by the Shenzhen Municipal Government

In August 2009, a reform of government institutions was implemented in Shenzhen to reduce the number of governmental departments from 46 to 31. One of the highlights of this institutional reform was the integration of the Municipal Administration for Industry and Commerce, the Municipal Bureau of Quality and Technical Supervision, and the Municipal Intellectual Property Office into the Market Administration of Shenzhen Municipality.

It was the first combined department to realize initial unified supervision of food production, distribution, and catering in China. However, edible agricultural products were not involved. In December 2011, the Shenzhen government decided to abolish the Municipal Bureau of Agriculture and Fisheries and assign the agricultural product safety supervision responsibilities to the Market Administration. At this point, the Market Administration of Shenzhen Municipality has taken total responsibility for food safety supervision and has begun to extend its responsibility to the source of food production.

In 2012, the Shenzhen government transferred the responsibilities for the comprehensive coordination of food safety from the Office of Food Safety Commission to the Market Administration to create the Food Safety Administration of Shenzhen Municipality. The Food Safety Administration is devoted to food safety management. As a significant adjustment following the institutional reform for supervision of food production, distribution, and catering in 2009, and the institutional reform for supervision of edible agricultural products in 2011, it has basically streamlined the food market supervision system.

8.5.3 Basic Features of the Innovation of Shenzhen

The innovation of the food safety management system of Shenzhen has certain features. They are described in Sections 8.5.3.1 through 8.5.3.3.

8.5.3.1 All-Out Building of a Food Safety Credit System In 2012, the Shenzhen government established a food safety credit information archive management system and prohibited owners of enterprises with revoked licenses due to food violations from engaging in the same industry. This was done to improve the punishment mechanism for breaking faith and to promote the lawful and credible operation of enterprises. Also, a public food credit information inquiry and disclosure platform was created. Moreover, a special column was set for disclosure of food safety credit information on the Shenzhen Credit Net and the first "blacklist" of food safety was exposed before the end of 2012. With this website, food safety credit information inquiry services have been provided to the public to strengthen social supervision.

8.5.3.2 Full Implementation of a Food Safety Manager System Shenzhen was the first to introduce the food safety manager system covering food production, distribution, and catering in China. The employment of food safety managers has become an essential condition for obtaining food-related licenses (Yu 2012). Food safety managers act as the first checkpoint of food safety. As of 2011, a total of 32,120 food safety managers (21,732 in catering services, 10,339 in food distribution enterprises, and 49 in food production enterprises) had been trained and obtained the Food Safety Manager Certificate in Shenzhen. It is expected that more than 120,000 food-related entities in Shenzhen will have at least one qualified food safety manager by 2015. An in-house food safety management team will be built in enterprises to improve the overall food safety management level in Shenzhen.*

8.5.3.3 Introduction of a Quantified and Classified Management System for Food Service Entities   In May 2011, a daily supervision information system based on a grid supervision platform was formally launched in Shenzhen. With the support of the supervision system combined with quantified and classified management measures, daily supervision and inspection has been strictly carried out for food enterprises to supervise the fulfillment of their corporate responsibility for food safety. Quantified and classified management has been implemented completely for food service entities. In 2012, there were 452 Class A entities, 20,811 Class B entities, and 39,207 Class C entities in Shenzhen. The number of Class A school canteens increased by 9.5% compared to the previous year. Furthermore, great efforts have been made to enhance unified tableware disinfection and sanitary inspection, and a quantitative risk assessment system has been established. In total, 32 registered tableware disinfection companies exist in Shenzhen, including 12 Class A, 12 Class B, and 8 Class C companies.

In fact, the reform of the food management system by the Shenzhen government has achieved significant results. In 2012, 20,495 batches were sampled in food production, distribution, and catering in Shenzhen, with an overall pass rate of 94.2%. Food safety there has been improved steadily. Overall, the Shenzhen model has produced positive effects (Dayoo.com 2013). The initial connection of supervision responsibilities between food production, distribution, and catering has been realized, and the Shenzhen model has overcome the institutional barriers in food traceability and linked supervision between different stages. It has also combined the complementary advantages in human resources and technological strength of the original regulatory bodies.

---

*White paper on the food safety management responsibilities in 2012 of Shenzhen, Food Safety Commission of Shenzhen Municipality.

8.6 Future Prospects for Deepening Reform

Based on China's experience of over 50 years in food safety management, the innovative model of Shenzhen with local characteristics, and the current super-ministry reform plan of food safety management, the Chinese central government and local governments must stand on a higher historical starting point to reflect on the institutional management system and government actions.

Theoretically, the Chinese food safety management system reform implemented in March 2013 can integrate regulatory power and achieve centralized management of food safety and has made a new step toward seamless management. Moreover, it has established a mechanism that has shifted the focus of work to lower levels and attached more importance to strengthening and improving grassroots management. However, it should be noted that the Chinese food safety management system still has many problems. For example, the source of food safety risks, that is, the safety of edible agricultural products, is not under the control of the CFDA but of the Ministry of Agriculture, which means that the quantity and quality are managed by the same department. Understanding how to manage the division of responsibilities in food safety management between the central government and local government, to deepen the role of local governments in food safety management while strengthening direct supervision by the central government, and to continue to explore innovation of local management systems will be the focus of the future reforms.

The exploration of reform and innovation of the food management system and mechanism by the Shenzhen government provides institutional experience in the reform and improvement of local food safety management systems and even in the reform of the national management system. Local governments should reform the local food safety management systems according to the general requirements established by the State Council in March 2013. However, it is not a justification for ceasing exploration and innovation of local food safety management. On the contrary, local governments should make pioneering and innovative efforts in the context that most local food safety management systems are basically the same. In particular, local governments should increase efforts in exploration and innovation of the specific food safety management mechanisms and enhance competition, sharing, exchange, and learning between local governments in food safety management mechanisms, to lay a solid foundation for innovation of the national food safety management mechanism and create a more high-quality, efficient, and professional food safety management system.

9

Construction and Development of a Food Safety Standard System

Before the implementation of the *Food Safety Law*, more than 2000 national standards, 2900 industry standards, and 1200 local standards, as well as a large number of enterprise standards, had already been developed for food, food additives, and food-related products. A preliminary food standard system based on national standards and supported by industry, local, and enterprise standards was thus established. However, the overall level of standards is low and there are contradictions, overlaps, and duplications among the standards. Moreover, some important standards or indicators are absent, and the scientificity and rationality of some standards need to be improved. In addition, some standards, including mandatory standards, are poorly implemented. With the implementation of the *Food Safety Law* on June 1, 2009, explicit provisions were made for the construction of the Chinese food safety standard system and other issues. This created conditions for the cleanup, integration, and unification of the food safety standard system. This chapter aims to summarize the construction and development of the Chinese food safety standard system after the implementation of the *Food Safety Law* and presents the author's insights.

9.1 Improvement of the Management System of the Food Safety Standards

Article 21 of the *Food Safety Law* provides that "the national food safety standards shall be formulated and announced by the health administrative department of the State Council, for which the standardization administrative department of the State Council shall provide the serial number of national standards." After the implementation of the *Food Safety Law* on June 1, 2009, the Ministry of Health and other relevant departments have done fruitful work in improving the management system of the safety standards.

9.1.1 Promulgation of Administrative Measures for National Food Safety Standards

To regulate the management of food safety standards and ensure smooth cleanup, integration, revision, and formulation of the food safety standards, the Ministry of Health drafted the *Administrative Measures for National Food Safety Standards* (Draft) based on the characteristics of the work related to national food safety standards under relevant provisions of the *Food Safety Law*. After revision in consultation with the relevant ministries and commissions, as well as provincial Health Departments, the *Administrative Measures for National Food Safety Standards* (hereinafter referred to as the *Measures*) were adopted at the executive meeting of the Ministry of Health on September 20, 2010, and promulgated on December 1, 2010. The *Measures* have four characteristics in content, which are provided in Sections 9.1.1.1 through 9.1.1.4.

9.1.1.1 Emphasis on Scientific Formulation The *Measures* require that the drafting of the national food safety standards shall be based on quality and safety risk assessment results of food and edible agricultural products, consult relevant international standards, and give full consideration to China's social and economic development level.

9.1.1.2 Principle of Openness and Transparency The *Measures* require that after the standards are drafted, the opinions of users of the standards, research institutions, industries, enterprises, consumers, experts, regulatory departments, and other walks of life shall be solicited and that the standards that have passed the preliminary examination by the Secretariat shall be published on the website of the Ministry of Health for public comments.

9.1.1.3 Encouragement of Broad Participation The *Measures* encourage citizens, legal persons, and other organizations to participate in the formulation and revision of the national food safety standards and put forward comments and suggestions, and state that any citizen, legal person, or any other organization may put forward comments and suggestions on the problems existing in the implementation of the standards.

9.1.1.4 Stress on Standard Examination The *Constitution of the National Food Safety Standard Review Committee* divides the examination of the drafts of the national food safety standards into three procedures, including preliminary examination by the Secretariat, examination by the meeting of the professional subcommittees, and deliberation by the directors' meeting. Moreover, contents and format of the materials on standards submitted to the Ministry of Health for approval shall be reviewed by the National Center for Health Inspection and Supervision of the Ministry of Health to ensure maximum standardization.

9.1.2 Promulgation of Administrative Measures for Local and Enterprise Food Safety Standards

Currently, numerous varieties of food exist, and new varieties are being regularly introduced. However, the national food safety standards are formulated and promulgated according to strict procedures, and therefore, national standards that cover all food varieties cannot be formulated in a timely manner. When there is no need to formulate national standards for products that are produced and consumed within a small area, such as local traditional food, for the time being, local or enterprise standards are required to regulate the production. According to the *Food Safety Law*, in the absence of national food safety standards, local food safety standards may be formulated; in the absence of national food standards or local standards for food produced by an enterprise, the enterprise shall formulate enterprise standards as the basis for organizing the production thereof. These provisions are important to the protection of the consumers' health and safety and are also reflective of respect to the autonomy of enterprises in food production and operation. The Ministry of Health promulgated the *Measures for the Archival Filing of the Enterprise Food Safety Standards* on June 10, 2009, which make provisions on the content, formulation principles, and archival filing deadline of enterprise food safety standards and place special emphasis on the legal liability of enterprises. In March 2011, the Ministry of Health issued the *Administrative Measures for Local Food Safety Standards* to make detailed provisions for conditions where local food safety standards could be formulated, as well as contents of the standards, prohibitions, and filing deadlines.

9.1.3 Establishment of the National Food Safety Standard Review Committee

The first general meeting of the National Food Safety Standard Review Committee was held in Beijing on January 20, 2010. This committee is primarily responsible for reviewing the national food safety standards, putting forward proposals for the implementation of national food safety standards, and providing consulting services for major issues of food safety standards. The first committee comprised of 10 professional subcommittees. These subcommittees are responsible for reviewing the national food safety standards in their respective professional fields, including food products, microorganisms, production and business operation regulation, nutrition and foods for special diets, test methods and procedures, contaminants, food additives and food-related products, pesticide residues, and veterinary drug residues. Establishment of this committee gives full play to the role of experts in different disciplines and is conducive to the scientific and rational formulation of national food safety standards.

9.2 Formulation and Revision of the New National Food Safety Standards

With China's rapid economic and social development in recent years, more strict food safety standards are required. In 2008, the State Council promulgated the *Regulation on the Supervision and Administration of the Quality and Safety of Dairy Products* and *Program Outline for Rectifying and Promoting the Dairy Industry*, which required the cleanup of safety standards for dairy products to be completed within 1 year. In February 2009, the State Council issued the *Food Safety Rectification Program*, which required the standards of pesticide and veterinary drug residues, toxic pollutants, pathogenic microorganisms, mycotoxins, and food additives to be revised and improved and relevant food safety standards to be integrated within approximately 2 years. Since 2010, the Ministry of Health has promulgated 269 new national food safety standards concerning the safety of dairy products, the use of food additives and compound food additives, mycotoxin limits, prepackaged food labeling and nutrition labeling, pesticide residue limits, and food additive products. During this time, the Ministry of Health has also revised and improved the standards of food packaging materials and improved the scientificity and practicability of the national food safety standards.

9.2.1 *National Safety Standards for Dairy Products*

Dairy products are not only directly consumed by consumers, but also widely used in the food industry as important raw materials. The safety of dairy products is very important to product safety in the entire food industry. On April 21, 2010, the Ministry of Health promulgated 66 new standards regarding the safety of dairy products, such as *Raw Milk* (GB 19301-2010), after examination by the first National Food Safety Standard Review Committee according to the *Food Safety Law*, *Regulation on the Supervision and Administration of the Quality and Safety of Dairy Products*, and *Program Outline for Rectifying and Promoting the Dairy Industry*. The new standards include 15 product standards, 11 of which are regarding general products and 4 are regarding infant food, 49 standards for test methods, with 39 regarding physical and chemical methods and 10 regarding microbiological methods, as well as 2 production standards. The new standards have improved the scientificity of the national safety standards for dairy products, established a unified system of national safety standards for dairy products, and solved most problems in dairy product standards, such as contradictions, duplications and overlaps of standards, and the unscientific setting of indicators.

Standards for dairy products are divided into standards for general products and safety standards for infant foods. Standards for general products include 11 categories: *Raw Milk*; *Pasteurized Milk*; *Sterilized Milk*; *Fermented Milk*; *Milk Powder*; *Condensed Milk*; *Watery Cream, Cream and Dehydrated Cream*; *Whey Powder and Whey Protein Powder*; *Cheese*; *Modified Milk*; and *Processed Cheese*. Safety standards for infant foods include four categories: *Infant Formula, Formula for Older Infants and Young Children, Cereal-based Supplementary Foods for Infants*, and *Canned Supplementary Foods for Infants*. Infant food is listed separately because infants are considered to be a special population. Products in this category have to meet the growth and physiological needs of infants. Nevertheless, most infant formulas contain milk, so the infant food safety standards are included in the dairy product standards. Standards for test methods are divided into standards regarding physical and chemical testing methods and those regarding microbiological testing methods. Among them, 39 standards pertain to physical and chemical testing methods, mainly for the determination of food ingredients, physical properties, vitamins, trace elements, and pollutants; 10 standards pertain to microbiological testing methods, specifying general requirements for microorganisms, indicator bacteria, and pathogens. The two production standards are *Good Manufacturing Practice for Dairy Enterprises* and *Good Manufacturing Practice for Infant Formula Powder Producers*. Sections 9.2.1.1 through 9.2.1.6 describe the six aspects where the progress of the new national safety standards for dairy products is mainly reflected.

9.2.1.1 Emphasis on Safety Requirements The new standards focus on safety management in production and other source stages. The production standards highlight the requirements for product recall, training, consumer communication, and records. In addition, some new indicators, such as *Enterobacter sakazakii*, have been set.

9.2.1.2 Scientific Setting of Indicators In microbiological testing standards, microbiological indicators are set using hierarchical sampling principles according to the biological characteristics of microorganisms, which yields more reliable testing results of indicators.

9.2.1.3 Enhancement in Consistency and Integrity The new standards integrate China's current food hygiene standards, food quality standards, and relevant industry standards and revise the duplicate and contradictory elements.

9.2.1.4 Initial Improvement in the Standard System of Dairy Products First, coverage of the standards has been expanded, and versatility of the food safety standards has been enhanced. Second, product standards are supported by testing methods, and some dairy product-related testing methods are divided or integrated based on parameter setting. For example, the *Method of Analysis of Hygienic Standard of Milk and Milk Products* is divided into *Determination of Fat in Milk and Milk Products*, *Determination of Impurities in Milk and Milk Products*, *Determination of Acidity in Milk and Milk Products*, and *Determination of Relative Density in Raw Milk*. Methods for determination of fat, acidity, protein, moisture, ash, relative density, impurity, and fatty acids have been integrated into one method. Finally, new testing methods, such as determination of *trans*-fatty acids, have been developed in reference to the international standards.

9.2.1.5 Standardization of Dairy Product Names Common types of liquid milk are clearly defined. For example, liquid products produced from raw milk by pasteurization are called pasteurized milk, liquid products produced from raw milk by sterilization with or without the addition of reconstituted milk are called sterilized milk, and liquid products produced from raw milk or reconstituted milk by sterilization with the addition of other ingredients, food additives, or nutritional supplements are called modified milk.

9.2.1.6 Emphasis on the Safety of Infant Food All products covered by the four new safety standards for infant foods, that is, *Infant Formula*, *Formula for Older Infants and Young Children*, *Cereal-based Supplementary Foods for Infants*, and *Canned Supplementary Foods for Infants*, are not dairy products.

Nevertheless, they are the focus of national safety standards for dairy products. Moreover, in the revision of infant food safety standards, not only are some standards integrated according to the characteristics of the products, but also contents of some nutrients are adjusted based on similar international standards, *Chinese Dietary Reference Intakes for Nutrients*, and data from Chinese nutrition surveys to better meet the dietary needs of Chinese infants while ensuring safety. In addition, though infant formulas for special medical purposes are prepared based on ordinary infant formulas, they have to be adjusted according to the special requirements of different diseases or medical conditions and thus have a complex composition and different nutritional requirements. Therefore, supporting management measures are still absent for the newly formulated *Standards of Infant Formulas for Special Medical Purposes*, and further intensive research efforts are required to develop, improve, and promulgate appropriate management measures.

The introduction of the safety standards for dairy products, which were the first national food safety standards issued after the enactment of the *Food Safety Law*, provided a specific understanding of food safety standards for the entire food industry. In particular, they turned the understanding of food safety of a large number of middle- and small-sized enterprises from abstract concepts to concrete representations, allowing them to fully understand the meaning and existence value of law. Moreover, their introduction is of great importance in establishing a sound system of food safety standards, protecting the health of the people, promoting the healthy development of dairy industry, and improving food safety.

9.2.2 Standards for Use of Food Additives

Standards for use of food additives are national mandatory standards. To meet the social and economic development needs, the new standard *National Food Safety Standard—Standards for Use of Food Additives* (GB 2760-2011) replaced the old standard *Hygienic Standard for Use of Food Additives* (GB 2760-2007) and has been formally implemented as of June 20, 2011. Compared with GB 2760-2007, the key changes in this new standard are as follows:

1. *A more reasonable name.* To further improve scientificity of the standard, the name of the standard was converted from "Hygienic Standard for Use of Food Additives" into "National Food Safety Standard—Standards for Use of Food Additives."
2. *Replenishment of contents.* The Food Additives Regulations announced in No.4 Proclamation of Ministry of Health 2007–2010 were added.

3. *Adjustment of usage regulation for some food additives.* For example, the new standard newly included the use of the commonly used preservative benzoic acid and its sodium salt in protein drinks, tea, coffee, and vegetable drinks and restricted its use in oyster sauce, shrimp sauce, fish sauce, and wine.

4. *A fundamental change in the principle of addition.* For example, "the additives allowed to be used in food and their dosage" in Table A.2 in GB 2760-2007 were replaced with "the list of food additives that can be used in all types of foods with appropriate dose as required in production."

5. *Adjustment of the classification system for some foods.* The classification system for some foods was adjusted, and the usage regulations for food additives were also adjusted according to the adjusted food classification.

6. *Addition of the usage principles for flavoring substances and essence used in food.* The classification of flavoring substances used in food was also adjusted. Moreover, the usage principles for food processing aids used in food industry were added, and the list of food processing aids used in food industry was adjusted.

Food additives are an important part of the modern food industry and an important driving force of technological progress and innovation in the food industry. However, they are also a main cause of the frequent food safety incidents in recent years in China. In general, the GB 2760-2011 *National Food Safety Standard—Standards for Use of Food Additives* focuses on cleanup and integration to solve the absence and the duplications and contradictions of standards. This is of great importance to the development of the Chinese food industry and the protection of public safety and health. On the basis of GB 2760-2011, 101 new national standards for food additives were issued in 2012. Among them, 11 focused on colorants, nine of which are natural plant pigments. This indicates that the Sudan red and malachite green incidents have improved the public awareness of food safety. Thus, food safety standardization must be in the forefront of the food safety system.

9.2.3 General Standard for Compound Food Additives

The Ministry of Health issued the GB 26687-2011 *National Food Safety Standard—General Rule on Compound Food Additives* on July 5, 2011, which came into force on September 5, 2011. It applies to all compound food additives, except food flavoring and chewing gum base. GB 26687-2011 provides a definition of compound food additives and excipients, the nomenclature of compound food additives, the basic requirements for the use of compound food additives, and the requirements for information that should be included on labels and in instructions.

The introduction of GB 26687-2011 finally addressed the outstanding standard issues for the food additive industry, especially for compound food additive producers. Industry insiders have pointed out that although GB 26687-2011 is still not perfect, it at least provides a unified management standard for the development of compound food additives, which is very helpful for the development of the industry. GB 26687-2011 has the following main features:

1. *Clear definition of compound food additives.* GB 26687-2011 defines the compound food additive as "a mix of two or more single food additives by physical methods with or without excipients to improve food quality and facilitate food processing." This provides a sound basis for the management of compound additives.

2. *Provision of rules of nomenclature of compound food additives.* It provides a basis for the nomenclature of compound food additives to avoid misunderstandings and confusions caused by the name.

3. *Emphasis on the transparency of the ingredients of compound food additives.* Article 2 of the information that should be included in labels and instructions of compound food additive products states that common names of single food additives and names of excipients shall be included and that the dosage of single food additives shall also be included for compound food additives that enter the stage of marketing or catering. This provision achieves transparency of the ingredients of compound additives, which facilitates the selection by consumers.

From the perspective of complex applied technology and the safety of food additives, the research and development of compound food additives is an important method for the healthy development of the food industry and an important bridge for connecting the monomeric food additive producers and food producers. In developed countries, compound food additives have become the major final products. Therefore, along with the promulgation of GB 26687-2011, compound additives will be gradually regulated by laws and standards in China. This will not only help improve the food safety, but also provides a legal guarantee for the healthy development of food industry.

9.2.4 Maximum Levels of Mycotoxin

On April 20, 2011, the Ministry of Health issued the GB 2761-2011 *National Food Safety Standard on Maximum Levels of Mycotoxins in Foods* to replace the indicators of maximum levels of mycotoxins in the GB 2761-2005 *Maximum Levels of Mycotoxins in Foods* and the GB 2715-2005 *Hygienic Standard for Grains.* GB 2761-2011 stipulates the indicators of maximum levels of six mycotoxins (i.e., aflatoxin B1, aflatoxin M1, ochratoxin A, patulin, deoxynivalenol, and zearalenone) in foods. Moreover, it defines 11 major categories of food and divides the major categories into several level-two or level-three subcategories, by which maximum levels of mycotoxins are set.

GB 2761-2011, formulated based on GB 2761-2005, deleted duplicate contents and filled in missing contents by summarizing and analyzing the mandatory indicators of maximum levels of mycotoxins in China's current quality and safety standards for edible agricultural products, food hygiene standards, food quality standards, and relevant industry standards in accordance with the food classification system described in appendix F of GB 2760-2011. Moreover, the maximum levels of mycotoxins in food are determined by comparative analysis of the standards implemented by the Codex Alimentarius Commission (CAC) and those implemented in the United States, European Union, Australia, New Zealand, Japan, Hong Kong, Taiwan, and other countries in reference to the monitoring results of mycotoxins in food in China combined with the Chinese dietary exposure. Therefore, the standard is much more scientific and practical. Compared with GB 2761-2005, the major modifications in this standard are as follows:

1. Modification of the standard name
2. Addition of the definition of edible parts, which improves the scientificity and rationality of the standard
3. Addition of the application principles to stipulate the applicable scope of the standard, special conditions, and control measures, which makes the standard more rigorous and practical
4. Addition of the indicators of ochratoxin A and zearalenone
5. Modification of the indicators of maximum level of aflatoxin B1—in particular, the value is reduced by 90% in infant foods compared with that in GB 2761-2005, thus better protecting the health of infants and young children
6. Modification of the inspection methods for aflatoxin B1, aflatoxin M1, and deoxynivalenol
7. Addition of a description of food categories (names) in appendix A, which facilitates the implementation of the standard

9.2.5 General Rules for the Labeling of Prepackaged Foods

The GB 7718-2011 *General Rules for the Labeling of Prepackaged Foods* was promulgated on April 20, 2011. It was formulated based on the revision of the prepackaged food labeling standards by the Ministry of Health in accordance with the *Food Safety Law* and its implementing regulations combined with the requirements for food safety supervision. It was reviewed and approved at the fifth directors' meeting of the National Food Safety Standard Review Committee. Compared with GB 7718-2004, the major changes of this standard are as follows:

1. *A clearer scope of application.* Taking into account the specific conditions in actual distribution, GB 7718-2011 specifies both the applicable and inapplicable conditions, with statements that "this standard applies to the labeling of prepackaged foods to be offered directly or indirectly as such to consumers" and that "this standard does not apply to the labeling of transportation packages providing protection for prepackaged foods during transportation, or the labeling of foods in bulk or produced and sold on the spot."

2. *Modification of some of the definitions.* Based on GB 7718-2004, modifications were made to the definitions of prepackaged food and date of manufacture. The definition of configuration was added and the definition of storage period was deleted, while the shelf-life requirements remained the same.

3. *Modification of the labeling method for food additives.* GB 7718-2011 provides detailed labeling requirements for food additives. It requires that the names of food additives be declared in general names in accordance with the *National Food Safety Standard—Standards for Use of Food Additives* (GB 2760-2011). On the label of the same prepackaged foods, general names of food additives can all be declared as the specific name of the food additives and also can all be declared as their class names combined with the specific name or international code of the food additives.

4. *Addition of labeling method for configuration.* The declaration of configuration consists of net weight of the prepackaged food and the number of inner individual food units, or only the number of inner individual food units. The word "configuration" may not be declared. For prepackaged food with no inner packs, the configuration means net weight.

5. *Further standardization of the labeling method.* Modifications were made to the labeling method for name, address, and contact information of the manufacturer and distributor. Name, address, and contact information of the manufacturer shall be declared. The contact information of the manufacturer or distributor, which by law bears independent statutory responsibilities, shall be declared including at least one of the following: telephone number, fax number, network contact information, or postal address. In addition, GB 7718-2011 provides that the name, address, and contact information of the manufacturer may not be declared for imported prepackaged food.

6. *Specification of the largest surface area of a package of prepackaged food.* Modifications were made to the largest surface area of a package (container) of prepackaged food, where the minimum size of the words, symbols, and numerals in the mandatory labeling information shall not be less than 1.8 mm in height. In addition, modifications were made to the calculation method of the largest surface area in appendix A, and a new provision that "the area of seal shall be excluded when determining the surface area for package bag and so on" was added so that calculation of the largest surface area is more reasonable.

7. *Addition of a recommended labeling requirement for food containing allergenic substances.* In reference to the Codex Alimentarius standard, a recommended labeling requirement for the food that may contain allergenic substances was added, which allows consumers to choose foods according to their physical conditions.

8. *Addition of appendixes.* Appendix B "Declared form of food additives in the list of ingredients" and appendix C "Recommended declared form of some label items" were added.

The changes in the scope of application, product definition, basic requirements, and labeling of food additives in GB 7718-2011 enhance the scientificity and operability of food labels, improve the convenience and accuracy for consumers' access to food information, and further legalize and standardize the judgment of labels by supervision and inspection agencies. With the promulgation of GB 7718-2011 *General Rules for the Labeling of Prepackaged Foods*, food labeling will play a more prominent role in food safety management.

9.2.6 Food Nutrition Labeling Standards

According to the national nutrition survey results, Chinese residents have both nutritional deficiencies and overnutrition. In particular, high intake of fat, sodium (salt), and cholesterol are the major factors contributing to chronic diseases. Food nutrition labeling is a mandatory national food safety standard to provide consumers with nutrition information and a description of characteristics of food. It is an effective way for consumers to gain an intuitive understanding of the nutritional components and characteristics of food. Therefore, improving nutrition labeling standards is conducive to the popularization of the nutritional knowledge of food and guides the public in the scientific selection of food to achieve a balanced diet and improve physical health. Moreover, it helps standardize the correct nutrition labeling by enterprises, publicizes nutritional knowledge, and promotes the healthy development of the food industry. In view of this, the GB 28050-2011 *General Rules for the Nutrition Labeling of Prepackaged Foods* was promulgated by the Ministry of Health on October 12, 2011, to guide and standardize Chinese food nutrition labeling, guide consumers in the reasonable selection of prepackaged foods, and protect consumers' right to know, choose, and supervise. It was formulated according to the relevant provisions of the *Food Safety Law* and in reference to the CAC and the international management experience.

By giving full consideration to the implementation of *Regulations on Food Nutrition Labeling Management* and drawing on the advanced experience of other countries, GB 28050-2011 further improves the nutrition labeling management system. The main improvements are summarized as follows:

1. *Simplification of nutrient classification and label format.* The category of "nutrients that should be declared" was deleted, declaring order of nutrients was adjusted, restrictions on the nutrition label format were reduced, and a basic format of text representation was added.
2. *Addition of mandatory labeling requirements.* Mandatory declaration of relevant information is required when nutritional supplements and hydrogenated oils are used. When the levels of energy and nutrients are lower than the threshold value of "0," "0" shall be declared as a mandatory labeling requirement.
3. *Simplification of allowable error.* Differences in the allowable error of contents of vitamins A and D between "fortified and nonfortified foods" were deleted. Additionally, requirements relating to the optional declared nutrients, chromium and molybdenum, and their nutrient reference values (NVA) were removed.

4. *Adjustment of requirements for nutrition claims.* The standard term and synonyms of nutrition claims, claim requirements and conditions for "0" trans fat (fatty acid), and claim conditions for the declaration of nutrients per 420 kJ were added.

5. *Adjustment of nutrient function claims.* Restrictions on the position of nutrient function claims were removed; terms of function claims for energy, dietary fiber, *trans* fat (fatty acid), and other ingredients were added; and terms of function claims for saturated fat, pantothenic acid, magnesium, iron, and other ingredients were modified.

9.2.7 Standard for Pesticide Residue Limits

Pesticides are a basic means of production. However, not only the inherent chemical toxicity of pesticide residues is a risk factor in the safety of edible agricultural products, but it also destroys the agricultural ecological environment. Therefore, the maximum residue limits (MRLs) of pesticides are the basis not only for ensuring food safety, but also for promoting producer compliance with good agricultural practices and eliminating unnecessary pesticide use to protect the ecological environment.

As of 2012, the Chinese food standard system included a total of 2319 pesticide residue limits. The pesticide residue standards were formulated in full compliance with the relevant standard of CAC based on the data of Chinese pesticide registration and dietary consumption and risk assessment. The standards were formulated in strict accordance with the following procedures: request for public comments, notification to the WTO, and examination by the Review Committee on National Pesticide Residue Standards. On November 16, 2012, the Ministry of Health and Ministry of Agriculture jointly issued the latest national food safety standard GB 2763-2012 *National Food Safety Standard—Maximum Residue Limits for Pesticides in Food.* Compared with the 2005 edition, the 2012 edition contained 100 more pages and the number of pesticides involved increased from 136 to 322. This standard replaces the GB 25193-2010, GB 26130-2010, GB 28260-2011, and Article 4.3.3 of GB 2715-2005 and abolishes 10 standards that are most relevant to pesticide residue limits issued by the Ministry of Agriculture, such as NY660-2003. Thus, it becomes the only mandatory national standard for pesticide residue regulation. GB 2763-2012 specifies 2293 MRLs for 322 pesticides, such as 2, 4-dichlorophenoxyacetate, in 10 categories of food, including cereals, oils and fats, vegetables, fruits, nuts, sugars, beverages, edible mushrooms, spices, and animal-derived food. Compared with the standards it replaced, GB 2763-2012 has the following main changes:

1. It verifies and revises the name of pesticides, residues, acceptable daily intake, food names, and other information in the original standards.
2. It modifies some of the MRLs according to the dietary exposure assessment and pesticide registration to eliminate the overlapped, repetitive, outdated, and uncoordinated elements.
3. It adds MRLs for pesticides in some foods.
4. It refines food varieties and categories.

9.2.8 Standard for Maximum Levels of Contaminants in Food

In recent years, China's National Center for Food Safety Risk Assessment recruited experts from scientific research institutes in agriculture, hygiene, quality inspection, food, and other fields to establish a Standards Drafting Team. The team analyzed and reviewed the specifications and requirements for the maximum levels of contaminants in more than 600 standards of agricultural product quality and safety and food quality and hygiene and relevant industry standards. Moreover, food safety risk assessment was performed based on data of Chinese food production and food contaminant monitoring. The standard *Maximum Levels of Contaminants in Food* (GB 2762-2005), issued in 2005, was revised in reference to the food safety standards of CAC, the European Union, the United States, Australia, New Zealand, and other international organizations and countries to formulate the new standard *Maximum Levels of Contaminants in Food* (GB 2762-2012). After notification to the WTO members and approval by the directors' meeting of the National Food Safety Standard Review Committee, the new standard was issued on November 13, 2012, and implemented as of June 1, 2013.

Before the implementation of the *Food Safety Law*, there were a total of 608 Chinese food standards involving maximum levels of contaminants in food, including 86 food hygiene standards, 35 quality and safety standards for edible agricultural products, 76 food quality standards, and 411 relevant industry standards. These standards covered 16 food contaminants, including lead, cadmium, total mercury and methylmercury, arsenic and inorganic arsenic, tin, nickel, chromium, nitrite and nitrate, benzo[*a*]pyrene, *N*-nitrosamines, polychlorinated biphenyl, 3-chloro-1,2-propanediol, rare earth elements, selenium, aluminum, and fluoride. Compared with the previous standards, the new standard GB 2762-2012 has the following main changes:

1. The definition of edible parts was added. First, it helps focus on strengthening the processing management of edible parts of food, preventing and reducing pollution, and improving relevance of the

standard. Second, the edible parts objectively reflect the actual dietary consumption of the general public, thus improving the scientificity and operability of the standard.

2. It cleans up all the previous provisions on maximum levels of contaminants in food standards. Provisions on maximum levels of 13 contaminants, such as lead, cadmium, mercury, arsenic, benzo[*a*]pyrene, and *N*-nitrosodimethylamine, in more than 20 categories of food, such as cereals, vegetables, fruits, meat, aquatic products, condiments, and beverages and liquor, are integrated and revised. Three indicators, including selenium, aluminum, and fluoride, were removed, and a total of more than 160 indicators were finally set.

3. Requirements for maximum levels of tin, nickel, 3-chloro-1,2-propanediol, and nitrate were added.

4. The indicator of maximum *N*-nitrosamine level was adjusted from a combination of *N*-dimethylnitrosamine and *N*-diethylnitrosamine to *N*-dimethylnitrosamine, and the name of the indicator of maximum *N*-nitrosamine level was changed to *N*-dimethylnitrosamine.

9.3 Substantial Efforts on the Implementation of National Food Safety Standards

Food producers and traders implement the food safety standards. Only by efficient implementation of the standards can food quality and safety be guaranteed to the greatest extent. To this end, in recent years, Chinese food safety regulators have made great efforts to promote the nationwide implementation of national food safety standards while vigorously cleaning up, integrating, and supplementing the food safety standards. Specifically, a series of publicity and training activities have been carried out.

9.3.1 Focused Publicity and Training

Since December 20, 2010, the Ministry of Health has promulgated a series of national food safety standards successively, such as GB 25596-2010 *General Rules for Infant Formulas for Special Medical Purposes*, GB 2760-2011 *National Food Safety Standard—Standards for Use of Food Additives*, and GB 7718-2011 *General Rules for the Labeling of Prepackaged Foods*. However, the standards were not properly understood or fully implemented by food producers and related enterprises after promulgation. To clear up doubts for enterprises, help them correctly understand the food safety standards, and thus achieve a better implementation of the standards, the China Dairy Industry Association took the lead in holding a training meeting on national food safety standards in Beijing in July 2011 (Qianlong News Network 2012). Experts from the Chinese Disease Control and Prevention Center were invited to give a detailed explanation of the three standards (GB 25596-2010, GB 2760-2011, and GB 7718-2011) and answer the questions. More than 120 people from over 70 producers and traders of dairy products or additives from all over the country attended the training meeting.

In March 2012, the National Center for Health Inspection and Supervision of the Ministry of Health held a national training meeting on the *General Rules for the Nutrition Labeling of Prepackaged Foods* in Beijing to promote the full implementation of the national standards for prepackaged food labeling and nutrition labeling (Health Standard Network 2012a). More than 160 representatives from 30 provincial health departments, Administration of Quality Supervision, Inspection and Quarantine (AQSIQ), State Food and Drug Administration (SFDA), State Administration for Industry and Commerce (SAIC), Chinese Nutrition Society, and other departments attended the training meeting. The training aimed to improve the supervisory staff's awareness of the importance of national food safety standards and deepen their understanding of the standard terms, thereby effectively enhancing the actual power of these two standards.

In April 2012, the National Center for Health Inspection and Supervision of the Ministry of Health held a seminar on the publicity and implementation of national food safety standards in Beijing to promote the implementation of national food safety standards (Health Standard Network 2012b). Experts provided a detailed and systematic interpretation of two key standards, *General Rules for the Nutrition Labeling of Prepackaged Foods* and *Standards for Use of Food Additives*, and answered questions from participants. More than 60 central media outlets, major mainstream media outlets, and mainstream network media outlets from all over China also attended the meeting.

In May 2012, the Ministry of Health and Ministry of Agriculture jointly ran the National Journalist Seminar on Safety Management of Genetically Modified Organisms (GMOs) and the Training Course for Publicity and Implementation of National Food Safety Standards in Beijing (China Network Television 2012). At the seminar and training course, not only were the *General Rules for the Nutrition Labeling of Prepackaged Foods* interpreted by experts, but also the research progress on safety of GMOs and relevant risk management measures were discussed. The aim was to help journalists to correctly understand the GMOs and timely and correctly comprehend the new national standards, to prevent false reports, which can cause unnecessary public panic.

In addition, the China National Center for Food Safety Risk Assessment of the Ministry of Health ran open days for face-to-face communication of food safety standards in July 2012 (Foodmate Net 2012). More than 70 people, including representatives of the food enterprises, media, industry associations, universities and other institutions, and the general public, took part in the activity. Experts explained the Chinese standards for food additives and contaminants and the *General Rules for the Nutrition Labeling of Prepackaged Foods* in an easy-to-understand way. Moreover, the public's comments and suggestions on formulation and revision of food safety standards were heard, to strengthen the rights of the public to participate in standardization.

First, this series of publicity and implementation activities played a positive role in promoting the implementation of national food safety standards and deepening the supervisory staff's understanding and consequent better enforcement of the standards. Second, they are conducive to furthering enterprises' proper understanding of the standards, which allows for better implementation. Third, they have helped the media, the public, and other social organizations to have a systematic and correct understanding of the purposes and basis of formulation of the food safety standards, thus providing rational and correct guidance of public opinion. Finally, such activities have provided a platform for public participation in the formulation and revision of food safety standards, which enables consumers to safeguard their legal rights.

9.3.2 *Follow-Up Evaluation of Standards*

Due to time and space constraints, it is difficult to take into account all aspects in the formulation of food safety standards. Moreover, a variety of problems will occur during the implementation of food safety standards due to the limitation of geographic, economic, and technological environments. The problems must be identified before they can be solved and effective implementation of the food safety standards can thereby be ensured. Therefore, follow-up evaluation of the standards is very important. It not only enables discovery of problems in the implementation of the standards to comprehensively evaluate their scientificity and applicability, but also provides an important reference for the revision of standards. After promulgation of the *Food Safety Law*, the Ministry of Health conducted follow-up evaluations of a series of new national food safety standards on dairy products and food additives to investigate their adaptability.

In September 2011, a kick-off meeting for a follow-up evaluation project for national food safety standards was held in Urumqi (National Center for Health Inspection and Supervision 2011). The importance of follow-up evaluation of food safety standards was emphasized and the *Rules for the Implementation of Follow-up Evaluation of National Food Safety Standards* were created at the meeting. The *Rules* make detailed provisions on the respondent, contents, methods, preparation, statistics, personnel training, and other aspects of the survey, to ensure smooth implementation of the follow-up evaluation.

In November 2011, the Health Department of Hebei Province issued the *Work Program for Follow-up Evaluation of Standards for Use of Food Additives and Enzyme Preparations for Food Industry Application* (Foodmate Net 2011), to understand the implementation of two national food safety standards, *National Food Safety Standard—Standards for Use of Food Additives* (GB 2760-2011) and *Enzyme Preparations for Food Industry Application* (GB 25594-2010), and problems in their application, as well as to collect comments and suggestions on their revision. The survey involved food safety regulators, food inspection agencies, and 60 users of food additives (including 30 food producers and 30 food service entities), as well as all producers of enzyme preparations for food industry application in Hebei Province and 30 users of enzyme preparations for food industry application. In June 2012, the Health Department of Inner Mongolia Autonomous Region issued *A Notice on Completing the Follow-up Evaluation of National Food Safety Standards* (Health Department of Inner Mongolia Autonomous Region 2012) and promulgated the *Implementation Program of Follow-up Evaluation of National Food Safety Standards of the Inner Mongolia Autonomous Region*. The program focuses on the evaluation of testing standards for food additives, including laboratory verification of some test items of the testing methods used in the food safety standards for nine food additives, such as agar, potassium dihydrogen phosphate, potassium hydrogen tartrate, sulfur, amaranth red, tartrazine aluminum lake, tartrazine, sunset yellow, and butylated hydroxy toluene.

In September 2011, follow-up evaluation of three standards, *Infant Formula*, *Sterilized Milk*, and *Modified Milk*, was conducted in Yunnan (Yunnan Health Inspection Information Network 2011), Xinjiang (National Center for Health Inspection and Supervision 2011), Hubei (Health Inspection Bureau of Health Department of Hubei Province 2012), and other provinces or regions. China's Public Union of Nutrition and Institute of Food Science and Technology also participated in the follow-up evaluation of the three standards. The two institutions were responsible for inspecting state-level food safety supervision departments, state-level inspection agencies, and infant formula producers across the country (except the 10 provinces participating in the project, Xinjiang Production and Construction Corps, Hong Kong, Macao, and Taiwan). Meanwhile, each province was responsible for inspecting food safety supervision departments and inspection agencies at the provincial-, prefecture-, and city-levels and all producers of infant formula, sterilized milk, and modified milk in that province. The survey was primarily aimed at understanding what the supervision departments, inspection agencies, and producers knew about the standards for infant formula, sterilized milk, and modified milk. Furthermore, the survey aimed to understand how they used the standards and how they implemented the main technical contents of the three dairy standards. Problems in their use of the standards and their comments and suggestions on the revision of standards were also collected.

The follow-up evaluations enable the departments formulating food standards to understand how food safety supervision departments, inspection agencies, and food producers across China have used the national food safety standards. Moreover, they allow for the collection of suggestions on the revision of the national food safety standards and discovery of problems of the relevant standards in the food safety chain. This facilitates timely communication of food safety risks and helps the food safety supervision departments to improve their work. The completion of these tasks and analysis and summarization of the evaluation results will provide many constructive suggestions for the formulation and revision of national food safety standards, which will play an important role in improving the scientificity and applicability of national food safety standards and protecting the health and food safety of consumers.

9.4 Active Participation in Affairs of the Codex Alimentarius

The Codex standards are formulated on the basis of scientific evidence. As the only international reference standard in the field of agricultural products and food, the Codex standards have been widely accepted by the international community. Therefore, the international food safety standards have developed from the recommended standards into widely accepted and commonly used food safety standards in the international community and mandatory standards in international food trade. A country's active participation in formulation and revision of international food safety standards is very important for improving its national food safety standards and safeguarding its legitimate interests in international food trade. In recent years, China has actively participated in affairs of the Codex Alimentarius and hosted a series of sessions of the Codex Committee on international standards for food additives and pesticide residues to further strengthen its position in the international standard-setting organizations.

In July 2006, China was selected as a new host country for two committees, one on food additives and the other on pesticide residues, by consensus at the 29th session of the Codex Committee in Geneva, Switzerland (China Agricultural Quality Standards Network 2007). It is the first time that China has served as a host country for the subsidiary committees of CAC since its establishment in 1963. Moreover, China is the only developing country in the 10 host countries for integrated committees of CAC. According to regulations, as the host country for the Codex committees on food additives and pesticide residues, China is responsible for the operation, costs, administration, and selection of the chairpersons of the committees. The selection of China as the host country for the committees on food additives and pesticide residues is important for promoting China's fair food trade and improving the food safety control. In recent years, as the host country for the committees on food additives and pesticide residues, China has presided over a number of sessions of the Codex Committee on Pesticide Residues and Codex Committee on Food Additives.

9.4.1 The Work of the Codex Committee on Food Additives

As the host country for the Codex Committee on Food Additives, China has presided over seven sessions (from 39th to 45th) of the Codex Committee on Food Additives up to 2013. The Codex General Standard for Food Additives (GSFA), quality specifications for food additives, draft guidelines for the use of spices, international numbering system for food additives (INS), and other issues were discussed at these sessions. After discussion at the 43rd Session of the Codex Committee on Food Additives, it was decided that the draft food additive regulations and proposed draft amendment of the GSFA, proposed draft amendment of the INS, proposed draft amendment of the GFSA food classification system, quality specifications for identity and purity of food additives proposed at the 73rd session of the Joint FAO/WHO Expert Committee on Food Additives (JECFA), proposed draft amendment of standard for food-grade salt, and revocation and termination of GSFA food additive regulations be submitted to the 34th session of the CAC for approval

(Fan et al. 2011). Furthermore, to accelerate the formulation and revision of standards, the Codex Committee on Food Additives decided to set up seven electronic working groups to continue the work in related areas. As part of the groups, the Chinese delegation undertook the creation and maintenance of a database for processing aids.

9.4.2 The Work of the Codex Committee on Pesticide Residues

Similarly, as the host country for the Codex Committee on Pesticide Residues, China has presided over seven sessions of the Codex Committee on Pesticide Residues up to 2013. These sessions focused deliberating on the drafts and draft proposals of MRLs of pesticides in food and feed, which involved a total of 511 MRLs of 53 pesticides in food, meat, vegetables, fruits, and other plant and animal products (China Agricultural Quality Standards Network 2008). In addition, a priority list of pesticides for deliberation on MRLs was developed. At the 41st Session of the Codex Committee on Pesticide Residues, the Chinese delegation recommended, based on the actual situation in China, that acephate be included in the priority list for deliberation to determine the MRLs of acephate in rice as early as possible. Moreover, the Chinese delegation proposed to retain the Codex MRLs for tea, which has delayed the scheduled cancellation of cypermethrin MRLs in tea for at least four years (Song et al. 2010). The setting of these standards on a priority basis or their retention is undoubtedly conducive to safeguarding the interests of China in international trade.

According to the ranking by International Organization for Standardization, the contribution of China has exceeded that of Canada, Italy, Australia, Switzerland, and other medium-developed countries and is rated sixth after the United States, Britain, Germany, Japan, and France (Foodmate Net 2008).

9.5 Preliminary Thinking on the Future Construction of Food Safety Standard System and the Major Measures

Since the implementation of the *Food Safety Law*, notable achievements have been made in the construction of the Chinese food safety standards system. In particular, the cleanup and integration of food safety standards have been accelerated, and overlap, repetitions, and contradictions among the current standards have been properly solved. However, many outstanding problems in the construction of the Chinese food safety standard system still exist. With the purpose of ensuring the physical health of the general public, as proposed in the *Food Safety Law*, the Chinese food safety standard system shall be constructed by adhering to the basic principles, setting goals of development, and implementing concrete measures to effectively support the improvement of the overall level of Chinese food safety and quality.

9.5.1 Basic Principles

The purpose of formulating food safety standards shall be to ensure the physical health of the general public. The food safety standards shall be scientific, reasonable, safe, and reliable. The formulation of food safety standards must always adhere to the principles provided in Sections 9.5.1.1 through 9.5.1.4.

9.5.1.1 The Principle of Obedience to Law   The *Food Safety Law* specifically devotes a chapter (Chapter 3) to defining the purpose and contents of food safety standards, entities responsible for the formulation, basis for the formulation, and entity responsible for the review and the review process, as well as entities responsible for formulation of local and enterprise food safety standards and the relevant conditions and reporting system. Moreover, the detailed rules and regulations of implementation of the relevant provisions have been specifically addressed in the subsequent *Implementing Regulations of the Food Safety Law* and *Implementing Rules of the Food Safety Law*. These regulations and rules provide not only a legal basis and guarantee, but also requirements for the formulation of food safety standards. In particular, they require that food safety standards shall be formulated with the purpose of ensuring the health of the general public, the standard-setting procedures shall be strictly followed, and technical requirements for food safety closely related to human health shall be contained.

9.5.1.2 The Principle of Grounding on Risk Assessment   Risk assessment has become a necessary technical tool for CAC to formulate food safety laws, regulations, and standards. Moreover, risk analysis was applied to the formulation of food safety regulations and standards as early as in the 1990s in western developed countries. This indicates that risk assessment plays an important role in improving food safety management, especially in identifying key (high-risk or nonnegligible) regulatory targets (fields or varieties), determining the risk of a certain factor, and setting a food safety standard. In particular, the risk assessment results provide an important basis for setting the limits of pesticide and veterinary drugs, environmental pollutants, food additives, and pathogenic microorganisms. The standards can be scientific and reasonable and recognized internationally only with the support of assessment results. Therefore, the formulation of Chinese food safety standards must be based on risk assessment to improve their scientificity.

9.5.1.3 The Principle of Combing the Realities and International Standards With advanced science and technology, a rational industrial structure and other advantages, the CAC and some developed countries have formulated more scientific and reasonable food safety standards, which should be drawn upon by China. However, the unique realities of China, such as low levels of industrial development and organization, should also be carefully considered. The standards may be inadaptable and unworkable in China if indiscriminately introduced from other countries without examination or investigation. Therefore, China should formulate scientific, reasonable, and workable food safety standards by both learning from the advanced technology and concepts of other countries in the formulation of standards and taking into full consideration its own realities.

9.5.1.4 The Principle of Openness and Transparency   After the promulgation of 66 new national standards regarding the safety of dairy products, such as the *Raw Milk* (GB 19301-2010), in March 2010, there has been an upsurge of discussion on the new dairy standards in China. The general public has many questions about the new dairy standards and believes that the new standards are regressive and even "hijacked by large enterprises." These concerns not only damage the credibility and image of administrative departments, but also hinder the implementation of the standards. As a matter of fact, it is because the standard-setting process is not transparent enough and the channel to request public comments is not effective or broad enough. Therefore, it is very important to adhere to the principles of openness and transparency during the formulation of national food safety standards. Moreover, consideration must be given to broadening the channel to publicize the standards and request public comments, so that more people and social organizations can participate in the formulation of standards. This can not only prevent unnecessary misunderstandings of the general public, but is also an effective supervisory mechanism for enhancing the legitimacy of the standard-setting process.

9.5.2 Key Tasks

The *12th Five-Year Plan for National Food Safety Standards* issued by the Ministry of Health and seven other departments explicitly requires that cleanup and integration of mandatory contents in the quality and safety standards for edible agricultural products, food hygiene standards, food quality standards, and industry standards be largely completed and the contradictions, duplications, and overlaps among the current standards be largely solved by 2015 to develop a more perfect system of national food safety standards. The joint efforts of the entire society are required to achieve these goals (Figure 9.1). Key tasks to be further performed are discussed in Sections 9.5.2.1 through 9.5.2.4.

9.5.2.1 Clean Up and Integrate the Current Food Safety Standards Since multiple departments are responsible for formulating Chinese food safety standards and sufficient coordination is not in place, there are contradictions, duplications, and overlaps among many of the current standards. To overcome this flaw of the Chinese food safety standard system, the contents of the current basic food standards, product standards, management and control standards, and standards regarding agricultural products must be carefully compared and analyzed to delete duplicate contents and annul unreasonable regulations, thereby establishing a coordinated food safety standard system.

FIGURE 9.1
Goals of the National Food Safety Standard System Construction during 2011–2015

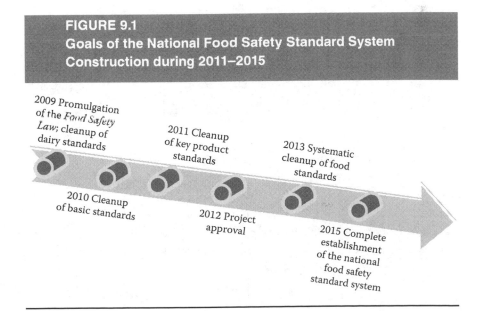

2009 Promulgation of the *Food Safety Law*; cleanup of dairy standards

2011 Cleanup of key product standards

2013 Systematic cleanup of food standards

2010 Cleanup of basic standards

2012 Project approval

2015 Complete establishment of the national food safety standard system

9.5.2.2 Accelerate the Formulation and Revision of Basic National Food Safety Standards   Basic food safety standards include important standards, such as limits of pesticides and veterinary drugs, environmental pollutants, food additives, and pathogenic microorganisms, and standards for labeling. The imperfection of these standards will directly lead to blind spots and gaps in food safety supervision and management, which goes against the protection of safety and health of the general public. Therefore, the *12th Five-Year Plan for National Food Safety Standards* emphasizes that the formulation and revision of national food safety standards shall be accelerated by cleaning up and improving the standards simultaneously and actively learning from the advanced standards of other countries.* Great efforts should be devoted to revising the standards for maximum levels of pesticide and veterinary drug residues, food contaminants, mycotoxins, and other substances, as well as the standards for use of food additives and nutritional supplements based on the follow-up evaluation results of food safety standards. Furthermore, focus should be placed upon setting the standards for maximum levels of pathogenic microorganisms in food and the control requirements for indicator microorganisms during food production and distribution, that is, guidelines for microorganism control in food and other basic standards, by the end of 2015. Moreover, the product standards for food packaging, containers, materials of processing equipment, and additives and the standards for the use of additives should be improved. In addition, in view of the absence of risk prevention and control measures during food production and distribution in China, emphasis should be placed on the formulation and revision of standards regarding food production and distribution. In particular, sanitary control requirements for raw material, planting and breeding, processing, sales, storage, and transportation shall be strengthened to standardize food production and distribution, so as to prevent and control food safety risks.

9.5.2.3 Improve the Administrative Mechanism for National Food Safety Standards   At present, China's management of national food safety standards is moving toward legalization and standardization, and it meets the basic requirements for formulation and implementation of national food safety standards. However, due to limitation of resources and time, appropriate management systems have been developed for some of the more specific and detailed aspects, while further improvement is required for others. In

---

*Refer to the *12th Five-Year Plan for National Food Safety Standards* issued by the Ministry of Health.

the next few years, China should further develop the follow-up evaluation management system for national food safety standards so as to standardize the follow-up evaluation of various standards. In addition, although a National Food Safety Standard Review Committee has been set up, a management system for the selection and tenure of committee members has not been established, and improvement should be made in this aspect. Furthermore, although an open and transparent standard-setting process is discussed in all of the above-mentioned regulations, the specific mechanisms for participation and communication have not been established; therefore, it is difficult to truly achieve openness and transparency. To this end, the regulators need to establish relevant mechanisms and improve relevant supporting measures to provide the necessary platforms and channels for the public to participate in the formulation of standards and provide comments.

9.5.2.4 Strengthen the Publicity and Implementation of Standards The insufficient efforts in publicizing and implementing the food safety standards will result in doubts and misunderstandings of the general public regarding the standards, which will affect the implementation of standards. Therefore, more efforts should be devoted to publicizing and implementing the standards, broadening the communication channels, enriching the forms of publicity, and reaching a wider audience. In particular, emphasis should be placed on the publicity of standards and popularization of relevant scientific knowledge, especially the publicity and explanation of standards with highly technical contents and of great public concern. The public's understanding of the significance and meaning of the standards can effectively avoid prevalence of the wrong public opinions. Furthermore, smooth implementation of a food standard requires the correct understanding and effective implementation of producers and traders. Therefore, the Ministry of Health and other regulatory authorities should further promote the food producers and traders to actively implement the national food safety standards according to laws, regulations, and standards. Moreover, follow-up evaluation should be performed to see how the standards have been implemented, to provide an important reference for the revision of food safety standards.

9.5.3 Supporting Measures

To achieve the goals of the food safety standard system construction, appropriate supporting measures should be implemented at least in the three aspects described in Sections 9.5.3.1 through 9.5.3.3.

9.5.3.1 Establish a Coordination Mechanism for National Food Safety Standards   As more than a dozen departments are involved in the management of food safety standards in China, an effective and reasonable interdepartmental coordination mechanism should be established in the future formulation and revision of food safety standards. This will help to achieve efficient and successful cleanup, integration, formulation, and revision of the standards. With the Ministry of Health as the lead department, a reasonable allocation of responsibilities should be made based on the characteristics and strengths of each department. Moreover, clear objectives and detailed tasks should be identified for each department to ensure orderly conduct of each task. All responsible departments need to actively cooperate with each other, break the boundaries between departments, provide the necessary monitoring data, collect and summarize new problems and conditions in a timely manner, and have active communications and discussions to ensure effective formulation, revision, and management of the food safety standards.

9.5.3.2 Increase Investment in the National Food Safety Standard System Construction   At present, there are more than 5000 national and industry food safety standards in China. Thus, it will not be easy to complete the cleanup and integration of food safety standards within the next few years. First of all, a large amount of money is required to support the daily operation. Therefore, more financial resources should be allocated to the formulation and revision of national food safety standards. In addition, the funds allocated to the formulation and revision of standards should be used under strict supervision to ensure that the funds are used as intended in an efficient and legal way.

9.5.3.3 Strengthen Professional Resources for Food Safety Standards   Professional resources are crucial, in addition to a reasonable and efficient working mechanism and adequate financial support, because the overall staff competence has a decisive impact on the efficiency and quality of work. Therefore, there is an urgent need to strengthen the professional resources in the field of risk assessment and standard formulation and evaluation in China. To this end, overall staff competence can be improved and the staffing levels for research, formulation, and management of standards can be increased by introducing leading professionals.

10

Research Report on Food Safety Information Disclosure

The essential characteristic of food safety risk is food safety information asymmetry. Disclosure of food safety information is of great significance for the prevention of food safety risks. The document *Arrangement of Key Work on Food Safety in 2013* (document number: [2013] No. 25), issued by the General Office of the State Council on April 7, 2013, clearly requires improvement of the food safety information release system, enhancement of intercommunication before information release to ensure the scientificity, accuracy and timeliness of the information, uniform release of important food safety information by specialized departments, timely release of authoritative information, objective, and accurate response to public concerns on hot issues concerning food safety (Website of the Central People's Government of the PRC 2013a). Food safety information can be disclosed by the government as well as by the concerned enterprise. As food safety information has the properties of public goods, government disclosure is particularly important. This book investigates voluntary disclosure of food safety information by the food safety supervision and management departments from June 2012 to July 2013 and related systems.

10.1 *Status Quo* of Food Safety Information Disclosure

This section analyzes the *status quo* of information disclosure of food safety and relevant incidents by the food safety regulators at the central government level.

TABLE 10.1
White Papers on Food Safety in 2012 Issued by the
Local Governments

Area	Name	Time of Release	Note
Qingdao	*2012 Report on Food Safety Situation in Qingdao (White Paper)*	April 16, 2013	Published for the first time
Fujian Province	*2012 White Paper on Work on Food Safety in Fujian Province*	January 17, 2013	Published for the first time
Shanghai	*2012 Report on Food Safety Situation in Shanghai (White Paper)*	February 5, 2013	Published for the second time
Shenzhen	*2012 White Paper on Food Safety Supervision and Management Responsibility in Shenzhen*	March 9, 2014	Published for the sixth time
Shunde District, Foshan	*2012 White Paper on Food Safety Supervision and Management in Shunde District*	April 2, 2013	Published for the first time
Qingpu District, Shanghai	*2012 Report on Food Safety Situation in Qingpu District (White Paper)*	March 1, 2013	Published for the first time

Source. Information collected from the Internet.

10.1.1 Improvement of the Food Safety Information Disclosure System

Three documents were issued and implemented by food safety regulators at the central government level from June 2012 to July 2013, one by the Ministry of Health and two by local governments. Table 10.1 also lists the white papers issued by some local governments describing the food safety within their administrative area.

10.1.2 Information Disclosure on the Web by Food Safety Regulators at the Central Level

Before the reform of the food safety management system, 13 departments, including the National Development and Reform Commission (NDRC), Ministry of Science and Technology (MOST), Ministry of Industry and Information Technology (MIIT), Ministry of Public Security, Ministry of Finance, Ministry of Environmental Protection, Ministry of Agriculture, Ministry of Commerce, Ministry of Health, State Administration of Industry and Commerce (SAIC), Administration of Quality Supervision, Inspection and Quarantine (AQSIQ), State Administration of Grain, and State Food and Drug Administration (SFDA), were responsible for food safety supervision and management. Of them, the Ministry of Health, Ministry of

Agriculture, Ministry of Commerce, SAIC, AQSIQ, and SFDA had direct responsibilities for administrative law enforcement and management. Other departments were only partially responsible for food safety management. For example, the salt industry was managed by the NDRC. This book analyzes the situation of food safety information disclosure based on search of the websites of the six departments with direct responsibilities for administrative law enforcement and management.

10.1.2.1 China Food and Drug Administration   The SFDA released government information on food and drug administration of general concern and information related to the vital interests of the general public via the government website and press conferences (Website of the Central People's Government of the PRC 2013b). For example, a special column was set up for "handling of medicinal capsules containing excessive chromium" on the website to release related work news and results of inspection and testing in a timely manner and to introduce relevant scientific knowledge; an official microblog account was specially registered to release relevant policies of the food and drug administration by microblogging. Through this, warning information on quality and safety of food, drug, health products, cosmetics, and medical devices was released, and food and drug safety problems of public concern were answered. According to the *Institutional Reform and Functional Transformation Plan of the State Council* adopted at the First Meeting of the 12th National People's Congress, the China Food and Drug Administration (CFDA) has been responsible for the release of important food safety information in place of the Ministry of Health since March 2013. However, in the section "Special Column" on the website of the CFDA, only the column "Food Safety Rectification in Catering" is related to food, while all other columns are about drug-related information. Also, the column "Warning Information" in the section "Public Service" is all about drugs. The CFDA has not reached the legal requirements for disclosure of food safety information in breadth or depth. Work on food safety information disclosure may be delayed due to the institutional reform.

10.1.2.2 National Health and Family Planning Commission   After the establishment of the CFDA, the former Ministry of Health was no longer in charge of the administration of food and drug safety. The newly created National Health and Family Planning Commission (NHFPC) is primarily responsible for food safety risk assessment and food safety standard establishment, and it has a relatively simple responsibility in the field of food safety information. Its internal organization Food Safety Standards and

Monitoring and Assessment Department is particularly responsible for food safety information disclosure. The former Ministry of Health issued *A Notice on Improving Disclosure of Food Safety Standards Related Information* (document number: [2012] No. 77) on November 30, 2012, which placed the focus of information disclosure on promoting the disclosure of food safety standards-related information. According to our search results, the website of the Food Safety Standards and Monitoring and Assessment Department of the NHFPC made more than 60 disclosures of important food safety standard information from June 2012 to July 2013.

Food safety standards-related information was the focus of the food safety information disclosure of the NHFPC from June 2012 to July 2013. A total of 60 disclosures of announcement, specifications, and replies directly related to food safety standards, excluding the notices on meetings, were made during this period. The *2012 Annual Report on Government Information Disclosure of the Ministry of Health* pointed out the existing problems of government food safety information disclosure and related future work to be "strengthening food safety information disclosure, clarifying the content and form of information disclosure and division of responsibilities for health administrative departments at all levels, closely integrating information disclosure with the popularization of science and public communication, and guiding the general public in correctly understanding the work on national food safety standards" (Website of the Central People's Government of the PRC 2013c).

10.1.2.3 Ministry of Agriculture   The Ministry of Agriculture is responsible for quality and safety supervision and management of edible agricultural products from farming until their entry into wholesale and retail markets or production and processing enterprises. Supervision and management of the use and quality of veterinary drugs, feed, feed additives, and pesticides, fertilizers, and other agricultural inputs are within the scope of its official duties, as are quality and safety supervision and management in livestock and poultry slaughter and raw milk procurement. According to the *Measures for Administration of Disclosure of Agricultural Product Quality and Safety Information (For Trial Implementation)* (document number: [2010] No. 10) (Website of the Central People's Government of the PRC 2010), the main contents of the disclosure of agricultural product quality and safety information include the following: (1) routine monitoring, sampling inspection, and special monitoring results of agricultural product quality and safety; (2) quality and safety incidents of agricultural products caused by problems in production and related handling information; (3) investigation and verification of quality and safety problems of agricultural products reported by consumers or the media and related handling information; and (4) other information on agricultural product quality and safety released by the agriculture department. Food safety information associated with the above functions is distributed in different sections of the website of the Ministry of Agriculture. The two hot topics, "Authoritative GMF Information" and "National Food Safety Awareness Week," release hot food safety information of general concern, which were the highlights of food safety information disclosure by the Ministry of Agriculture from June 2012 to July 2013.

10.1.2.4 Administration of Quality Supervision, Inspection and Quarantine The AQSIQ is responsible for supervision and management of production and processing of food-related products, such as packaging materials, containers, and tools. It is also responsible for the safety and quality supervision and inspection of imported and exported food. AQSIQ shall collect safety information of imported and exported food and promptly inform the CFDA. If food safety incidents occurring outside the territory of China may have effects within China, or serious food safety problems are found in imported food, AQSIQ shall promptly take risk warning or control measures and inform the CFDA. Upon notification, the CFDA shall promptly take appropriate measures to address the situation. Before the reform in 2013, AQSIQ comprised the Food Production Supervision Department and the Import and Export Food Safety Bureau, which were responsible for food safety management in production and in entry and exit, respectively. The webpage of the Food Production Supervision Department on the then AQSIQ website (http://spscjgs.aqsiq.gov.cn/) could be searched for "food production license information" and "information on certified companies for food additives and food-related products." Before the reform, the webpage released a large amount of information on the excessive use of food additives in food production and processing detected by sampling inspection, including food producers using excessive food additives. Because of the institutional reform and functional adjustment, the Food Production Supervision Department is no longer included on the AQSIQ website. Currently, the website of the Import and Export Food Safety Bureau on the AQSIQ website (http://jckspaqj.aqsiq.gov.cn/) includes a section of risk warning for imported and exported food and cosmetics, where one can search for information on substandard imported food, including food additives-related information. Moreover, this section has provided a report almost every month.

10.1.2.5 State Administration of Industry and Commerce The food and drug supervision and administration department is responsible for advertising content censoring for drugs, medical devices, and health food, while the industry and commerce administrative department is responsible for supervision and inspection of advertising campaign for drugs, medical devices, and health food. Based on this division of function, the responsibility of SAIC that is directly related to food safety is the "supervision and inspection of advertising campaign for health food." In the section "Regulation and Law Enforcement" on the SAIC website, two columns are related to food safety, that is, "Food Safety Supervision" and "Advertising Regulation and Law Enforcement." The former mainly releases information on food safety supervision in circulation. After the institutional reform, the function of food safety supervision in circulation was transferred to the CFDA. Apparently, it will take some time to complete the transfer of this function. According to the document *Notable Effects Have Been Achieved in Food Safety Rectification in the National Industry and Commerce System, with Accumulated Experience and Achievement in Food Safety Management* issued by the SAIC on June 25, 2013, as of the end of March 2013, 520,000 cases related to food safety in circulation were investigated in the entire system, with a total involved value of 2.188 billion yuan; 509 of these cases were transferred to judiciary. After the completion of the function transfer, such information will no longer appear on the SAIC website, and only food safety information related to health food advertising regulation will be presented. As for food safety information disclosure by SAIC from June 2012 to July 2013, information disclosure was well executed for "advertising regulation and law enforcement" (Chapter 3). However, information disclosure on food safety in circulation was relatively insufficient, at least during the period from June 2012 to February 2013, before the institutional reform. Important information, such as "sampling qualified rates of food in circulation," cannot be found on its website as of the writing of this book.

10.1.2.6 Ministry of Commerce　According to the plan of institutional reform, the reformed Ministry of Commerce is responsible for creating development plans and policies to promote catering and liquor circulation, while the CFDA is responsible for supervision and management of food safety in catering and liquor safety. In addition, the responsibilities of the Ministry of Commerce for pig slaughter supervision and management were transferred to the Ministry of Agriculture. The Ministry of Commerce has a relatively simple function in food safety supervision and management, which focuses on creating development plans and policies and does not involve administrative law enforcement and practical regulation. The development plans and policies are made in the form of notices, provisions, and other normative documents in practice. This type of information is government information that an administrative authority shall disclose actively and can be searched for in the special column of government information disclosure.

The Ministry of Commerce issued the *2013 Working Points on Food Safety in the Commercial System* on May 22, 2013. It is the only central state organization that has issued a document on working points on food safety. The document proposed the following working points for the management of imported and exported food:

1. Strengthen information service system construction for imported and exported products and release relevant information in a timely manner
2. Strengthen public services for agricultural exports
3. Enhance public opinion monitoring and information release and respond to public concerns about hot issues in various forms and channels in a timely, objective, and accurate manner
4. Improve communication before information release to ensure the scientificity and accuracy of the information released

10.1.3 Government Information Disclosure of Food Safety Incidents

Disclosure of information on food safety incidents has a positive effect for the prevention of public scares over food safety. However, information disclosure is currently seriously insufficient with respect to this and cannot produce a beneficial effect on preventing food safety risks.

10.1.3.1 Normative Documents on Food Safety Incidents   According to the *National Food Safety Incident Response Plan*, food safety incidents are incidents that are harmful or potentially harmful to human health and caused by food, such as food poisoning, food-borne illness, and food contamination. They are classified into four levels, namely special major food safety accident, major food safety accident, large food safety incident, and ordinary food safety accident. However, the *National Food Safety Incident Response Plan* does not clearly define the actual conditions of these different levels of food safety incidents. The local governments have issued over 100 food safety incident emergency plans, which provide similar definitions of different levels of food safety incidents. The most recent one is the *Food Safety Incident Response Plan of Xi'an Municipality* (document number: [2013] No. 78) issued by the Office of the Xi'an Municipal People's Government on June 17, 2013.

According to the *Food Safety Incident Response Plan of Xi'an Municipality*, a food safety incident is defined as a special major incident (level I) if one of the following conditions occurs:

1. Distribution of contaminated food to two or more provinces or outside (the territory of) China (including Hong Kong, Macao, and Taiwan regions), which has caused or is assessed as having the potential to cause very serious harm to health
2. Identification as other level I food safety incidents by the State Council

A food safety incident is defined as a major incident (level II) if one of the following conditions occurs:

1. Distribution of contaminated food to two or more districted cities, which has caused or is assessed as having the potential to cause food poisoning or food-borne diseases that could result in serious harm to the public health
2. A food-borne disease caused by a new contaminant detected in China for the first time that has resulted in serious harm to people's health and has the tendency to spread
3. An incident of food poisoning with more than 100 people poisoned or one that has caused the death of more than 10 people
4. Identification as other level II food safety incidents by a government at or above the provincial level

A food safety incident is defined as a large incident (level III) if one of the following conditions occurs:

1. Distribution of contaminated food to two or more counties, which has caused serious harm to health
2. An incident of food poisoning with more than 100 people poisoned or one that causes the death of any person
3. Identification as other level III food safety incidents by a government at or above the city level

A food safety incident is defined as an ordinary incident (level IV) if one of the following conditions occurs:

1. The presence of contaminated food, which has caused serious harm to health
2. An incident of food poisoning with less than 99 people poisoned and causing no deaths
3. Identification as other level IV food safety incidents by a government at or above the county level

10.1.3.2 Information Disclosure of Food Safety Incidents According to the *Measures for Administration of Food Safety Information Disclosure* (document number: [2010] No. 93) issued by the former Ministry of Health, Ministry of Agriculture, Ministry of Commerce, SAIC, AQSIQ, and SFDA on November 3, 2010, in case of a major food safety incident, information should be disclosed for the major food safety accident and its handling, including location, basic information of the responsible units, number of casualties, medical treatment, cause of the incident, investigation of liability, and emergency response measures. Many "food safety scandals" were reported by the media between June 2012 and July 2013. However, these scandals can only be referred to as "hot food safety issues," not "food safety incidents" or even "food safety events" (China Economic Net 2013). Official and authoritative data are not available with regard to which hot food safety issues of public concern constitute "food safety incidents." This lack of information in itself demonstrates a deficiency in food safety information disclosure.

The incident of rice contaminated with cadmium is a food safety incident that occurred in the first half of 2013. A serious problem of dereliction of duty of government regulators in food safety information disclosure was demonstrated by this incident. According to the report of the *Southern Metropolis Daily*, the overall pass rate of food and related products in catering in the first quarter of 2013 released by the Guangzhou Food and Drug Administration was 92.92%. In this sampling inspection, the Guangzhou Food and Drug Administration inspected a total of 367 batches of food and related products in catering, with a pass rate of 92.92% (341 batches). Of them, 310 batches were food and its raw materials, with a pass rate of 92.58% (287 batches); 57 batches were tableware, with a pass rate of 94.73% (54 batches). The lowest pass rate was recorded in rice and rice products. A total of 18 batches of rice and rice products were inspected and 10 of them were found to be up to standard, with a pass rate of 55.56%. The other eight batches failed the test for cadmium contamination. As for the specific catering units using rice contaminated with cadmium, an official from the Guangzhou Food and Drug Administration said that only data could be published for the time being, instead of the specific list (Ma 2013). After exposure of the incident, the public repeatedly requested disclosure of specific information from the relevant government departments. After being questioned by the public, the Guangzhou Food and Drug Administration announced the units involved. However, the relevant departments have not given a clear response as to the distribution of the substandard food. Food safety information concerning the public's health is often regarded as a secret by the relevant governmental departments. After the food and drug supervision and administration departments of Guangzhou published the news about food safety on May 16, 2013, the information disclosure process has been like slowly squeezing toothpaste out of a tube under constant questioning by the public, from the nondisclosure of enterprises involved at the beginning to the conditional disclosure of part of the producers of unsafe food, all the while stating that it is inappropriate to disclose some producers and brands. This way of information disclosure has not eliminated the worry of consumers and has also sparked a wider range of questions (Guo 2013). Due to the unavailability of official information, the sources, areas of influence, affected populations, and results of rice contaminated with cadmium are not known, which further increases the public's fear.

10.2 Main Characteristics of Food Safety Information Disclosure

In recent years, food safety information disclosure has undergone new changes in China and has some new characteristics, which will be discussed in Sections 10.2.1 through 10.2.4.

10.2.1 Transition from Legal System Construction to Behavioral Regulation

The *Food Safety Law, Implementing Regulations of the Food Safety Law,* and *Measures for Administration of Food Safety Information Disclosure,* as well as many normative documents on local food safety information disclosure management, constitute the basis of the legal system on food safety information disclosure. These laws, regulations, rules, and documents were mainly enacted during 2010–2012, providing a basic normative framework for food safety information disclosure. Since the second half of 2012, the focus of relevant work on food safety information has moved from "creation of rules" to the stage of "concrete action." In terms of the number of norms issued, only three central and local norms on food safety information were issued from June 2012 to July 2013. However, in terms of action, the NHFPC, for example, disclosed relevant information almost every month during this book period, rather than disclosing information for a long period in single month after not disclosing information for a long period of time.

10.2.2 Change in Departments Responsible for Information Disclosure

The CFDA was created according to the *Institutional Reform and Functional Transformation Plan of the State Council* adopted at the First

Meeting of the 12th National People's Congress. Its key responsibilities are to perform unified supervision and management of food safety and drug safety and efficacy in production, circulation, and consumption. The plan has adjusted the food safety management system. Accordingly, the responsibility for food safety information disclosure has also been transferred to new departments with a transfer of function. Formerly, "major food safety information" was uniformly disclosed by the Ministry of Health and "routine food safety information" was disclosed by the other respective departments. Since the institutional reform, major food safety information is now disclosed by the newly created CFDA. The *Key Responsibilities, Internal Structure, and Staffing of the China Food and Drug Administration* (document number: [2013] No. 24 of the General Office of the State Council of the People's Republic of China) clearly provides that the Press and Publicity Department, as a division of the CFDA, is responsible for developing a system for uniform disclosure of food safety information and carrying out popularization of science, and news and information releases regarding food and drug safety (Website of the Central People's Government of the PRC 2013d). Other information on food safety supervision and management remains to be disclosed by the Ministry of Agriculture, NHFPC, AQSIQ, SAIC, Ministry of Commerce, and Ministry of Public Security in their respective areas of responsibility.

The newly created CFDA assumes primary responsibilities in food safety information disclosure. However, because the CFDA is newly created, its functions have not yet been fully organized. Much emphasis has been placed on drug information disclosure, while too little is done for food information disclosure. Hopefully, this situation will be improved in the next report period. After readjustment of the functions, the NHFPC plays a relatively simple role in food safety supervision and management and assumes more responsibilities in information disclosure than before. Its work on formulation, cleanup, and disclosure of food safety standards was one of the highlights in 2012. Of all the central ministries and commissions involved in food safety supervision and management, the NHFPC has done the most significant work in food safety information disclosure.

10.2.3 Emphasis on Information Disclosure of Standards

China has established nearly 1900 national standards regarding food, food additives, and food-related products. After implementation of the *Food Safety Law*, relevant departments have cleaned up and integrated the standards in a timely manner. As a result, 185 national food safety standards have been published and 2193 indicators of pesticide residue and veterinary drug limits have been cleaned up. However, a uniform food safety standard system has not been fully developed and some standards of food hygiene, quality, and edible agricultural products quality and safety, as well as industry standards, are absent, lagging, or overlapping or conflicting with one another (Website of the Central People's Government of the PRC 2012). Moreover, food safety standards are published in a disordered and decentralized manner and thus are often difficult to locate. The document *A Notice on Improving Disclosure of Food Safety Standards Related Information* (document number: [2012] No. 77), issued by the Ministry of Health on November 30, 2012, emphasizes that the health administrative departments at all levels shall voluntarily disclose the following food safety standard information within their scope of responsibilities in accordance with the *Food Safety Law* and its implementing regulations:

1. National food safety standards, local measures for standard administration, and measures for the archival filing of the enterprise food safety standards
2. Planning for national food safety standards
3. Annual plan of formulation/revision of food safety standards
4. Draft of food safety standards open for comments and explanation for its formulation
5. Texts of the national and local food safety standards and standards
6. Explanatory materials of the national and local food safety standards
7. Other information on food safety standards that should be voluntarily disclosed in accordance with the relevant laws, regulations, and provisions

The methods of disclosure include the department website, department microblog, information disclosure column, electronic screen, announcements, special public issues, printouts, food safety awareness weeks and open days, other major food safety publicity and education activities, newspapers, radio, television, internet, mobile phones, and other media. The 12320 health hotline also exists, as well as other accessible ways allowing the public to get information. As previously mentioned, the NHFPC has made nearly 60 disclosures of information on food safety standards.

10.2.4 A Trend to Disclose Safety Information by Special Topic

Due to the large quantity, disorder, and many different types of food safety information, both the discloser and the recipient may get caught up in a large amount of information, not knowing how to start. Collection of food safety information by special topic is one of the important ways of solving this problem. This method provides clear information at a glance and makes it easy to look up and use. Food safety information was mostly dispersed on various sections of the government website, which made it time-consuming and laborious to look up in 2011. The situation has since been improved. A greater number of special columns on food safety information now exist, such as the special columns "Risk Assessment Information" and "Risk Warning Information" on the NHFPC website, and the special column "GMF Safety" and the "National Food Safety Awareness Week" on the website of the Ministry of Agriculture.

10.3 Main Problems in Food Safety Information Disclosure

Many problems remain to be solved in the field of food safety information disclosure in China. Information asymmetry is still the essential feature of food safety information disclosure in China.

10.3.1 Disorder of Disclosers

The institution responsible for disclosure of certain food safety information shall be the government department supervising food safety in the relevant area. According to the *Food Safety Law*, *Implementing Regulations of the Food Safety Law*, and *Measures for Administration of Food Safety Information Disclosure*, such departments include the food and drug supervision and administration department, health department, agriculture department, Ministry of Commerce, quality inspection and quarantine department, and industry and commerce administrative department. These departments bear the responsibilities for collection, preparation, and disclosure of food safety information and the public has a right to access the relevant information. When these departments do not adequately perform their duties, the absence of authoritative food safety information results in a disorder of information. The hot issues on food safety in 2012 have truly demonstrated this problem.

As shown in Table 10.2, among the 12 hot issues on food safety between March and November 2012, the only true food safety incident, according to the judgment of experts, was the Coca-Cola fluoride scandal. Other

TABLE 10.2
Hot Food Safety Issues of Public Concern in 2012

Month	Hot Issue	Expert	Expert Interpretation
March	Lead contamination in Spirulina	Feng Chen	The media misused the standards, made improper judgment, and announced the news without verification; not a food safety incident
April	Gelatin scandal	Ying Sun	The addition of industrial gelatin to set style yogurt and jelly was unsubstantiated; not a food safety incident
April	Pesticide scandal of Lipton	Wei Chen	The judgment was made on a wrong basis; not a food safety incident
April	Coca-Cola fluoride scandal	Baoguo Sun	Management and control was not in place during production; a food safety incident
June	Detection of ethyl carbamate in Guyue Longshan rice wine	Zhenghe Xiong	There was no basis to judge whether the ethyl carbamate (content) in rice wine posed a potential carcinogenic risk to humans; not a food safety incident
July	Acid value scandal of Jinmailang	Xiaomei Yuan	The inspection agency inspection withdrew the report and declared it invalid; not a food safety incident
August	Fluorescent substance out of limits	Tianzu Wang	Paper containers conforming to the relevant Chinese standards can ensure food safety
August	Excessive pesticide residues in Changyu Wine	Xianming Shi	The media coverage was not comprehensive; pesticide residues were not found to be excessive in wines tested; the products were up to the standard
August–September	Issue of borax	Yimin Wei	Utilization of public opinion due to improper law enforcement; not a food safety incident
September	Inclusion of a banned substance in a cheese product of Bright Dairy for baby	Ning Li	The use of the substance requires approval under the relevant provisions
October	Excessive bacteria in KFC burger	Xiumei Liu	An improper standard was quoted; the judgment was unfounded; not a food safety incident
November	Excessive plasticizer in Chinese liquor	Yan Xu	The judgment was unfounded; not a food safety incident

Source. China Economic Net, 2013, http://www.ce.cn/cysc/sp/info/201301/06/ t20130106_21313898.shtml.

scandals caused huge losses to the enterprises, though many of them were ultimately proved to be not food safety incidents. Do institutions or individuals other than the statutory government departments have the right to publish information on food safety? What legal liability shall be borne for illegal disclosure? These questions have not been answered by the current laws. Food safety information should be disclosed following the legal, scientific, accurate, comprehensive, objective, and fair principle, as well as strict procedures. The ultimate goal should be to help strengthen food safety supervision, safeguard the right to know and supervise consumers and producers, and guide market consumption and the food industry in a healthy direction (Ma 2012).

10.3.2 The Lack of an Accurate Grasp of Food Safety Information

The *Food Safety Law* and the *Measures for Administration of Food Safety Information Disclosure* clearly define food safety information. However, the definition of food safety information is not accurately understood in practice. Food safety information has been confused with news reports, communications, and comments and has been published as the same type of information, making it difficult to search for and use food safety information. The extension of food safety information should cover the following contents: (1) information on the overall food safety situation, analysis and forecasting of the overall trend of food safety, and early warning; (2) information on routine monitoring of food safety, primarily including information on the local food safety and food safety of the relevant food system collected through routine monitoring by various food safety supervision departments; (3) information on food safety supervision and management, including policy measures, planning and deployment, program implementation, local legislation, standards development and revision, supervision and inspection, testing and examination, risk monitoring and evaluation, special rectification, supervision of case investigation, important meetings, important instructions from the leadership, and work news related to food safety management; (4) information on food safety incidents and risk warning, including report and complaints about food quality and safety problems, major food poisoning incidents, food contamination emergencies, zoonotic diseases, and other information related to food safety incidents and their handling; (5) license and credit information on food production and business operations, including information on food production, food circulation, food service, and business licenses, and relevant certifications and credits; (6) information on food safety system construction, including institutions, publicity, training, credit system construction, construction of important projects, and scientific achievements and their applications; and (7) public opinion on food safety, including food safety-related information that has been reported by the media or has caused widespread public concern and other important food safety information.

10.3.3 Incomprehensive Disclosure

The document *Arrangement of Key Work on Government Information Disclosure in 2012* (document number: [2012] No. 26 of the General Office of the State Council of the People's Republic of China) explicitly defines "promoting food safety information disclosure" as one of the key fields of government information disclosure in 2012. The document stresses that more efforts should be devoted to food safety supervision information disclosure. Specifically, food production and business operation licenses, special inspection and rectification, investigation of illegal production and business activities, and other daily supervision information and risk assessment and warning information should be disclosed pursuant to the law in principle. Moreover, the health administrative department should further strengthen food safety standards disclosure, broaden the channels for the public to participate in drafting of food safety standards, and improve public consultation and participation. The work arrangement of the State Council explicitly defines the six categories of information as key food safety information to be disclosed and requires that implementation of this task should be led by the Office of the Food Safety Committee and the Ministry of Health. Broadly speaking, the "overall food safety situation" is an overall evaluation of the food safety situation within a certain area and is an important part of food safety information. However, thus far, this vital information is not provided from the central to the local level. Although the six categories of food safety information have been disclosed, they are far from sufficient to meet the requirements of the State Council, except the information on food safety standards. The information disclosed is fragmented and not comprehensive in terms of both a single point and a plane. The nonperformance and inadequate performance of regulators result in the lack of authoritative information. From the point of view of specific incidents, food safety information on substandard food, such as rice contaminated with cadmium, remains incomprehensive. "In face of unexpected food safety problems, such as the cadmium contamination, the regulators should not keep silent, but actively respond to questions of all parties and provide as much information as possible for reference. For example, the brand, producer, trader, and distribution of the substandard food should be made public objectively and truly" (Liao 2013).

10.3.4 Failure to Make Timely Disclosure

Both the *Food Safety Law* and the *Measures for Administration of Food Safety Information Disclosure* clearly require the food safety supervision departments to disclose information in an accurate, timely, and objective manner. In practice, however, many food safety supervision departments do not strictly comply with this requirement. In the incident of rice contaminated with cadmium, for example, the Guangzhou Food and Drug Administration released monitoring results on May 16, 2012, but stated that it was not appropriate to disclose the list of specific catering units using the rice contaminated with cadmium at that time. Nevertheless, the Guangzhou Food and Drug Administration finally announced the relevant units in response to the subsequent strong questioning from the public. This incident ended with information disclosure by the relevant departments (China Food Safety Net 2013). However, the food safety information disclosure was made under pressure from the public instead of being made immediately after an incident. In the face of this incident, the public expects the regulators to investigate the incident quickly, destroy the rice, trace the source of the rice, and severely punish catering units using the rice to prevent the recidivity. Relevant information must be disclosed in a comprehensive and timely manner to satisfy the public's right to know and choose, and regulators/relevant departments must make efforts to prevent incidents that will do harm to the public health (Lian 2013).

10.3.5 A Huge Gap Exists between the Amount of Information Disclosed and the Public Demand

Food safety information is one category of key government information to be disclosed specified by the State Council in 2012. The *2012 Annual Report on the Chinese Government Transparency* issued by the Chinese Academy of Social Sciences on February 25, 2013, warns of a grim situation of food safety supervision information disclosure in many cities in China. A survey of the quality and technical supervision websites of 43 large cities revealed that only 10 of the websites could provide information on the supervision and sampling inspection of local food production in the last month, representing an information transparency of only 23%. In addition, the disclosure of food enterprise safety credit file information is not optimistic. The book revealed that only the quality and technical supervision departments of 16 large cities had provided food enterprise safety credit files, not to mention providing various categories of basic information in detail. In addition, the food safety information transparency of some key tourist cities, such as Luoyang and Haikou, was also not satisfactory. In the evaluation of food safety information transparency, Luoyang scored only 1.5 and Haikou actually scored 0 on a scale of 10. Another city with a score of 0 in terms of food safety information transparency was Xining (China E-government Network 2013).

10.4 Focus on Food Safety Information Disclosure System Construction

Great efforts should be made, based on the current situation, to disclose critical information in a timely manner, respond to the public concern, eliminate public scares regarding food safety, solve food safety information asymmetry to the best extent possible, and promote the construction of a food safety information disclosure system.

10.4.1 Refine the Classification of Food Safety Information

Food safety regulators have disclosed a certain amount of food safety information; however, the large quantity and disorder of information makes it difficult to search/find. The information classification is unscientific, and it does not always correspond exactly to the relevant management departments. The food safety information disclosure system should be first improved by refining the information classification and the information should be matched to the corresponding food safety regulator. Regulators should clearly define the scope of food safety information to be disclosed by them and disclose the information to the public in a timely and legal manner. From the geographical point of view, food safety information should include national and local food safety information. From the point of view of content, food safety information should include the following categories: (1) information on the overall food safety situation, including the annual national overall food safety situation, implementation of national food safety risk monitoring plans, and formulation and revision of national food safety standards; (2) information on food safety risk assessment and warnings, including warnings of the presence or potential presence of toxic and harmful factors in food, and risk warnings of food with a high level of safety risk; (3) information on major food safety accidents and their handling, including location, basic information of the responsible units, number of casualties, medical treatment, cause of the incident, investigation of liability, and emergency response measures; and (4) other important food safety information and information that should be disclosed uniformly as decided by the State Council.

A food safety information disclosure mechanism should be established and improved to disclose relevant hot food safety information on commonly consumed food varieties with city and county as the basic unit and food safety government regulator as the discloser. Moreover, relevant information of food producer and trader should be disclosed in a timely manner; periodically targeted food safety inspections should be performed on the basis of enterprise reporting, and the inspection results should be disclosed in an authoritative, standard, and publicly accessible way. A food producer and trader credibility evaluation and publicity system should also be established as soon as possible (Liu 2012b).

10.4.2 Organize the Information Disclosure Function of Regulators

The basic principle is that the regulator is the discloser of corresponding information, that is, the department that is responsible for food safety supervision is also responsible for the collection, preparation, and disclosure of the relevant food safety information. It is critical to avoid "buck-passing" and unclear division of information disclosure responsibilities between the regulators. All the food safety supervision departments above the county level are authorized to disclose daily supervision information. Because of the large number and multiple levels of disclosers, it is necessary to strengthen interdepartmental, interlevel, and interregional coordination and cooperation in information disclosure.

Specifically, emphasis should be placed on the following aspects.

1. The relevant departments should fulfill their respective duty in disclosing information on daily food safety supervision and disclose whatever should be disclosed.
2. In case the information to be disclosed involves more than two supervision departments, the relevant departments should closely communicate with each other. In case the sensitive information to be disclosed involves other provinces or cities, the relevant areas should be informed before the disclosure to ensure proper communication.
3. Food safety risk assessment and warning information that need be disclosed shall be managed and disclosed by specialized health administrative departments at or above the provincial level.
4. A uniform disclosure system should be implemented for important information; information on major food safety incidents and their handling, overall national food safety situation, and other important information should be well planned and coordinated before uniform disclosure (Zhang 2012).

Because of the large quantity and disorder of food safety information, it is difficult to achieve high-quality, high-efficiency, and full-coverage disclosure in the short term. Therefore, it is imperative to perform a rough classification of food safety information and define the specialized management responsibilities of specialized departments. Based on the classification of food safety

TABLE 10.3
Matching of Important Food Safety Information with Food Safety Regulators

Regulator	Category of Food Safety Information
Food and drug supervision and administration department	Overall food safety situation
	Food safety incidents
	Administrative licenses
	Hot issues of food safety
Health department	Information on food safety risk assessment
	Information on food safety standards
Agriculture department	Routine monitoring, sampling inspection, and special monitoring results of agricultural product quality and safety
Commercial department	Development plans and policies to promote catering and liquor circulation
Inspection and quarantine department	Food packaging materials, containers, and related tools
Industry and commerce administrative department	Information on advertising campaign of health food

information, important food safety information that should be managed by a certain food safety regulator should, at least, be first disclosed by that regulator. The food safety information disclosure system can thus be gradually improved on this basis. Table 10.3 lists the basic food safety information that should be managed by each food safety regulator. It should be taken as the outset of the improvement of food safety information disclosure. Among the various regulators, the food and drug supervision and administration department is an important department responsible for food safety information disclosure; however, it has fallen down on the job in this area, at least for now.

10.4.3 Strengthen the Construction of Provincial Food Safety Information Disclosure Platforms

In the longitudinal direction, the people's governments at or above the county level take full responsibility for food safety information disclosure. As to particularly important food safety information involving the entire province, the provincial people's government shall decide the disclosure details. If the food safety information involves two or more districted cities, or if the food safety information involves one districted city but the problem presents a prevailing tendency and needs to be disclosed at the provincial level, then the relevant department of the provincial people's government shall decide the disclosure details. For food safety information concerning only a single stage or department, the disclosure shall be made by the relevant competent department at the provincial level according to the division of responsibilities. Food safety information should be disclosed through government websites, government gazettes, press conferences, newspapers, radio, television, government affairs disclosure columns, electronic screens, electronic touch screens, and other convenient ways for the public to access the information. Because of the large number of disclosers and the diversity of disclosure channels, a major information disclosure platform should be developed as the basic way for the public to get the information. The provincial food safety information net and information conferences (press conference, briefing, etc.) held by the provincial food safety commission and its office are the main platforms for provincial food safety information disclosure. Because the provincial administrative region has the role of connecting the upper levels with the lower levels, the provincial food safety information net should serve as a local food safety information sharing platform. Currently, most provinces have built a provincial food safety net, but it has not been fully positioned as "provincial information disclosure platform" in terms of content. For example, the "Guangdong food safety information" is prepared by the Office of the Guangdong Provincial Food Safety Commission at regular intervals and disclosed on the Guangdong Provincial Food Safety Net (http://www.gdfs.gov.cn/). Although the information content is not comprehensive, this form of disclosure allows the public to obtain information in a relatively centralized manner. More efforts should be made to construct provincial information platforms, which are the basis of the central food safety information disclosure.

10.4.4 Improve the White Paper System of Food Safety

A white paper is an official report issued by the government that often addresses problems and how to solve them. These papers are applicable in many fields. Since the first white paper was published in 1991, China has published 85 white papers related to democracy, legal construction, political party system, human rights, arms control, national defense, nonproliferation, religion, population, energy, environment, intellectual property, food and drug safety, internet, and Tibet and Xinjiang as of May 2013 (Baidu Encyclopedia 2013). On August 17, 2007, the Information Office of the State Council published the white paper *Food Quality and Safety in China*, which reports the food quality and safety in China in various aspects, such as production, general quality situation, management system and related work, and supervision of imported and exported food. It is the first time the Chinese government provided a comprehensive report on the food safety situation in China in the form of a white paper. However, a national food safety white paper was not published annually after that, and the food safety white paper system has not yet been popularized nationwide. In fact, a member of the Chinese People's Political Consultative Conference (CPPCC) has suggested implementing an annual food safety white paper system. Specifically, it is to publish a white paper on the situation of food safety and food market regulation annually by the name of the government, with provinces, autonomous regions, and municipalities directly under the central government as the unit. Publishing in this way will help the people understand the food safety situation and supervision every year and subsequently lead to confidence in the Chinese food market (Liu 2012b). As a comprehensive description of food safety supervision efforts and results by the government, the food safety white paper contains a large amount of food safety information and should be used as a regular form of food safety information disclosure. Therefore, this book suggests that the central government and the provincial governments shall regularly publish a white paper on food safety in the first quarter of each year and that administrative regions below the provincial level may voluntarily publish a white paper on local food safety.

11

Food Safety Incidents
of Concern in 2011 in Chinese
Mainland and a Brief Review

Food safety is a long-standing, complex, and mixed social and global problem. Every year, a large number of consumers throughout the world are faced with different food safety risks (Sarig et al. 2003). In China, as intricate factors are intertwined, food safety risk has become a particularly prominent problem in recent years. Frequent food safety incidents are becoming a major social risk affecting social stability and consumer health. Taking into account the integrity and reliability of data, this chapter analyzes and reviews the food safety incidents of concern that occurred in China in 2011, as well as the major controversies on food safety standards.*

11.1 Background and Incident Selection

In Chapter 5, food safety risks in China during 2006–2012 were assessed, and the actual state of food safety risks was analyzed to provide a panoramic description of the real changes of food quality and safety in China, with the conclusion that China's food safety has been relatively safe. During that time, the food safety situation was generally stable and continued to

---

*Generally speaking, in the narrow sense, food safety incidents are incidents where consumption of food has directly or indirectly resulted in a consumer health problem due to various causes. In the broad sense, food safety incidents also include incidents that receive extensive attention and are closely related to food safety and possibly without any report of human health problems as a direct or indirect result (at least at that time). In the cases discussed in this chapter, use of shoddy and false advertising, for example, are food safety incidents in the broad sense.

gradually improve through 2011. However, some food safety incidents of concern still occurred in mainland China in 2011. During the 21st Meeting of the Standing Committee of the 11th National People's Congress on June 29, 2011, the inclusion of food safety, such as financial security, grain security, energy security, and ecological safety, was proposed in the "national security" system. This indicated that the issue of food safety was becoming one of the most important focuses in China and was considered to be a threat to national security. In-depth exploration of the deep-seated causes of major food safety incidents that occurred in China will not only help the government to strengthen food safety supervision and the society to understand the truth and harm of food safety incidents, but also help food producers and traders to better understand the situation and identify preventive measures.*

On December 10, 2011, Food & Beverage Online summarized and published 24 influential food safety incidents that occurred in China in 2011 (Food & Beverage Online 2011), which will be covered in this chapter. In addition, this chapter focuses on select influential food safety incidents from the related media reports, such as heavy metal pollution, tainted Chinese chives, exploding watermelons, and counterfeit green pork, as well as controversies regarding food safety standards of widespread concern. The primary aim of this chapter is to analyze these food safety incidents and controversies on food safety standards of public concern.†

11.2 Description and Analysis of Food Safety Incidents

Although the major food safety incidents exposed by the media have complex causes, they can be classified by the most important causes, including environmental pollution, excessive residues of agricultural chemicals such as pesticides and veterinary drugs, microorganisms, microbial toxins and parasites, illegal or improper use of food additives, use of nonfood raw materials, use of shoddy and false certification, and false advertising.

---

*In this book, if not otherwise noted, the scope of China does not include Hong Kong, Macau, or Taiwan.

†It should be noted that other incidents, such as the nitrate poisoning of several children in Pingliang, Gansu, and the Coca-Cola Minute Maid poisoning in Changchun, Jilin, also occurred in China. However, such incidents, in essence, are criminal offenses relating to food safety and not simply food safety incidents and therefore are not discussed in this book.

11.2.1 Food Safety Incidents Caused by Environmental Pollution

Among the food safety incidents caused by environmental pollution that occurred in 2011, the most notable one was the incident of cadmium-contaminated rice. In the process of industrialization and urbanization, ubiquitous mining and other industrial activities have resulted in the release of cadmium, arsenic, mercury, and other harmful heavy metals previously present in the form of compounds into the natural world. This has led to land contamination through water and air, thus causing contamination of agricultural products. After lettuce, rice has the second strongest ability to accumulate cadmium; therefore, rice planting in cadmium-contaminated soil will produce cadmium-contaminated rice with high potential food safety risks. On February 14, 2011, the *Century Weekly* published an article that gained widespread attention, titled "Murder by cadmium-contaminated rice." The article reported that a dozen elderly people in Si Village, Xingping Town, Yangshuo County, Guangxi Province, suffered from Itai-itai disease due to long-term consumption of rice containing excessive cadmium and were unable to walk normally.* In essence, the incident of cadmium-contaminated rice is a typical food safety incident caused by environmental pollution.

In fact, research on rice with excessive heavy metals has been carried out for more than 10 years in China. In 2002, sampling inspection of rice was conducted by the Rice Product Quality Supervision and Inspection Center of the Ministry of Agriculture in markets all over China. The highest over standard rate of 28.4% was recorded for lead and the second highest rate of 10.3% was recorded for cadmium.† In 2007, the Institute of Resource, Ecosystem and Environment of Agriculture of Nanjing Agricultural University randomly purchased over 100 samples of rice from the markets above the county level in six regions (East, Northeast, Central, Southwest, South, and North China). Test results revealed that 10% of the rice samples contained excessive

---

*Cadmium mainly accumulates in the human liver and kidneys and cannot be eliminated naturally. Several years or even decades of chronic accumulation may lead to significant symptoms of cadmium poisoning, which will inhibit bone growth and metabolism and cause various bone diseases. The dreaded Itai-itai disease results in the most severe cases. In the 1960s, the farmland along the Jinzu River in Toyama Prefecture, Japan, was seriously polluted by cadmium due to mining. Some local rice farmers were poisoned after long-term consumption of cadmium-contaminated rice. The disease is named for the severe pains (Japanese: 痛い itai) caused in the joints and spine. Poisoning resulting from consumption of cadmium-contaminated rice requires long-term accumulation. Generally, a health hazard to a 60-kg person is possible only with daily intake of at least 100 g of cadmium-contaminated rice for 50 consecutive years.

†In China, lead was the main cause of heavy metal pollution for quite some time. However, the situation changed after the prohibited use of leaded gasoline. Most incidents of excessive blood lead levels are not caused by food contamination. Excessive levels of lead are no longer the most critical problem in the food contamination incidents caused by heavy metals in recent years.

cadmium. In April 2008, the Institute randomly collected 63 samples from the farmers' markets in Jiangxi, Hunan, Guangdong, and other provinces and found that more than 60% of the samples exceeded the national limit of cadmium in rice. A research group from the School of Environmental Science and Engineering of Guangzhou University established stations in 10 major basins, such as Minjiang, Jiulongjiang, Jinjiang, and Mulanxi, in the Fujian coastal area from south to north to collect 185 rice samples. Test results revealed that 16.8% of the samples contained excessive lead, 11.4% contained excessive cadmium, and 0.5% contained excessive mercury and arsenic. Most samples with high levels of cadmium and lead were collected from the surrounding areas of Zhangzhou, Fuzhou, Fuqing, and other industrially developed cities. In January 2006, an alarming cadmium pollution incident occurred in Xinma Village, Zhuzhou City, Hunan Province, which caused two deaths and induced mild chronic cadmium poisoning in 150 villagers. The source of the cadmium pollution of farmlands in Xinma Village was 1 km away from the Xiangjiang River, a river that contained the most serious heavy metal pollution ever found in China. According to a survey of the Institute of Resource, Ecosystem, and Environment of Agriculture of Nanjing Agricultural University, the cadmium contamination rate was the highest in rice produced from Hunan and Jiangxi, and the major cause was soil polluted by mine wastewater in these two areas.

Chinese governments at all levels attach great importance to food safety problems caused by excessive heavy metals and have devoted more efforts to the relevant institutional improvement. However, it is still a difficult task to completely eradicate rice with excessive heavy metals. Under the strict supervision of the government, it is now quite difficult for rice contaminated with excessive heavy metals to find its way into large supermarkets in cities. However, with the expanding scope of soil contamination, the improvement of people's living standards, and increased health awareness, farmers may tend to sell rice containing heavy metals or even excessive heavy metals and purchase other rice to consume for themselves. Therefore, urban residents will be more likely to buy rice containing excessive heavy metals. It is also possible that a small amount of rice containing excessive heavy metals could evade supervision and find its way to the urban rice markets. Nevertheless, as citizens will generally change the rice varieties they eat from time to time, the occasional consumption of rice containing excessive heavy metals will lead to a relatively low accumulation of heavy metals, thus relatively little harm is done. Currently, the largest potential risk of rice containing excessive heavy metals is present in farmers' markets in the villages, towns, and counties, where frequent food safety incidents caused by rice containing excessive heavy metals have occurred. For example, rice produced by Qionglai City Ruitai Rice Co., Ltd. and Sichuan Province Wenjun Rice Co., Ltd. was found to contain excessive cadmium in a food sampling inspection performed by the Bureau of Quality and Technical Supervision of Chengdu, Sichuan, in February 2008 (China Economic Net 2011).

11.2.2 Food Safety Incidents Caused by Improper Use of Agrochemicals

Food safety incidents caused by improper use of agrochemicals that were reported by the media causing widespread concern in 2011 mainly include the incidents of tainted Chinese chives and exploding watermelons.

11.2.2.1 Tainted Chinese Chives   On March 25, 2011, 10 people from four families vomited continuously after eating scrambled eggs with Chinese chives and Chinese chive dumplings in Nanyang City, Henan Province. When they met in the hospital emergency room, they found out that they all developed the symptoms after eating Chinese chives from the same mobile vendor. The diagnosis was poisoning with highly excessive organophosphorus pesticide residues in the Chinese chives.

Because Chinese farmers are accustomed to using organophosphorus pesticides to kill maggots in Chinese chives, poisoning by tainted Chinese chives occurs occasionally. The main symptoms of organophosphate poisoning include a slowed heart rate, nausea, vomiting, sweating, miosis, and in severe cases, amyostasia and even coma. After investigation, the Nanyang Municipal Public Security Bureau confirmed that the tainted Chinese chives came from the same mobile greengrocer. The day after the incident, the Nanyang Inspection Center for Agricultural Product Safety and Quality conducted a dragnet investigation on Chinese chives sold in wholesale markets, farmers' markets, supermarkets, and by mobile vendors in the downtown area. A total of 106 samples were taken the same day, of which two samples contained excessive phorate residues. Moreover, the sources of the positive samples were identified to be two Chinese chive plantations in Luying Town, Wolong District, and Wadian Town, Wancheng District, Nanyang City. The Nanyang Inspection Center for Agricultural Product Safety and Quality confiscated and destroyed all Chinese chives containing excessive phorate residues and eradicated all Chinese chives in the two plantations. On this basis, relevant departments of the Nanyang municipal government further investigated the Chinese chive plantations, strengthened market admittance management for agricultural products, expanded the inspection scope, and increased the inspection frequency to prevent entry of tainted Chinese chives into the market. Furthermore, it was required that more efforts be devoted to managing agricultural commodity markets and guiding the farmers in scientific and rational pesticide use to ensure the safety and quality of agricultural products at the source and prevent similar incidents in the future (Sun 2011b).

11.2.2.2 Exploding Watermelons   Ripeners and swelling agents are plant growth regulators that are also widely used in developed countries. Ripeners can facilitate remote sales of fruits and swelling agents can promote the volume growth of fruits. Although they may reduce the fruit quality, no toxins will be produced.* The widespread concern over these plant growth regulators in 2011 was mainly due to the incident of exploding watermelons in Jiangsu Province.

In May 2011, farmer Mingsuo Liu in Dalv Village in Jiangsu Province was left perplexed after his watermelons began to explode one by one in his over 40-mu farm. Dozens of mu of watermelons also burst in the farms of other farmers. The overuse of a swelling agent was blamed in a report by China Central Television. Overdoses of swelling agent will lead to accelerated cell division and consequently to the explosion of the watermelon. After the exaggerated media reports, this opinion caused national concern regarding early maturing watermelons and resulted in poor watermelon sales in the early summer of 2011, causing significant economic losses to melon farmers.

In fact, agricultural experts investigating the incident were unable to offer a consistent explanation. Farmer Mingsuo Liu was reported to be the only farmer from the affected farmers who used swelling agents. Therefore, most experts did not agree that the use or overuse of swelling agents should be blamed. With regard to the impact on human health, one consistent view of the experts has been that the use of plant growth regulators, including swelling agents, under certain quotas should be safe. Large melons and fruits on the market may be a result of breeding and technological improvements, or they may be a result of the use of a swelling agent. However, the melons and fruits receiving a proper dose of swelling agent are safe. In fact, in addition to watermelons, tomatoes, kiwis, strawberries, other fruits and vegetables often receive plant growth regulators. With a proper and scientific usage, the residues of plant growth regulators in melons and fruits should be within the allowable range and cause no side effects (Sina Finance 2011).

---

*The chemical name of the swelling agent is forchlorfenuron. It is also called KT30 or CPPU. It was first developed in Japan in the 1980s and then introduced to China as a state-approved plant growth regulator. The swelling agent has been widely used in China and has been proven harmless according to its long-term use.

Agrochemicals mainly refer to chemical fertilizers, veterinary drugs, pesticides, and plant growth regulators. As an important means of agricultural production, they have played an important role in pest control and increasing agricultural production. For example, previous studies have shown that the use of pesticides has greatly contributed to increased production in the past 60 years (Fernandez-Cornejo 2004). The amount of grain recovered by pesticides around the world accounts for up to 30% of the global grain production (Liu et al. 2002). Pesticide use also has the added benefits of reducing labor intensity and improving agricultural productivity (Padgitt et al. 2000).

China is seriously affected by agricultural pests and diseases. Without any prevention or control methods, it is estimated that an average annual reduction of approximately 15% would occur in grain production due to crop pests. Zeng et al. (2002) pointed out that the loss would be much larger than previous years if a severe pest infestation occurred and that control with chemical pesticides could recover most of the losses. However, long-term heavy use of pesticides gradually kills predators of pests (Tang et al. 2010a) and also leads to a gradual increase in pest resistance (Gut et al. 2007). Moreover, if the resulting pesticide residues exceed the maximum limits, it will pose a serious threat to the quality and safety of agricultural products, resulting in significant food safety (Bourn and Prescot 2002) and human health (Koleva and Schneider 2009; Guler et al. 2010) problems and causing agricultural nonpoint source pollution. Currently, only a very small proportion of agricultural products are produced completely without agrochemicals. In fact, a large number of fruits and vegetables may contain agrochemical residues. Agrochemicals are widely used in developed countries, and tests conducted by the US Department of Agriculture and Food and Drug Administration detected pesticide residues in a large proportion of fruits and vegetables produced in the United States. Excessive pesticide residues are also often found in agricultural products imported to China from developed countries, including the United States. In recent years, because of the improvement of public awareness of food safety and popularization of scientific knowledge, the number of food safety incidents caused by agrochemical residues has been greatly reduced in China. However, food safety incidents caused by improper or even illegal use of chemicals by producers are not yet extinct.

The media and the parties concerned should learn a lesson from the watermelon explosion incident. The explosion of immature watermelons was not necessarily caused by the use of a swelling agent. In the case of Dalv Village, all farmers had exploding watermelons, and none of them, except Mingsuo Liu, used swelling agents, indicating that the use of swelling agents may not have been the main reason for the burst. Standard use of swelling agents itself does not harm human health. However, exaggeration by the media and consumers' unscientific knowledge of swelling agents resulted in major concerns over eating watermelons during this time, consequently affecting the income of watermelon farmers.

11.2.3 Use of Nonfood Raw Materials in Cooking

The use of nonfood raw materials in cooking can endanger the health of consumers and, in severe cases, can even cause diseases and deaths. The most influential food safety incident caused by nonfood raw materials in 2011 in China was the incident of swill-cooked dirty oil. Swill-cooked dirty oil itself has a certain economic value, as it can be used for the industrial production of soap, biodiesel, and other products, as well as a collector for beneficiation. The recycling of waste animal and vegetable oils itself complies with the principle of a circular economy and has a positive effect in reducing environmental pollution. Aflatoxin, one of the major hazardous substances in swill-cooked dirty oil, is a strong carcinogen with a 100 times higher toxicity than arsenic (Sun 2011a). Long-term intake of swill-cooked dirty oil will significantly harm human health and lead to developmental disorders, susceptibility to colitis, and even enlargement or lesions of the liver, heart, and kidneys.

On June 28, 2011, the Xinhua News Agency exposed the clandestine production of swill-cooked dirty oil in Beijing, Tianjin, and Hebei. Processing dens of swill-cooked dirty oil of an unexpectedly large scale were found in these areas (Jinghua News 2011). These illegal dens have begun using more and more advanced processing technologies and equipment after years of "technical transformation." As a consequence, it is increasingly difficult to distinguish the swill-cooked dirty oil from normal cooking oil. Swill-cooked dirty oil has been continually introduced into food processors and grain and oil wholesale markets through clandestine channels and has even found its way into the supermarkets in the form of small packages. Professor Dongping He from the School of Food Science and Engineering, Wuhan Polytechnic University, and the leader of the Fats and Oils Working Group of the National Oil and Grain Standardization Committee estimated that every year 2 to 3 million tons of swill-cooked dirty oil could have snuck onto our dining tables. In China's restaurants, approximately 1 in 10 meals are cooked with dirty oil, a calculation based on China's annual oil consumption of 22.5 million tons. After the incident of swill-cooked dirty oil in Beijing, Tianjin, and Hebei, departments in all the regions concerned took actions quickly to crack down on crimes relating to swill-cooked dirty oil. According to the information published online by the Ministry of Public Security on

December 12, 2011, the local departments of public security had detected 128 cases, in which swill-cooked dirty oil was produced and sold as edible oil, arrested more than 700 criminal suspects, verified more than 60,000 tons of oil involved, and destroyed 60 criminal networks involving 28 provinces in three months. These actions have effectively contained the hazards of swill-cooked dirty oil. The relevant circumstances of crimes relating to swill-cooked dirty oil are specifically discussed in Chapter 7, "Evolutionary development of the Chinese food safety legal system."

11.2.4 Use of Shoddy and False Certification

In China in 2011, the most common food safety incidents caused by use of shoddy and false certification were the Wal-Mart green pork scandal and false organic food certification.

11.2.4.1 "Green Pork" Scandal   On August 24, 2011, the Chongqing industrial and commercial department received reports from the public that a Wal-Mart store in Fengtian, Chongqing, was labeling ordinary pork as "green pork." Subsequent investigation by the relevant departments found three Wal-Mart stores in Chongqing selling incorrectly labeled "green pork" products. Through on-site assessments, examination of electronic records of purchases and sales, and investigation of its suppliers, the purchased volume of "green pork" was found to be clearly smaller than the sales volume. In October 2011, the Chongqing Municipal Industrial and Commercial Bureau informed the media that three Wal-Mart stores in Fengtian, Songqinglu, and Ranjiaba were suspected of selling 1,178.99 kg of counterfeit green pork since 2011, involving more than 40,000 yuan. The counterfeit green pork had been sold at a price 4–10 yuan/kg higher than ordinary pork. Further investigation found the same situation in other Wal-Mart stores in Chongqing, as well as two Trust-Mart stores acquired by Wal-Mart. Relevant departments verified that the Wal-Mart stores and other stores involved had sold a total of 59,049 kg of counterfeit green pork from January 2010 to August 2011. The Wal-Mart and Trust-Mart stores involved were ordered to close the stores and pay a fine of five times the illegal income (Food Industry Net 2011).

Wal-Mart has been known for its efficient supply chain and has been an example of standardized operation in the retail industry. However, as a well-known international brand, the company has been sanctioned by the Chongqing government 20 times in a five-year period for various violations, including false advertising and selling expired and substandard food. The major root cause is inadequate supervision and extremely low cost for violation of laws. In addition, the low awareness of rights and blind worship of international well-known enterprises of Chinese consumers lead to their unawareness of product quality supervision and right protection. These are the deep-seated causes of the Wal-Mart "green pork" incident. In recent years, food safety incidents involving foreign-branded food companies have been common in mainland China for various reasons.

11.2.4.2 False Organic Food Certification   In November 2011, the Xinhua News Agency and other media outlets reported on false organic food certification. Reporters' in-depth investigation and tracking of the industrial chain of some organic foods in Shandong, Guangxi, and other regions found use of organic food labels at will, purchase of organic food certifications, and use of fake organic food labels to be common occurrences. What the reporters had uncovered was the fact that the organic food market needed to be remedied immediately (Zou and Yu 2011).

Organic certification is the highest level of food quality certification. The use of pesticides, fertilizers, herbicides, and other synthetic materials is prohibited during production and processing of organic food. Organic certification also has a high requirement for soil, air, water, and other environmental factors. To ensure safety, consumers are often willing to pay a relatively high price for organic produce. Thus, the global organic food market has experienced an annual growth rate of 20%–30%. Chapter 1 of this book introduced the development of green and organic agricultural products in China. The sales of Chinese organic agricultural products (food) reached 14.539 billion yuan in 2010. Because of serious information asymmetry in the food market and the temptation of large profits, the organic food certification, which should represent the highest level of agricultural products, performs practically no function. Some certification bodies have turned the rigorous review process of organic food certification into "paid certification." Thus, the "organic" label is often unworthy of the term "organic" (China Certification and Accreditation Information Network 2010). At present, there are 24 organic product certification bodies in China. These certification bodies set up offices in various provinces and can independently assess and award organic food certifications (Shanghai Government Website 2011). A large number of agencies also exist in the market. "Any product can be labeled as organic, as long as you are willing to pay," an agency spokesperson in Beijing told a reporter who pretended to be a tea merchant. According to normal procedures, the first organic certification of a tea plantation must go through a land transition period of 2–3 years since the date of application. However, the agency promised that the whole certification process could be completed in only six months, at a cost of over 40,000 yuan. All the enterprise needed to do was to provide business licenses and seal and sign the documents. In addition, many certification bodies only focus on certification while neglecting management.

11.2.5 Controversies on Microbial Standards in Food

No major food safety incidents involving personal injury caused by microorganisms, microbial toxins, or parasites were reported in 2011 in China. However, the microbial standards in food, especially the dairy standards and the *Staphylococcus aureus* (*S. aureus*) standard in frozen rice and flour products, aroused widespread concern in 2011.

11.2.5.1 Controversy on Dairy Standards
At the Southern Pasteurized Milk Development Forum held in Fuzhou, China, on June 15, 2011, experts from the China Dairy Association suggested that political and economic support should be provided to producers of pasteurized milk to restore the leading position of pasteurized milk in the market. Dingmian Wang, the President of the Guangzhou Dairy Association, said that the Chinese dairy industry standards had been hijacked by some large enterprises and were the worst standards in the world. He went on to state that the government should support the development of pasteurized milk. In the Chinese dairy industry standards, the maximum allowed number of bacteria in raw milk is 2 million colony-forming units (CFU)/ml, 20 times higher than the EU standards; the minimum protein content per 100 g of raw milk decreased from 2.95% in the old national standard to 2.8%. Statements like "the dairy industry standards have record low requirements," "the worst milk standards in the world," and "the shame of the world dairy industry" immediately caused uproar in the country.

The working group of experts for Chinese dairy safety standards explained "the worst milk standards in the world." First, although there is a large gap between the new national standard for number of bacteria in raw milk and the international level, it is, in fact, stricter. Based on many years of experience in application of the *Raw Milk Purchasing Standards* enacted by the agricultural department in 1986, the total bacterial count in raw milk was adjusted from 4 million CFU/ml to 2 million CFU/ml in the newly released national food safety standard GB 19301-2010 *Raw Milk*, which increased the requirement for the purchasing of raw milk from farmers. Second, protein content is a quality indicator of raw milk, but not of the final product to be consumed by consumers. The protein content in raw milk ranges from 2.8%

to 3.2% in China, with an average of 2.95%. Because the protein content in raw milk is affected by dairy breeds, feed, feeding and management, lactation, climate, and other factors, it is lower than the average of 2.95% in a considerable part of raw milk, for example, during the lactation period from late May to late August. Therefore, the minimum protein content per 100 g of raw milk is decreased from 2.95% to 2.8%.

On July 13, 2011, the Ministry of Health held a press conference to explain the lower standards for raw milk safety in the new standard GB 19301-2010. Heping Wu, the Secretary-General of Heilongjiang Dairy Industry Association, told a reporter that the new standard was in line with China's national conditions. Professor Liguo Yang from Huazhong Agricultural University and Vice President of the China Dairy Association said that the Chinese dairy industry was still in a backward state, that it would be vain to blindly improve the national standards, and that we should realistically recognize the gap. Foreign expert, Deborah Gray, spokesperson of the Ministry of Agriculture and Forestry of New Zealand, who is responsible for the formulation of food standards, also believes that China's new national standard for raw milk was developed based on China's national conditions and is in line with international practice. The standards of each country are different, and thus, the most important thing is to be able to address the specific circumstances of a country (Yin and Yang 2011).

However, the explanation given by the working group of experts for Chinese dairy safety standards has not been universally accepted. Consumers believe that, although the standard for total bacterial count has indeed been improved in the new national raw milk standard compared with the *Raw Milk Purchasing Standards* (GB/T 6914-86), it is much lower than the relevant standards in many other countries. Some Chinese consumers have urged the government to make the Chinese dairy standards for colony count and protein content consistent with international standards.

11.2.5.2 Controversy on S. aureus *Standard in Frozen Rice and Flour Products* On November 25, 2011, the new standard, GB 19295-2011 *National Food Safety Standard—Frozen Rice and Flour Products*, in which the maximum microbial count was modified, was officially announced. The new national food safety standard changed the requirement that no detectable *S. aureus* was allowed in the old standard GB 19295-2003 *Hygiene Standard for Prepackaged Frozen Rice and Flour Products* into limitation in colony count. According to the Information Office of the Ministry of Health, as a bacterium widely distributed in the atmosphere and soil, heat-sensitive, *S. aureus* can be killed by heat treatment at 80°C for 30 minutes. It can usually produce pathogenic enterotoxin at a level greater than 10^5 CFU/g. Generally, dairy products, egg products, and meat products are suitable habitats for the growth of *S. aureus*. Therefore, maximum levels of *S. aureus* are provided in the relevant safety standards of these products.

The Ministry of Health held a news briefing on the day of formal announcement. A researcher from the Institute of Nutrition and Food Safety, Chinese Center for Disease Control and Prevention, said at the news briefing that no quantitative determination requirements of pathogenic bacteria were provided in GB 19295-2003, only the concept of qualitative determination, which means that products with any detectable level of pathogenic bacteria are substandard. However, this concept has been changed accordingly since the Codex Alimentarius Commission (CAC) made a qualitative change in the principles of control of microbial hazards in food in 1999. All pathogenic microorganisms will not have the same hazards in different foods. Specific pathogens should be controlled as a key target in certain foods. The provision that no pathogenic bacteria should be detected in the previous standard lacks a scientific basis. From a scientific perspective, the higher the temperature, the shorter the time required to kill bacteria. When frozen rice and flour products are heated to 100°C, the proteins will be destroyed, and the bacteria will thus be inactivated. "In other words, the proteins will coagulate at 100°C within seconds, thus inactivating the bacteria. It should be certain that the food is thoroughly cooked in the boiling process and the bacteria will thus be thoroughly destroyed," an expert said (Alibaba China 2012). "The requirement that no detectable *S. aureus* is allowed in the old standard is a one-size-fits-all approach, which, in fact, is unattainable and unnecessary. It will not cause disease in the intestine within a certain level. Limited *S. aureus* involves no food safety problems," an expert from the Department of Nutrition and Safety, College of Food Science, China Agricultural University, pointed out, and "the maximum level of *S. aureus* in GB 19295-2011 is consistent with the international standards" (China Business News 2011).

11.3 Brief Review

Analysis of the above incidents or controversies reveals that the influential food safety incidents of concern that occurred in 2011 in China have complex deep-seated causes. Accordingly, the controversies on food safety standards should be resolved in a coordinated manner, taking into account the entire current situation and the interests of all parties.

11.3.1 Main Causes of Food Safety Incidents

The deep-seated causes of major influential food safety incidents that occurred in 2011 included at least four aspects. These causes have been comprehensively analyzed in the relevant chapters of this book and, therefore, will be briefly reviewed here in combination with the specific incidents.

11.3.1.1 Environmental Pollution   Along with the progresses of industrialization and urbanization in China, unreasonable, unscientific, exhaustive, and extensive exploitation has severely damaged the local agricultural ecological environment in some mineral-rich places. The incident of cadmium-contaminated rice is a typical incident in which the risk of quality and safety of agricultural products is caused by environmental pollution resulting from economic development. It is the result of long-term extensive industrialization and urbanization.

The incident of cadmium-contaminated rice is only a microcosm of food contamination caused by soil heavy metals. Because situations are different from one contaminated area to another, excessive heavy metals may not only include cadmium, but also include arsenic, mercury, and lead. Moreover, crops with excessive heavy metals may not only include rice, but also include wheat, corn, and other crops. At present, environmental health crises caused by environment pollution are becoming common in China. In the area of food safety, heavy metal contamination may replace pesticide residues as a major food safety risk factor. A lag exists in the emergence of food safety caused by environmental pollution. Moreover, it is difficult to solve such problems. The lag means that a certain period of time exists between the production of hazards and the emergence of consequences. It is sometimes too late to solve problems when consequences have already emerged. The difficulty in solution lies in the fact that, not only must further pollution be prevented, but it will also take a long time to alleviate or eliminate the existing pollution. In addition, an institutional reason for the heavy metal pollution in agricultural products exists in China. The self-sustained production and consumption manner has exacerbated the heavy metal pollution of agricultural products to some extent. Therefore, preventive measures must be taken as soon as possible. The government should investigate heavy metal pollution in soil in severely polluted areas. The planting of crops that can accumulate heavy metals should be prohibited on contaminated land. Guidance and support should be provided to farmers involved on planting crops with low accumulation capacity. Moreover, special studies should be conducted to breed crop varieties with low accumulation capacity to prevent heavy metal pollution of crops at the source. In addition, detection of heavy metal pollution should be further strengthened in relevant agricultural markets in key areas. A warning shall be given to consumers upon discovery of agricultural products with excessive heavy metals, and products involved should be promptly recalled to prevent entry of these products into the market and safeguard the quality and safety of edible agricultural products.

11.3.1.2 Violation of Laws by Producers The above-mentioned incidents, including swill-cooked dirty oil, tainted Chinese chives, counterfeit green pork, and false organic food certification, in essence, were caused by violation of laws and regulations by producers who were motivated by pursuit of economic profit to distribute foods with serious quality problems or unqualified foods due to use of nonfood raw materials or shoddy food raw materials. This not only endangered the health of consumers, but also seriously disrupted the food market order. Therefore, the fundamental problem lies in the fact that illegal activities by producers and traders motivated by pursuit of economic profit cannot be eliminated despite repeated prohibitions. For example, the main cause of the frequent incidents of swill-cooked dirty oil may be that the mechanism of the profit chain in production and selling of swill-cooked dirty oil has not been destroyed. To capture more profits, the dens for production and selling of swill-cooked dirty oil have been developed from small workshops into large factories, and the production and selling process has been divided into many stages, including collection, primary processing, reselling, deep processing, wholesale, and sales. The production and selling of swill-cooked dirty oil is becoming a bigger and bigger business and more profits are being derived from this business.

11.3.1.3 Absence of Government Regulation The absence of food quality and safety regulation by the government also creates a loophole for the various violations by producers. False organic food certification demonstrates, yet again, the extreme importance of reforming the food safety regulation manner in China. According to the *Administrative Measures for Organic Product Certification*, organic food manufacturers must be examined and approved by a relevant approving body to obtain organic certification. Moreover, a national uniform organic food certification label and logo of the certification body must be placed on each package of the certified organic food to distinguish between organic, green, or pollution-free food. Due to the lack of effective regulation, some organic food certification bodies provide organic food certification to ineligible manufacturers for their own benefits. Even though it may be impossible to regulate numerous organic food manufacturers with limited regulatory power, it is absolutely possible to effectively regulate a limited number of organic food certification bodies. However, no effective regulation has been ever implemented. Due to the lack of regulation, some organic food certification bodies do not have the necessary financial resources to carry out effective follow-up examination or establish a problem discovery mechanism, a mechanism of withdrawal and information disclosure of certified manufacturers, or a problem tracing mechanism, which has eventually led to the disorder of organic food certification.

11.3.1.4 Inadequate Legislation   The case of swill-cooked dirty oil will be discussed here as an example. The problem of swill-cooked dirty oil is not unique to China. It was also common in the United States, Western Europe, Japan, and Taiwan decades ago. For example, in the 1960s, swill-cooked dirty oil was provided by Japanese manufacturers to Taiwanese manufacturers to be sold as cooking oil in Taiwan after processing, resulting in the incident of Japanese swill-cooked dirty oil in Taiwan. Taiwan was targeted by Japanese manufacturers because it was too difficult to sell the dirty oil in Japan where strict food regulations were implemented. Fortunately, the problem of dirty oil quickly disappeared in the United States, Western Europe, Japan, and Taiwan because of the very strict legal environments. However, incidents involving swill-cooked dirty oil have happened frequently in China due to inadequate legislation and legal punishment.

11.3.2 Essential Problems in Controversies on Food Standards

Food safety standards are an important component of the food safety supervision and management system. The level of scientificity and rationality of the standards directly determines the level of food quality and safety and profit distribution among many stages of the industrial chain. The controversies on dairy standards and *S. aureus* standards in frozen rice and flour products demonstrate the imperfection of the Chinese food standard system. For example, some standards are absent, some lag behind the international standards, and some are limited by special conditions in China. This is one of the important reasons for common concerns regarding the current status of food safety in Chinese society.

Essential problems in the controversies on dairy standards and *S. aureus* standard in frozen rice and flour products are given in Sections 11.3.2.1 through 11.3.2.3.

11.3.2.1 What Should Be Taken as Supreme Interests in the Formulation of Food Safety Standards   Standards are regulations. The formulation of standards is a process of public health decision making, with the purpose of minimizing possible risks to public health. Only on this basis, can maximum flexibility be given to producers to reduce social costs. Some experts believe that the formulation of the new dairy standards just ignores or even denies the interests of consumers and farmers and protects the interests of some large enterprises. Of course, this view is not entirely correct. However, the first priority and supreme interests must be the public health in the formulation of food standards.

11.3.2.2 How to Coordinate and Balance the Benefits of All Parties in the Formulation of Food Safety Standards   Bacterial counts are not related to farmers' farming technology. The total bacterial count is very low after milk is drawn from the breast. However, the bacterial count will increase 100 times due to backward transportation and processing technologies, which, in fact, are the results of the cost-saving strategy of enterprises. Therefore, while no technical barriers exist, a high cost in the production of high-standard milk does exist. In fact, the interests of dairy farmers are entirely determined not by the cost of production, but by the difference between cost of production and selling price. If relatively high quality standards are implemented in a fair and consistent way, milk companies will raise prices to promote the production of acceptable raw milk by farmers. Moreover, the increased costs will not entirely be covered by the milk companies, but eventually be included in the prices paid by consumers under market competition. Therefore, the controversy on dairy safety standards, in fact, is about the coordination of economic interests among dairy farmers, producers, and consumers, under the premise that public health is taken as the supreme interest.

11.3.2.3 How to Bring Chinese Food Safety Standards in Line with International Standards   The Chinese food standards are gradually brought in line with international standards, indicating that the Chinese food safety risk management is becoming more mature. The controversies on dairy standards and *S. aureus* standards in frozen rice and flour products indicate that some think that the two standards have been lowered. However, the modification of the two standards indeed takes into consideration the consistency with international standards. If high standards are blindly implemented without considering China's reality, some enterprises may cheat or be forced to stop production. However, if emphasis is only put on the specific national conditions and more consideration is given to the backward production technologies of enterprises, there will be endless controversies on food safety standards and it will be difficult to achieve transformation of the mode of food production and the guarantee of food safety. Chinese food safety standards must actively learn from the advanced international food safety standards, as they are the result of progress of human civilization.

Therefore, the current and future formulation and revision of Chinese food safety standards should be carried out based on reality and under the premise that public health is taken as the supreme interest, while trying to coordinate the interests of all parties, as well as be in congruence with international standards.

12

Research on Agricultural Production Mode Transformation

Vegetable Farmers' Willingness to Adopt Biopesticides in Cangshan County, China

The research in Chapter 1 of this book reveals that the abuse of chemical pesticides causes deterioration of the ecological environment and endangers the safety and quality of edible agricultural products, which is a major challenge for the safety of Chinese edible agricultural products. Reducing the use of chemical pesticides and encouraging and supporting the use of biopesticides have thus become important ways of promoting the transformation of the agricultural production mode in China. This chapter presents a case study on vegetable farmers' willingness to adopt (WTA) biopesticides in Cangshan County, Shandong Province, China.

12.1 Introduction

The presence of pests in agricultural products is a major biological problem. Since the Swiss chemist Paul Mueller discovered dichloro-diphenyl-trichloroethane (DDT) in 1945, farmers have utilized chemical pesticides to control pests (Agnew and Baker 2000). Total grain loss saved by the application of chemical pesticides accounts for 15% of the total production in the world (Hou and Wu 2010). On the downside, the application of chemical pesticides leads to a series of negative effects such as the disturbance of the natural ecosystem (Palikhe 2001) and the serious threat to human health owing to the existence of pesticide residues through the food chain and the cumulative effects of bioconcentration (EEA 2005).

Biopesticides are products used to control pests in agricultural products that take the form of living organisms or their preparations, and they are highly compatible with the environment and are relatively safe for humans/animals. They have gradually been accepted for use over time (Chen 2003). Compared with chemical pesticides, biopesticides are not only more effective in pest control, but also more advantageous to the environment, the safety and quality of agricultural products, and human health (Laird et al. 1990; Hokkanen and Hajek 2003; Mandal et al. 2003; Bhalla and Devi-Prasad 2008).

At the World Conference on Environment and Development in Brazil in 1992, it was proposed that the application area proportion of biopesticides must reach 60% of the global pesticide application area by the end of the twentieth century to replace chemical pesticides. Until 2004, the biopesticide application area was only ≈400–500 million mu in China, accounting for ≈9% of the country's total pesticide application area (Zheng 2006). Historically, the application quantity of chemical pesticides has been high in China, and as early as the 1990s, the application quantity of chemical pesticides per unit area of rice in Eastern China was twice that of the Philippines (FAO 2007). Statistics showed that the average growth rate of the amount of China's chemical pesticide application during 2001–2008 was as high as 4.05%, and even reached a record high (167.23×10^7 kg) in 2008, ranking first in the world at that time.

Farmer households are the basic unit of agricultural production in China and are the subjects of pesticide application. Therefore, investigating farmer households' acceptance of biopesticides and the main influencing factors is of great practical significance for the Chinese government. Understanding these issues will provide guidance on deepening the reform of the pesticide management system and ensuring the quality and safety of agricultural products.

12.2 Survey Design

12.2.1 Selection of Agricultural Product Varieties

Among the wide varieties of Chinese agricultural products, vegetables are the most basic and important agricultural product in China. Vegetable consumption occupies an extremely important position in the food consumption of Chinese residents.

As shown in Table 12.1, between 1995 and 2009, the proportion of per capita vegetable consumption was 25%–36% in China. In 2009, the per

TABLE 12.1

The Consumption and Proportion of Vegetables and Food of Urban and Rural Residents in China

	Urban Residents			Rural Residents		
Year	Vegetables Consumption (Kg)	Food Consumption[a] (Kg)	Proportion[b] (%)	Vegetables Consumption (Kg)	Food Consumption[c] (Kg)	Proportion (%)
1995	116.47	312.75	37.24	104.62	400.37	26.13
2000	114.74	321.08	35.73	106.74	411.13	25.96
2005	118.58	335.20	35.37	102.28	370.06	27.64
2008	123.15	–	–	99.72	359.60	27.73
2009	120.45	–	–	98.44	351.26	28.02

Source.   China Statistical Yearbook, 2010.

[a] According to the statistical standard, food consumption in urban China consists mainly of 10 categories: cereals, fresh vegetables, edible vegetable oil, pork, beef and mutton, poultry, eggs, aquatic products, milk, and fruits.

[b] As 2008 and 2009 food consumption data are lacking, the proportion of vegetable consumption in the total food consumption is not given.

[c] According to the statistical standard, food consumption in rural China consists mainly of nine categories: cereals, vegetables, cooking oil, meat and their products, poultry and their products, aquatic products, milk and milk products, melons and fruits and their products, and nuts and nut products.

capita vegetable consumption of China's urban and rural residents was 120.45 and 98.44 kg, respectively (National Bureau of Statistics of China 2010). In China, the frequency of vegetable quality and safety incidents caused by pesticides has been high in recent years. Some of these incidents included the detection of a banned pesticide, isocarbophos, in cowpeas in Hainan Province (2010) and the poisoning of many citizens in Qingdao by leeks containing pesticide residues far exceeding the standard (2010). As vegetables occupy an important position in the food consumption, more focus should be placed on promoting the use of biopesticides. Therefore, studying Chinese vegetable farmers' willingness to adopt biopesticides is an important, forward-looking endeavor.

12.2.2 Study Area

In 2009, the vegetable planting area in China was 18,414,300 hm², with a total production of 618 million tons; in the same year, the planting area in Shandong Province, China's largest vegetable planting province, was 1,756,000 hm², with a total production of 89 million tons (Ministry of Agriculture of China 2009). Cangshan County, under the jurisdiction of

Shandong Province, is an agricultural county (population 1.2 million). Currently, the vegetable planting area in Cangshan County is more than 1 million mu. There are about 630,000 farmers in Cangshan County engaged in vegetable planting, mainly common vegetables, which are not only sold in China's Southeastern coastal cities, but also exported to many countries/regions. For the most part, these vegetable farmers use fungicides, acaricides, herbicides, and other chemical pesticides to control pests. Therefore, the vegetable farmers of Cangshan County were examined in the current study.

12.2.3 Questionnaire Design and Survey Subjects

In this study, only ordinary farmer households with pesticide application experience were examined. The questionnaire was composed of demographics, knowledge of biopesticides, and adoption willingness. Twenty-one mainly closed questions were included in the questionnaire, and it was revised and established according to experience gained by a preparatory survey; samples were selected by random stratified sampling. Four townships, Zhuangwu Town, Xiangcheng Town, Xingming Town, and Lanling Town, in Cangshan County were randomly selected. Three natural villages were randomly selected from each township, and 30 farmers were randomly surveyed in each. In total, 360 farmers were interviewed, and 294 valid questionnaires were collected. The cultural level of the vegetable farmers surveyed may have been low; thus to avoid bias in understanding that may affect the authenticity of the questionnaire answers, the survey was carried out using one-on-one interviews. During this process, the respondents were asked to answer on the spot, and the questionnaires were filled out by the investigators. The survey was completed in March 2011.

12.2.4 Sample Analysis

12.2.4.1 Household Demographics As shown in Table 12.2, household demographics obtained from the survey were as follows:

1. The farmers were older with a lower level of education. In total, 66.30% of the respondents were older than 40 years, and 76.6% had ≤9 years of education. The age and education demographics of the farmers interviewed were consistent with the basic demographics of agricultural workers in rural China.
2. The household income level was low. For 84.70% of the respondents, the total household annual income was under 30,000 yuan, slightly higher

TABLE 12.2
Relevant Demographics of the Households Surveyed

Demographics	Classification Index	Number (Person)	Effective Proportion (%)
Gender	Male	195	67.00
	Female	99	33.00
Age	≤29 years	15	5.10
	30–39 years	84	28.60
	40–49 years	108	36.70
	50–59 years	63	21.40
	≥60 years	24	8.20
Household per capita	≤1 mu	102	34.70
cultivated area	1–2 mu	102	34.70
	≥2mu	90	30.60
Nonagricultural income	Yes	150	51.00
	No	144	49.00
Years of education	≤6 years	96	32.70
	7–9 years	129	43.90
	10–12 years	63	21.40
	≥13 years	6	2.00
Household total annual	≤30,000 yuan	249	84.70
income	30,000–50,000 yuan	42	14.30
	≥50,000 yuan	3	1.00

than the average level of farmer households in Shandong Province. Although the absolute income continued to increase, the increase in the relative income of the respondents was limited due to inflation and other factors.

3. The farmers displayed differentiation of occupation. A larger proportion of farmers had part-time jobs, and 51.00% of the respondents had nonagricultural income. Vegetable planting was not the most important source of income for the farmers.

4. Typical small-scale planting habits were identified. The average per capita cultivated area of Chinese rural households is 1.39 mu (Fan 2007). For 69.40% of the farmer households surveyed, household per capita cultivated area (household cultivated area divided by the number of farmers in the household) was 2 mu and below.

12.2.4.2 Farmers' Knowledge and Adoption Willingness of Biopesticides   Sixty-six percent (192) of the respondents had no idea about, were not clear about, or had never heard about biopesticides; 28.57% (84) of them were using or had used biopesticides, and 69.05% (203) were skeptical of their effects.

12.3 Modeling and Econometric Analysis

12.3.1 Modeling and Variable Setting

Agricultural production purposes (PPs) vary; when the PP is to meet market supply, allocation of production factors is cost-effective in agricultural production motivated by profit maximization (Schultz 1964). However, in a society extremely self-sufficient in agricultural products, the PP of farmers is household utility maximization, and the allocation of production factors is not optimized. Farmers pursuing profit maximization are more concerned about the potential economic gains of biopesticides, while farmers in pursuit of household utility maximization are more concerned about the satisfaction of family needs when using biopesticides. China's per capita vegetable consumption has reached 120 kg; however, farmers do not selectively use pesticides according to family consumption or selling.[*] Vegetable production is more of a market behavior, with the direct PP of increasing income. This is also demonstrated by the fact that vegetable planting was the only source of income for 49% of the farmers in the survey area of this study. Therefore, this study assumed that vegetable farmers[†] satisfied the rational agent assumption. In addition, it was assumed that their selection of biopesticides and chemical pesticides in the production process was based on the difference in their expected utilities.

Assumptions: (1) Vegetable farmers faced the same macro/natural risks regardless of the use of biopesticides or chemical pesticides; (2) Application of different pesticides in planting vegetables did not affect other relevant inputs in their agricultural production; (3) The utility function of the vegetable farmers was the "mean–standard deviation" (Dillon and Scandizzo 1978). Thus, the increase in the expected utility of farmer i by applying biopesticides relative to chemical pesticides is expressed as follows:

---

[*] The farmers in the survey site of this chapter, Cangshan County, Shandong Province, mainly grew vegetables, most of which were for sale, with only a very small part reserved for family consumption.

[†] "Peasant" and "farmer" are two terms with similar meanings. Ellis (1998) argued that a "peasant" only partially integrated into the market that was not defective, while a "farmer" completely integrated into an ideal market. Economic activity of the former was not intended to seek maximum benefits, but to seek conditional maximum benefit with a primary purpose of meeting the family consumption demand. In China, the agricultural workers do not have ownership of the land, but they have a complete land use right in the household contract responsibility system. In addition, they can sell their agricultural products directly to the market. Therefore, Chinese agricultural workers more closely fit the term "farmer."

$$EΔU_i = EΔTR_i + βV_i - ΔTC_i \qquad (12.1)$$

where:

$EΔTR_i$ is the increase in average return of the vegetable farmers by applying biopesticides over chemical pesticides

V_i is the specific standard deviation of return of the vegetable farmers due to biopesticide application, including both market and technology risks

$β$ is the standard deviation coefficient, representing the farmer's attitudes toward risk

$EΔTR_i + βV_i$ is the risk return by applying biopesticides

$ΔTC_i$ is the increase in procurement costs of the vegetable farmers by applying biopesticides over chemical pesticides[*]

Therefore, the economic factors affecting farmers' WTA biopesticides can be divided into two categories given as follows:

1. *Factors affecting the risk return by applying biopesticides, that is,* $EΔTR_i + βV_i$. $EΔTR_i$ in Equation 12.1 depends on differences in vegetable yield and market price by applying biopesticides and chemical pesticides. As biopesticides are new pesticide products in China, biopesticide application in vegetable production actually has two risks affecting the standard deviation of return. The first is the market risk. Vegetables produced by applying biopesticides should have more market demand compared with those produced by applying chemical pesticides; however, it is difficult for consumers to identify the quality of vegetables produced by applying different pesticides, and thus, the market risk exists. If biopesticide application is promoted by the agricultural cooperative organizations, and the vegetables produced by farmers are purchased by such organizations, consumers may be able to identify the type of pesticide used, and the market risk will be reduced. The second risk is the technology risk.[†] Currently, in China, biopesticides are not widely used, and the variety is limited; there are also risks in their effects of killing pests and the vegetable farmers' biopesticide application skills. If the government can provide relevant pesticide application skills training

---

[*]Biopesticides are already available on the market in China, so the difference in procurement cost of biopesticides and chemical pesticides is definite.

[†]Technology risk is the possibility of the deviation of actual return from expected return in the application of agricultural technology (Liu 2010b).

or implement agricultural insurance,[*] the technology risk can be reduced.

2. *Factors affecting the procurement costs of the pesticides* ΔTC_i. At present, in China, the procurement cost of biopesticides is higher than that of chemical pesticides; therefore, in Equation 12.1, $\Delta TC_i > 0$. However, if the government could provide proper subsidies to vegetable farmers for biopesticide procurement, the difference in procurement costs between the two types of pesticides could be reduced,[†] which may help increase vegetable farmers' WTA biopesticides.

The above-mentioned model follows the assumption that farmers are economic, but studies on farmers in developing countries revealed that they faced many constraints in their pursuit of benefit maximization. In China, the farmers' input of agricultural production factors and market return are disproportionate. Their agricultural production behavior is affected more by their individual characteristics (level of education, age, gender, and cognitive level of pesticide residues), as well as household characteristics (cultivated area, household size, and household income level) (Peng 2002). These farmers are in pursuit of maximization of limited benefit, and thus, the factors affecting their WTA biopesticides should include individual characteristics, household characteristics, and economic factors.

Based on the brief analysis given earlier and the existing domestic and foreign literature, the factors affecting farmers' WTA biopesticides are summarized in Table 12.3.

The vegetable farmers' WTA biopesticides was selected as a proxy variable for reflecting their pesticide application behavior. The discrete choice variables were set as follows:

$$E\Delta U_i = B'X_i + \varepsilon_i$$

$$\begin{cases} Y_i = 1 & \text{if } \Delta U_i > 0 \\ Y_i = 0 & \text{if } \Delta U_i < 0 \end{cases} \tag{12.2}$$

where X_i = the factor affecting the vegetable farmers' WTA biopesticides, and $X_i' = (X_{i0}, X_{i1}, \ldots, X_{i12})$ and $X_{i0} = 1$; according to the preceding

---

[*]Carlson (1979) demonstrated that government insurance was an alternative to additional pesticides, that is, the insurance contributed to farmers' reduced use of pesticides. However, Horowitz and Lichtenberg (1993) reached the opposite conclusion. It should be noted that these studies did not relate to biopesticides. The impact of agricultural insurance on the use of biopesticides has not been reported.

[†]Some Chinese local governments provide farmers who sell vegetables to the city with direct subsidies to encourage them to apply less toxic and more efficient pesticides.

TABLE 12.3
Factors Affecting the Farmers' WTA Biopesticides

Class	Influencing Factors	References
Individual characteristics	Gender of the farmer	Michael et al. (2001); Kishor (2007); Fu and Song (2010)
	Age of the farmer	Binswanger (1980); Sule and Ismet (2007)
	Years of education of the farmer	Binswanger (1980); Hruska and Corriols (2002); Li et al. (2007); Zhou and Hu (2009); Abhilash and singh (2009)
	The farmer's knowledge of the vegetable price level	Hoogeveen and Oostendorp (1996)
	The farmer's knowledge of pesticide residues	EEA (2005); Rother (2005); Waichman and Eve (2007); Jolankai et al. (2008); Wu et al. (2011)
Household characteristics	Nonagricultural income of the farmer household	Binswanger (1980); Dasgupt et al. (2007); Dariush et al. (2009)
	Vegetable cultivated area of the farmer household	Feder (1980); Sanzidur (2003); Zhou (2006); Ngow et al. (2007)
	Farmer household population	Binswanger (1980); Wang and Xu (2004); Zhang et al. (2004)
Factors of risk return	Whether the farmer participates in cooperative organizations	Zhang et al. (2004); Zhou (2006)
	Whether the farmer has ever received skill training organized by the government	Hruska and Corriols (2002), Zhou (2006); Zheng (2006)
	Whether the farmer participates in agricultural insurance	Carlson (1979); Horowitz and Lichtenberg (1993)
Factors of procurement costs	Whether the farmer is satisfied with the government's related subsidies for pesticides, etc.	Hruska and Corriols (2002); Zhou (2006)

TABLE 12.4
Variable Definition and Sample Statistics

Variable	Definition and Assignment	Mean Value[a]	Standard Deviation[a]
Adoption willingness of the farmer (Y_i)	Whether the farmer is willing to adopt biopesticides; yes = 1, no = 0	0.2755	0.4491
Gender of the farmer (GEN)	Dummy variable; male = 1, female = 0	0.6633	0.4750
Age of the farmer (AGE)	Dummy variable; >40 years of age = 1, otherwise = 0	0.6653	0.4348
Years of education of the farmer (EDU)	Dummy variable; >9 years = 1, otherwise = 0	0.6735	0.4714
The farmer's knowledge of pesticide residues (RUD)	Dummy variable; understand the hazards of pesticide residues = 1, otherwise = 0	0.6939	0.4633
Vegetable cultivated area of the farmer household (ARE)	Dummy variable; >5 mu = 1, otherwise = 0	0.3387	0.4850
Farmer household population (POP)	Dummy variable; >5 = 1, otherwise = 0	0.3367	0.4750
Nonagricultural income of the farmer household (INC)	Dummy variable; >30,000 yuan = 1, otherwise = 0	0.2653	0.4438
Whether the farmer participates in cooperative organizations (COO)	Dummy variable; yes = 1, no = 0	0.2143	0.4124
The farmer's knowledge of the vegetable price level (PRI)	Dummy variable; thinking that vegetables produced by applying biopesticides are of higher prices = 1, otherwise = 0	0.3469	0.4784
Whether the farmer has ever received skill training organized by the government (TRA)	Dummy variable; have received skill training organized by the government = 1, otherwise = 0	0.3571	0.4816
Whether the farmer participates in agricultural insurance (INS)	Dummy variable; yes = 1, no = 0	0.1735	0.3806
Whether the farmer is satisfied with the government's direct subsidies for biopesticides (SUB)	Dummy variable; yes = 1, no = 0	0.6735	0.4714

[a] Mean value and standard deviation are the results after rounding.

analysis, assignment and interpretation of the relevant variables are shown in Table 12.4. B = the parameter vector to be estimated, $B' = (B_0, B_1,..., B_{12})$, and ε_i = the error term. If $DU_i > 0$, the vegetable farmers are more willing to use biopesticides, that is, $Y_i = 1$; otherwise $Y_i = 0$. Thus, Equation 12.3 can be given as follows:

$$\text{prob}(Y_i = 1) = \pi_i = \text{prob}(\varepsilon_i > -B'X_i) = 1 - F(-B'X_i) \qquad (12.3)$$

Assuming that the cumulative distribution of the error term is logistic, it is represented as follows:

$$\text{prob}(Y_i = 1) = \pi_i = \frac{e^{B'X_i}}{1 + e^{B'X_i}} \qquad (12.4)$$

Due to the nonlinear nature of the logistic model, it is not applicable to ordinary least squares (OLS) or weighted least squares (WLS). Although the nonlinear form of OLS and WLS can be adopted, maximum likelihood estimation (MLE) is commonly used in the parameter estimation shown in Equation 12.4 due to its features of consistency, asymptotic normality, and asymptotic efficiency. The corresponding log-likelihood function and score equation are, respectively, as follows:

$$\ln L = \sum_{i=1}^{n} [Y_i \ln \pi_i - (1 - Y_i) \ln(1 - \pi_i)] \qquad (12.5)$$

$$U(B_r) = \partial \ln \frac{L}{\partial B_r} = \sum_{i=1}^{n} [(Y_i - \pi_i) X_{ir}] = (r = 0, 1, \ldots, 12) \qquad (12.6)$$

For small samples, the bias[*] (Bartlett 1953; Shenton and Wallington 1962; Cordeiro and McCullagh 1991) and separation problem (Albert and Anderson 1984) commonly exist in direct parameter estimation from Equation 12.6. Thus, to avoid the iterative failure caused by the separation problem due to the single use of the binary logistic regression model with a small sample size, while ensuring the objective validity of the model results, this study adopted the suggestions of Firth (1993) and Heinze and Schemper (2002), that is, the penalized log-likelihood function, and let Equation 12.7 be represented as follows:

$$\ln L^* = \ln L + \frac{1}{2} \ln |I(B)| \qquad (12.7)$$

---

[*] Spicer (2005) holds that the basic sample size of the logistic model is 100. An increase of 50 samples is required for each additional variable, that is, the number of samples is 50 times the number of variables. Corpas estimated the bias of a hypothetical single-variable logistic model of which the true value was zero and concluded that if the sample size was 10, the bias of the estimated value was ±3%–4%.

In Equation 12.7, $I(B)$ is the information matrix,[*] and $|I(B)|$ is the determinant of the information matrix. The corresponding score equation is as follows:

$$U(B_r)^* = U(B_r) + \frac{1}{2}\mathrm{trace}\left[I(B)^{-1}\left(\frac{\partial I(B)}{\partial B_r}\right)\right] \quad (r = 0,1,\ldots,12) \quad (12.8)$$

12.3.2 Model Analysis Results and Discussion

The parametric solutions of Equation 12.8 were obtained using the MATLAB® (R2009a) analysis tool through the N–R iterative method with the iterative equation as follows: $B^{(s+1)} = B^{(s)} + I^{-1}(B^{(s)})U(B^{(s)})^*$, where $S =$ the Sth iteration. The relevant parameter estimation results are shown in Table 12.5.

Gender, knowledge of pesticide residues, household vegetable cultivated area, and household nonagricultural income were significant at the 5% level; age, years of education, whether the farmer participated in cooperative organizations, and the constant term were significant at the 1% level (Table 12.5).

TABLE 12.5
Penalized Binary Logistic Regression (LR) Model Results ($n = 294$)

Independent Variable	B	Penalized LR	p value	Odds Ratio
GEN[a]	−0.7372	4.8630	0.0274	0.4784
AGE[b]	−0.9107	8.8677	0.0029	0.4022
EDU[b]	2.7092	33.6238	<0.0001	15.0165
RUD[a]	0.7860	5.4481	0.0196	2.1793
ARE[a]	−0.6771	4.9313	0.0264	0.5081
POP	0.2129	2.1369	0.1438	1.2372
INC[a]	0.7219	4.7432	0.0294	2.0583
COO[b]	1.3016	7.8307	0.0051	3.6751
PRI	−0.4801	2.8909	0.0891	0.6187
TRA	0.7498	3.6213	0.0570	2.1165
INS	0.4277	3.1346	0.0766	1.5337
SUB	−2.1235	12.7270	0.0004	0.1196
Constant term[b]	−0.7372	4.8630	0.0274	0.4784

Note. Comprehensive test: Penalized LR = 208.9560; degree of freedom = 12; Goodness of fit: Pseudo $R^2 = 0.3380$.
[a] Denotes significance at the 5% level.
[b] Denotes significance at the 1% level.

[*] The first approximation of I(B) is $-\partial^2 \ln L^* / \partial B^2$.

Gender, age, and household vegetable cultivated area were negatively correlated with the WTA biopesticides, while the level of education, knowledge of pesticide residues, household nonagricultural income, and whether the farmer participated in cooperative organizations were positively correlated with the WTA biopesticides. The possible reasons for the above-mentioned results are listed as follows:

1. The odds ratio of the WTA biopesticides of vegetable farmers with 9 years of education or above was 15.0165 times than that of those with <9 years of education. The effect of years of education on their biopesticide application has two sides. One is the negative effect. The research of Zhang et al. (2004) and Fu and Song (2010) demonstrated that farmers with a higher level of education better understood the rapid and effective insecticidal effect of chemical pesticides and that when the agricultural product traceability system was not yet perfect, vegetables produced by applying biopesticides were not able to achieve a higher market price. Thus, it can be concluded that such farmers' WTA biopesticides is often lower due to the pursuit of economic interests. The other side is the positive effect. Based on the analysis that those with a high level of education could easily master the application skills of biopesticides and could relatively easily control risk, a high level of education should enhance farmers' willingness to apply biopesticides. The empirical results of this study revealed that, in the survey area, the positive effect dominated.

2. The odds ratio of the male vegetable farmers' WTA biopesticides was significantly lower than that of females. The general consensus is that the difference in a farmer's gender affects his/her pesticide application behaviors and adoption level of new agricultural technology (Doss 2001; Doss and Morris 2001); the related research of Fu and Song (2010) supported the above-mentioned research conclusion. However, the results of this study were inconsistent with other studies. For example, the study of Rizwan et al. (2005) in Pakistan and the study of Kishor (2007) hold that women's educational level was generally relatively low; therefore, their pesticide selection ability was poor, leading to a higher risk. The study of Wu et al. (2011) also pointed out that, when selecting varieties of pesticides, compared with that of female farmers, the probability of standard and rational behaviors of male farmers was higher. Risk averters are generally more inclined to use chemical pesticides with relatively stable technology and income; however, women's and men's attitudes toward

risk have demonstrated no significant differences (Binswanger 1980). Thus, the difference in risk preference cannot be the main explanatory basis for this conclusion. Farmers with more years of education should more easily be able to acquire new technology, but in general, the level of education of women in rural areas was revealed to be lower than that of men*; therefore, this conclusion cannot be explained from the perspective of years of education. Studies have shown that, in the process of pesticide application, women tended to pay more attention to health and safety (Khwaja 2001; Xing et al. 2009). The difference in psychological characteristics may be the main reason for the significantly lower odds ratio of male farmers' WTA biopesticides. This explanation was supported by Kruger and Polanski (2011).

3. The willingness of vegetable farmers aged ≥40 to adopt biopesticides was significantly lower than that of vegetable farmers who were <40. This conclusion was similar to that of Isina and Yildirim (2007). Binswanger (1980) demonstrated that young farmers were more inclined to venture a small stake. This means that, compared with the application of chemical pesticides, if biopesticide application allows farmers to obtain higher expected net profit, but the difference is not significant, young farmers are more willing to run a "risk" to apply biopesticides. In addition, younger farmers may be more highly educated and may more easily master biopesticide application skills, thus reducing the technology risk of biopesticide application. Older farmers have accumulated experience by applying chemical pesticides in long-term agricultural production practices. Due to behavioral inertia, older farmers have been dependent on chemical pesticides; therefore, they will inevitably have a low WTA biopesticides.

4. Vegetable farmers with a cultivated area of ≥5 mu were significantly less willing to apply biopesticides than those with a cultivated area of ≤5 mu. One explanation is that the main income source of farmers with a large vegetable planting area is agricultural production, and they have been dependent on chemical pesticides to control pests and obtain stable income in long-term vegetable planting practices. Understandably, if farmers adopt biopesticides, they will inevitably take high market and technology risks; therefore, they are more inclined to use familiar chemical pesticides. This conclusion was supported by the conclusions

---

*The proportion of male and female farmers with more than nine years of education was 69.23% and 63.64%, respectively, in the survey sample of this chapter.

of Wang and Xu (2004) and Ngow et al. (2007), that is, for the farmers with a larger cultivation area, the degree of agricultural commercialization is higher and their WTA biopesticides is lower.

5. Vegetable farmers with a household nonagricultural income of ≥30,000 yuan were more willing to apply biopesticides compared with those with <30,000 yuan. Farmer household income is one of the most important factors influencing pesticide application behavior. A high nonagricultural income indicates that ≥1 household member is a migrant worker. Since household members interact with each other, they are more open-minded and more receptive to new things, and thus, they are more concerned about quality/safety of agricultural products (Wu et al. 2011). In addition, for vegetable farmers with a high nonagricultural income, agricultural production, including vegetable planting, is no longer the main income source; with good economic conditions, they have a stronger ability to take the potential market and technology risks caused by biopesticide application. Therefore, farmers with a high nonagricultural income are relatively more willing to use biopesticides, which is consistent with the conclusions of Kong et al. (2005) and Amaza and Ogundari (2008) regarding the influence of the economic characteristics of the farmer households on their adoption of new pesticide technologies.

6. The farmers with a higher cognitive level of chemical pesticide residues were more willing to use biopesticides. The reason for this is obvious. Farmers who are aware of the hazards of chemical pesticide residues are usually more highly educated and also younger. They are more aware of the negative impact of chemical pesticide application on their own health and the related potential losses caused by vegetable safety risk; therefore, they have a higher WTA biopesticides.

7. Vegetable farmers participating in agricultural cooperative organizations were more willing to use biopesticides than the individual farmers. Agricultural cooperative organizations can provide the vegetable farmers with preproduction services like special production materials, such as improved varieties, related pesticides, and production services. Cooperative organizations can also provide required management techniques during the planting process and other supporting activities, and postproduction services such as packaging, storage, transport, purchase, and sales for vegetable farmers. This assistance can not only stabilize the purchase price of vegetables and reduce the market risks arising from biopesticides, but also provide technical support to vegetable farmers, thus reducing technical risks for vegetable farmers in biopesticide application.

12.4 Main Conclusions and Policy Implications

According to data published by the Medicine Inspecting Institute of the Ministry of Agriculture of the People's Republic of China, 97 varieties of registered active ingredients of biogenic pesticides existed in 2008 in China, accounting for 13.8% of total varieties of active pesticide ingredients; 3071 biogenic pesticide products also existed during this time, accounting for 10.2% of registered pesticide products. In 2008, the Chinese biopesticide market size was approximately 6 billion yuan, accounting for 10% of the total sales of pesticides, which was far below the international level of 20%. Currently, three problems exist in popularizing biopesticides in China. First, the low agricultural production income results in farmers' insufficient demand. Second, most of the Chinese biopesticide enterprises have backward production technology, a small sales volume, and low independent innovation ability, and the industry concentration is low. Third, the use of biopesticides is only being promoted in limited varieties of agricultural products, such as leafy vegetables, melons and fruits, tea, and Chinese herbs, and they are promoted on a small scale. Therefore, the Chinese government must gradually implement the most stringent agricultural environmental policies and take strong measures to promote biopesticides for pest control in agricultural production.

In this study, using 294 vegetable farmers in Cangshan County, Shandong Province, China, as the study subjects, the major factors affecting vegetable farmers' WTA biopesticides were analyzed. The results demonstrated that vegetable farmer households with individual characteristics, such as females, age <40 years, perceived hazards of chemical pesticide residue, and ≥9 years of education, and with household characteristics, such as a household vegetable planting area of <5 mu, a nonagricultural income of ≥30,000 yuan, and participation in agricultural cooperatives, were more willing to apply biopesticides. These findings bear a wealth of policy implications. According to these findings, the promotion of biological pesticides in China's agricultural production transformation both now and in the future may be adopted by the following policies:

1. The government should promote biopesticides by means of agricultural cooperative organizations, and by taking into account that more highly educated, young, female farmers with a small planting scale and high nonagricultural income are more likely to adopt biopesticides, the government can rely on agricultural cooperative organizations to select vegetable farmers with the above-mentioned characteristics to promote biopesticide use.

2. The government must increase efforts to support biopesticide application. Our results demonstrated that government subsidies for biopesticides, skill training in pesticide application organized by the government, and other government-related variables did not show significant correlation with WTA. Therefore, the current biopesticide application in Shandong Province mainly depends on the farmers' internal drive, and the external driving force from the government is not sufficient. The basic point of a series of policies supporting agriculture and benefiting farmers issued by the government in recent years, which are traditional agriculture supporting policies, was to ensure the increase of agricultural production, rather than improvement in quality. The skill training provided for farmers by the government was not directed at biopesticides, and no subsidies are currently provided to reduce the relative prices of biopesticides to chemical pesticides. Hence, it is difficult to significantly increase the farmers' WTA biopesticides. Therefore, the government's policy on agricultural production should be adjusted to promote biopesticides in a large area with a more explicit policy orientation.

3. The government should guide the agricultural cooperative organizations in focusing on building sales networks for high-quality agricultural products and related works. Because it is difficult for the consumers to distinguish vegetables produced by different pesticides in the general vegetable market in China, vegetable farmers expecting a higher price for vegetables produced by applying biopesticides do not show high WTA biopesticides. Therefore, to encourage more consumers to buy and effectively increase consumer confidence, agricultural cooperative organizations should establish a clear and definite standard, build a credible certification system for agricultural products produced by applying biopesticides, develop sales networks for high-quality agricultural products, and implement proper marketing strategies such as pricing strategy.

This chapter presented a case study on vegetable farmers. There are numerous factors affecting food safety in China. However, it is difficult to analyze all the factors due to space limitations. Currently, at least 200 million farmers exist in China, and they are distributed in areas with different and even very different economic and social levels. The significance of the case study presented in this chapter was to demonstrate that the prevention of food safety risks is a very complex issue and will be a long and arduous process in China.

13

Main Conclusions
and Research Prospects

To facilitate the readers' general understanding of main conclusions of this book, this chapter summarizes all the research conclusions and provides a preliminary indication of future research emphases.

13.1 Main Research Conclusions

This book mainly describes the changes in Chinese food safety during 2006–2011 (Chapter 5 assesses the food safety risks during 2006–2012). The major changes in China's food safety can be summarized by 14 research conclusions given in Sections 13.1.1 through 13.1.14.

13.1.1 Guaranteed Quantitative Safety of Edible Agricultural Products and Food

With the increasing production and adequate market supply of major edible agricultural products, the quantitative safety of edible agricultural products has been effectively guaranteed in China. Moreover, the internal structure of major edible agricultural products has been undergoing continual adjustments and will continue escalating long into the future, effectively meeting the requirements of dietary upgrades resulting from the increasing standard of living of Chinese urban and rural residents.

In addition, in China, the production of rice, wheat flour, edible vegetable oil, fresh frozen meat, biscuits, juices and juice drinks, beer, instant noodles, and some other foods ranks first or very high on the list in the world.

The food industry has become one of the most important economic pillar industries for the Chinese national economy. The great development of food industry has guaranteed quantitative food safety.

13.1.2 Gradual Improvement of Qualitative Safety of Edible Agricultural Products

With the gradual implementation of agricultural production policies, continual transformation of the production mode of edible agricultural products, effective development of the production and market of emerging edible agricultural products, gradual construction of the standardization systems, and in particular, the government's increasing efforts in safety supervision of edible agricultural products and the qualitative safety of major edible agricultural products closely related with people's daily lives, such as vegetables, fruits, and livestock and aquatic products, advanced to a new level to varying degrees during 2006–2011 in China. The qualitative safety of edible agricultural products has been effectively guaranteed for urban and rural residents and overall has maintained a generally stable development trend with gradual improvement.

13.1.3 A High Level of Food Quality and Safety in Processing

Since 2005, national food quality checks by the Administration of Quality Supervision, Inspection and Quarantine (AQSIQ) have covered all main food categories, involving thousands of different types of food. Although the pass rates of the same food in different years, as well as those of different foods in the same year, differed to varying degrees in national food quality checks, the overall pass rate increased from 80.1% in 2005 to 96.4% in 2011, representing an increase of 16.3% during the 6 years. According to the data obtained from the AQSIQ spot checks in 2011, the pass rates of most food categories closely related with people's daily lives, including carbonated beverage products, instant noodles, wheat flour, milk powder, meat products, sugar, and edible vegetable oil, were higher than 90%, and only those of soy products and oolong were lower than 90%.

13.1.4 Gradual Improvement of Food Quality and Safety in Circulation

Over the years, the Chinese government industrial and commercial management system has achieved positive results in strengthening special law enforcement inspection of food safety in circulation to actively and effectively prevent and manage food safety incidents, and in enhancing regular supervision of food quality and safety in circulation and in the food market. At present, the sampling scope and varieties in the Chinese food market have been expanding, more effort has been devoted, and classified supervision systems have been gradually established. Chinese industrial and commercial administration departments sampled 285,200, 208,300, and 332,000 sets of food in 2009, 2010, and 2011, respectively, and observed a general increasing trend in pass rates in circulation. Consumer safety has been basically guaranteed in the Chinese food market.

13.1.5 A High Level of Quality and Safety of Imported and Exported Food

Since the 1990s, both the total volume and quality of imported and exported food have continued to increase in China. China plays an indispensable role in regulating the global food supply and demand and in guaranteeing the food supply of the world in the context of economic globalization. Moreover, the pass rates of exported Chinese foods have remained at a stable level of 99%. The quality and safety of exported food have been fully guaranteed. However, unacceptable food quality, excessive pesticide and veterinary drug residues, noncompliance with quarantine regulations, unacceptable food additives, micro-biological contamination, and unacceptable certificate remain the major causes of quality deterioration in exported Chinese food. In addition, the overall quality and safety of imported Chinese food have also been relatively high. However, most quality failures are caused by microbial contamination, unacceptable food additives, unacceptable food quality, and heavy metal pollution. Not surprisingly, as the demand for imported food has been growing in China, safety risks in imported food have also increased. Therefore, imported food safety standards must be further improved, substandard imported food should be disposed of according to laws and regulations, and detection techniques should be gradually improved. Also, efforts should be devoted to guiding consumers in the rational consumption of imported food. Furthermore, establishment of technical trade measures integrating technology, regulations, and culture against imported food in the World Trade Organization (WTO) framework by drawing on international experience may need to be considered as an important agenda.

13.1.6 A Relatively Safe Range of Food Safety Risks

China's food safety risks during 2006–2012 were assessed at the macrolevel using catastrophe models based on the national statistics of relevant departments. Results demonstrated that overall food safety risk values showed a continuous downward trend since 2006 and remained stable in a relatively safe range in 2012. Of course, safety is a relative concept. No absolutely safe food exists. Thus, although China's food safety risks are in a relatively safe range in general, potential risks still exist. The possibility of rebound or even great fluctuations cannot be excluded in some food industry sectors or local production areas. Nevertheless, the overall basic trend of "general stability and gradual improvement" is unlikely to change, provided no irresistible, sudden, catastrophic event occurs at a large scale. This conclusion provides a reference for correctly understanding and assessing China's food safety risks.

13.1.7 Particularity of China's Food Safety Risks

As a global issue, food safety risks have common characteristics throughout the world. However, particular characteristics have also been observed in China's food safety risks. In developed countries, food safety incidents are mainly caused by biological, physical, chemical, and other nonhuman factors, such as environmental pollution and food chain contamination. However, in China, although some of the recent major food safety incidents were the result of technical deficiencies and environmental pollution, most of them were caused by human factors, that is, misconduct, noncompliance or partial compliance to existing food technical specifications and standard systems, and other illegal acts by producers and operators. This feature is closely related to the fact that China is at the economic and social development stage. Therefore, to improve China's food safety, it is important to enhance the integrity of food producers and operators by the combination of legal, economic, and administrative means.

13.1.8 Urban and Rural Residents' Low Evaluation of Food Safety

According to the survey data collected from 4289 urban and rural residents in 96 sites in 12 provinces, autonomous regions, and municipalities, including Fujian, Guizhou, Henan, Hubei, Jilin, Jiangsu, Jiangxi, Shandong, Shaanxi, Shanghai, Sichuan, and Xinjiang, less than 30% of the respondents believed that local food was "safe" and "very safe." More than 70% of the respondents' confidence in food safety has been affected by frequent food safety incidents, and 45.44% of the respondents had confidence in the improvement of food safety. Respondents believed that the most important food safety risk factors were, in descending order of importance, abuse of food additives and intentional addition of nonfood substances, excessive pesticide and veterinary drug residues, heavy metal contamination, and bacterial and microbial contamination. Moreover, 60.65% of the respondents considered the main cause of food safety risks to be exploitation of the inability of consumers and government regulators to gain complete relevant information by unscrupulous food producers and operators, irresponsible profiteering of food-related enterprises, or a weak sense of corporate social responsibility. Although urban and rural respondents had a different awareness of food safety risks and consumer safety, as well as food consumption habits, they had consistent opinions on the same issue. In particular, it should be noted that both urban and rural respondents were unsatisfied with the government's efforts in food safety supervision and law enforcement.

13.1.9 Basic Establishment of a Food Safety Legal System

Along with the implementation of the *Food Safety Law* on June 1, 2009, China has basically established an integrated food safety legal system. It centers on the *Food Safety Law*, is supported by relevant special laws, and connects with the laws on environmental protection, product quality, import and export commodity inspection, and quarantine of animals and plants, as well as other relevant aspects. It has played a very large role in ensuring the stable improvement of food quality and safety in China. However, the Chinese food safety legal system and regulation model essentially maintains a top-down approach. The *Food Safety Law* has strengthened the administration pattern led by government regulation. In the system, emphasis has been placed on the allocation and implementation of regulatory responsibilities of administrative organs in the field of food safety. In practice, overreliance has been placed on the regulation approach of special rectification. Although the special rectification approach can manage disorders using heavy penalties, it lacks universality and continuity, cannot fundamentally solve food safety problems, and can only provide stop-gap measures in most cases. Therefore, to complement government regulation with the social power to address food safety problems should become the general trend and orientation of the construction of the Chinese food safety legal system.

13.1.10 Significant Progress in the Food Safety Management System Reform

This book describes the evolution of the Chinese food safety management system according to the history of China's development and reform with the founding of New China in 1949 as the starting point. In this way, a panoramic view of the development process and the actual state of the Chinese food safety management system is provided. Before the reform in February 2013, a food safety management system was implemented in China. In this system, the Food Safety Commission of the State Council served as the coordinating agency. In addition, segmented regulation involving a number of regulatory authorities was implemented and local governments maintained full responsibilities. Despite its gradual development after several reforms, this system failed to fundamentally address the overlap of and gap in responsibilities caused by the multidepartmental segmented regulation. In March 2013, the Chinese food safety management system was reformed again to restructure regulatory bodies. This important reform has made a new step toward the exploration and final solution of overlapped and segmented food safety management, "buck-passing," and unclear division of powers and responsibilities among different departments. It has also had a positive effect on developing an integrated, extensive, specialized, and efficient food safety management system, and comanagement of food safety by the government and civil society. However, the new food safety management system fails to develop an effective incentive mechanism to give full play to the role of civil supervisors. Moreover, a long process will be required to establish good coordination mechanisms among the multiple food safety regulators led by the China Food and Drug Administration, and seamless management is yet to be tested in practice.

13.1.11 New Progress in Construction of a Food Safety Standard System

A preliminary food standard system based on national standards and combined with industry, local, and enterprise standards has been established. However, the overall level of China's food safety standards is low and there are contradictions, overlaps, and duplications among the standards. Moreover, some important standards or indicators are absent, and the scientificity and rationality of some standards should be improved. In addition, some standards, including mandatory standards, are poorly implemented. With the implementation of the *Food Safety Law* on June 1, 2009, explicit provisions were made for construction of the Chinese food safety standard system and other issues. This created conditions for the cleanup, integration, and unification of the food safety standard system. Great efforts have been devoted, not only to improving the food safety laws and regulations and promoting the implementation of national food safety standards, but also to revising and improving the food standard system in a targeted manner. Since 2010, the Ministry of Health has promulgated 269 new national food safety standards concerning the safety of dairy products, the use of food additives and compound food additives, mycotoxin limits, prepackaged food labeling and nutrition labeling, pesticide residue limits, and food additive products. During this time, the Ministry of Health has also revised and improved standards of food packaging materials and improved the scientificity and practicability of national food safety standards. However, it will be a long-term task to largely solve the contradictions, duplications, and overlaps among the current standards to develop a more perfect system of national food safety standards in China.

13.1.12 Seriously Insufficient Disclosure of Food Safety Information

The essential characteristic of food safety risk is food safety information asymmetry. Disclosure of food safety information is of great significance for the prevention of food safety risks. Although the Chinese government is committed to, and has made some progress in, the legal system construction for food safety information disclosure, in general, food safety information disclosure by the government is clearly unsatisfactory. Prominent problems are disorder of disclosers, incomprehensive disclosure, and a huge gap between the amount of information disclosed and the public demand. Therefore, the public's right to know and choose cannot be satisfied.

13.1.13 Difficult Transformation of Production Mode of Edible Agricultural Products and Food

Very complicated factors affect the quality and safety of Chinese agricultural products at this stage, but the most realistic threat is excessive chemical pesticide residues. This book presents a case study on vegetable farmers' willingness to adopt biopesticides in Cangshan County, Shandong Province, China. Our study demonstrated that vegetable farmers' gender and knowledge of pesticide residues, vegetable cultivated area of the farmer household, and nonagricultural income of the farmer household significantly affected the farmers' willingness to adopt biopesticides. Currently, there are at least 200 million farmers in China who are distributed in areas with different and even very different economic and social levels. Therefore, it will take quite some time to regulate farmers' application of pesticides, reduce the use of chemical pesticides, and gradually popularize the use of biopesticides in China. In addition, the analysis of typical food safety accidents that occurred during 2002–2011 in China revealed that 68.2% of food safety incidents were the result of food quality and safety problems knowingly caused by stakeholders in the supply chain that were motivated by a desire for personal gain or profit.

This fully demonstrates that "knowing violations" by food producers and operators are currently the main cause of food safety problems; that is to say, most Chinese food safety incidents are caused by the moral failure of food producers and operators in the face of interests. The ethical quality of food producers and operators cannot be improved in an action. In light of this, a very long journey lies ahead with respect to improving China's food safety.

13.1.14 A Long and Arduous Road to the Prevention and Control of Food Safety Risks in China

In short, the gap between the decentralized, small-scale food production and the inherent requirements of the modern food industry and consumers' growing demand for food safety has become the principal contradiction in the prevention and control of food safety risks in China. This contradiction will remain for a very long period of time in the future and cannot be fundamentally changed in the short term. Thus, prevention of food safety risks will be a long-term arduous process in China. Optimizing and restraining the behaviors of food producers and operators and changing the food production mode are the only ways to prevent food safety risks in China at present and in the foreseeable future. Therefore, the limited regulatory power and the major social resources in food safety risk monitoring should be centralized to solve the principal contradiction.

13.2 Research Prospects

Food safety risk management is a very complex issue for China and for the world. Based on the understanding of China's food safety problems, the future research priorities in this regard should be as given in the subsequent sections.

13.2.1 The Particularity of Food Safety Management in China

The fundamental cause of frequent occurrence of food safety incidents in China is the noncompliance of farmers, food producers, and operators. Furthermore, the contradiction between the backward agricultural production and operation and decentralized, small-scale food production, and the growing demand for food safety occupies an important position among China's current social contradictions. Therefore, future research should be focused on investigating the factor set affecting food producers' safe production and the relationship between the factors, as well as identifying the key factors, and analyzing the producer set in preventing food safety risks to identify target groups for governmental supervision based on China's reality.

13.2.2 The Relationship between China's Food Safety and Food Standards

Food standards are the basic norms of food production and the basic means of ensuring food safety. At present, considerable controversy exists regarding Chinese food standards, both domestically and internationally. Determination of the food standards that are lagging behind, lead to blind spots, are repeated or overlapped, or are mutually contradictory is worthy of further study. Moreover, compliance of food producers and operators to existing food standards, as well as the implementation results, should be intensively investigated.

13.2.3 The Relationship between Food Safety Management and the Market in China

Both market and government failures occur in food safety governance in China. Deepening the reform and improvement of social governance was an extremely important topic of the Third Plenary Session of the 18th Communist Party of China (CPC) Central Committee, which was just closed when the drafting of this book was completed. Food safety governance is an important content of social governance. Both an effective policy system and market mechanism that provide clear boundaries between the government and market functions and are mutually integrated should be further designed and implemented in the Chinese food safety management system. This is related to the overall food safety governance in China and should be regarded as a priority in future research.

13.2.4 Rural Food Safety Governance in China

Improving rural food safety is an important part of food safety governance in China. Currently, counterfeit and expired foods are commonly seen in rural food markets in China. The outlook on rural food safety is not optimistic at this time. Therefore, emphasis should be placed on deepening the reform of the rural food safety management system and investigating the new features of rural food safety governance in China in future research.

13.2.5 Summary of Experience in Food Safety Governance in China

Although food safety needs to be further improved in China, it should be noted that great progress has been made in food safety management over the 30 years of reform and opening up. Relevant experience, particularly with respect to ensuring the supply of safe food, should be summed up. In 1994, the American scholar Lester Brown asked, "Who will feed China?" The nine successive increases in grain output and a self-sufficiency rate of 95% in 2012 have eliminated Brown's "concern" and some of the "worry" in the world. These nine successive increases in grain output are rare, not only in China's history, but also in the world. This is a microcosm of the universally acknowledged great achievements made by China over the 30 years of reform and opening up. China's high grain output also provides a useful reference to the majority of developing countries in agricultural policy-making and production practices. Therefore, emphasis should also be placed on providing an accurate, in-depth, and objective description and summary of the practical experience in China's food safety governance reform and development, answering the major questions of common concern of the international community, describing the "Chinese characteristics" in preventing food safety risks, sharing the experience in food safety governance with developing countries, and strengthening exchange and cooperation with the international community, especially the academic community, to contribute to the global governance of food safety risks.

Food safety risk management is a global issue, and no country can stand aloof. Food safety risks must be comanaged by all countries, and international cooperation is mandatory because of their global characteristics. Some of the basic characteristics and problems of China's food safety governance are global; however, most of them are unique to China. As Chinese food safety risks are of global concern, China plays an important role in defending against the global challenge of food safety. To this end, we are willing to cooperate with researchers worldwide to devote persistent efforts to promoting the research on China's and global food safety governance.

References

Abhilash, P.C. and Singh, N. 2009. Pesticide use and application: An Indian scenario. *Journal of Hazardous Materials* 165: 1–12.

Administration of Quality Supervision, Inspection and Quarantine of the PRC. 2002. *GB/T 7635.1-2002 Categories and Codes of Major Products in China.* Beijing, People's Republic of China: China Standard Press (in Chinese).

Administrative Law Office of the Commission of Legislative Affairs of the Standing Committee of the National People's Congress. 2009. *Interpretation and Application of the Food Safety Law of the PRC,* ed. Y. Li. Beijing, People's Republic of China: China Legal Publishing House (in Chinese).

Agnew, G.K. and Baker, P.B. 2000. Arizona cotton pesticide use data: Opportunities and pitfalls. *Proceedings of the Beltwide Cotton Conferences,* January 4–8, San Antonio, TX. Memphis, TN: National Cotton Council, pp. 1239–1246.

Albert, A. and Anderson, J. 1984. On the existence of maximum likelihood estimates in logistic regression models. *Biometrika* 71: 1.

Alibaba China. 2012. The Ministry of Health says the *Staphylococcus aureus* standard is not lowered in the questioning of new national standard for frozen food. http://info.china.alibaba.com/news/detail/v0-d1023918977.html (accessed June 1, 2012) (in Chinese).

Amaza, P.S. and Ogundari, K. 2008. An investigation of factors that influence the technical efficiency of soybean production in the Guinea savannas of Nigeria. *Journal of Food, Agriculture & Environment* 6: 92–96.

AQSIQ Research Centre for International Inspection and Quarantine Standards and Technical Regulations. 2012. Analysis report on detention/recall of Chinese exports. WTO/TBT-SPS Notification and Enquiry of China. http://www.tbt-sps.gov.cn/riskinfo/riskanalyse/Pages/riskreport-cn.aspx (accessed November 10, 2012) (in Chinese).

Baidu Encyclopedia. 2013. White paper. http://baike.baidu.com/view/1778.htm (accessed June 28, 2013) (in Chinese).

Bartlett, M. 1953. Approximate confidence intervals: More than one unknown parameter. *Biometrika* 40: 306.

Beneficiation Technology Network. 2012. Millions and millions of tons of food are polluted by heavy metals every year in China. http://www.mining120.com/show/1203/20120323_82731.html (accessed October 10, 2012) (in Chinese).

Bhalla, R.S. and Devi-Prasad, K.V. 2008. Neem cake-urea mixed applications increase growth in paddy. *Current Science* 94: 1066–1070.

Binswanger, H.P. 1980. Attitudes toward risk: Experimental measurement in rural India. *American Journal of Agricultural Economics* 62: 395–407.

Bourn, D. and Prescott, J.A. 2002. Comparison of the nutritional value, sensory qualities, and food safety of organically and conventionally produced food. *Critical Reviews in Food Safety and Nutrition* 42(1): 1–34.

Carlson, G.A. 1979. Insurance, information, and organizational options in pest management. *Annual Review of Phytopathology* 17: 149–161.

Chen, J.G, Huang, Q.H., Yu, Q. et al. 2009. *China's Industrialization Report*. Beijing, People's Republic of China: Social Sciences Academic Press (in Chinese).

Chen, M.Z. 1995. An explanation for the Food Hygiene Law (Revised Draft). *Gazette of the Standing Committee of the National People's Congress* 7: 92–94 (in Chinese).

Chen, P. 2008. An empirical Study on relationship between food trade and economic growth in food industry of China. PhD dissertation, Jiangnan University, Jiangsu, People's Republic of China (in Chinese).

Chen, Q.L., Zhang, Q., and Xiao, L. 2010. The indexes, model and method for urban safety assessment under emergent situation base on the catastrophe model. *Chinese Journal of Management* 7(6): 891–895 (in Chinese).

Chen, W.Y. 2003. A brief comment on biopesticides. *Pesticide Science and Administration* 2: 3–6 (in Chinese).

Chen, X.H. 2011. A speech in the forum on the National Rural Land Contract Management: Recognize the situation and identify the tasks to further strengthen the management and service of transfer of land use rights. http://www.gov.cn/gzdt/2011-08/19/content_1928920.htm (accessed August 6, 2012) (in Chinese).

Chen, X.Z. 1990. Food poisoning in Xuhui District during 1960–1989. In *Shanghai health and epidemic prevention*. Health and Epidemic Prevention Station of Shanghai Municipality. Shanghai, People's Republic of China: Shanghai Peoples Publishing House, pp. 233–235 (in Chinese).

Chen, Y. 2007. Issues and countermeasures on imported food security in China. *Food Research and Development* 6: 183–187 (in Chinese).

China Agricultural Quality Standards Network. 2008. Report on participation in the 39th session of the Codex Committee on Pesticide Residues. http://www.caqs.gov.cn/News/Detail/?ListName=食品法典专栏-02会议动态&UserKey=8SS2ZAM99NXMB4E7 (accessed August 6, 2012) (in Chinese).

China Business News. 2011. New national standard for dumplings includes limits for *Staphylococcus aureus*. http://www.ccgp.gov.cn/jrcj/mrshts/201111/t20111121_1883838.shtml (accessed June 1, 2012) (in Chinese).

China Certification and Accreditation Information Network. 2010. Organic food becomes popular after the hormone scandal. http://www.cait.cn/spnc_1/xwdt/zxbd/201012/t20101203_69000.shtml (accessed June 1, 2012) (in Chinese).

China Economic Net. 2011. A survey finds that 10% of Chinese rice is contaminated by cadmium and can lead to Itai-itai disease. http://www.tianjinwe.com/hotnews/gnxw/201102/t20110214_3414801.html (accessed June 1, 2012) (in Chinese).

China Economic Net. 2012. National Development Exchange Conference on the 12th Five-Year Plan of Food Industry. http://news.163.com/12/0614/14/83VGPMV400014JB5.html (accessed October 18, 2013) (in Chinese).

China Economic Net. 2013. An expert states that most hot food safety issues in 2012 are not food safety incidents. http://www.ce.cn/cysc/sp/info/201301/06/t20130106_21313898.shtml (accessed March 12, 2013) (in Chinese).

China E-government Network. 2013. Key points of the 2012 annual report on the Chinese government transparency. http://www.e-gov.org.cn/news/news007/2013-02-27/138925.html (accessed June 30, 2013) (in Chinese).

China Food Machinery and Equipment Network. 2006. Five major gaps faced by food packaging machinery industry in China. http://www.foodjx.com/Tech_news/Detail/261.html (accessed October 18, 2013) (in Chinese).

China Food News. 1983. *Compilation of Food Hygiene Laws.* Beijing, People's Republic of China: China Food News (in Chinese).

China Food Safety Net. 2013. 21 cities in Guangdong publish the list of substandard rice. http://www.cfsn.cn/news/content/2013-05/27/content_125118.htm (accessed May 28, 2013) (in Chinese).

China Food Safety Strategy Research Group of Development Research Center of the State Council. 2005. Study on China's national food safety strategy. *Agricultural Quality and Standards* 1: 5–6 (in Chinese).

China Low-Carbon Economy Network. 2012. State Council officially has approved the 12th Five-Year Plan of Heavy Metal Pollution Prevention and Control. http://www.lowcn.com/xinnengyuan/zhengce/201204/2853685.html (accessed October 10, 2012) (in Chinese).

China Network Television. 2012. The Ministry of Health and Ministry of Agriculture jointly runs a training course on national food safety standards. http://nongye.cntv.cn/20120511/106707.shtml (accessed August 6, 2012) (in Chinese).

China.org.cn. 2013. Work report of the Supreme People's Procuratorate by Jianming Cao (complete record). http://www.china.com.cn/news/2013lianghui/2013-03/10/content_28191919_2.htm (accessed April 23, 2013) (in Chinese).

Cordeiro, G. and McCullagh, P. 1991. Bias correction in generalized linear models. *Journal of the Royal Statistical Society Series B (Methodological)* 53: 629–643.

Dariush, H., Bijan, A., Reza, M., and Maryam, H. 2009. An empirical model of factors affecting farmers' participation in natural resources conservational programs in Iran. *Food, Agriculture & Environment* 7: 201–207.

Dasgupta, S., Meisner, C., and Huq, M. 2007. A pinch or a pint? Evidence of pesticide overuse in Bangladesh. *Journal of Agricultural Economics* 1: 91–114.

Dayoo.com. 2013. Food safety has been improved steadily in Shenzhen, with a pass rate of food sampling over 94% last year. http://news.dayoo.com/finance/201302/07/54401_109806776.htm (accessed July 2, 2013) (in Chinese)

Den Ouden, M., Dijkhuizen, A.A., Huirne, R., and Zuurbier, P.J.P. 1996. Vertical cooperation in agricultural production—Marketing chains, with special reference to product differentiation in pork. *Agribusiness* 12(3): 277–290.

Dillon, J.L. and Scandizzo, P.L. 1978. Risk attitudes of subsistence farmers in northeast Brazil: A sampling approach. *American Journal of Agricultural Economics* 60: 425–435.

Dong, Y.G. and Chu, X. 2013. Institutional cause of the Chinese food safety crisis. *Journal of Northwest Agriculture and Forestry University (Social Science Edition)* 4: 127–132 (in Chinese).

Doss, C.R. 2001. Designing agricultural technology for African women farmers: Lessons from 25 years of experience. *World Development* 29: 2075–2092.

Doss, C.R. and Morris, M.L. 2001. How does gender affect the adoption of agricultural innovations? The case of improved maize technology in Ghana. *Agricultural Economics* 1: 27–39.

Du, Q. 2008. Inspection and control of nitrite in meat. *Meat Research* 3: 55–58 (in Chinese).

Ellis, F. 1998. *Peasant Economics: Farm Households and Agrarian Development.* Cambridge: Cambridge University Press.

European Environment Agency. 2005. The European environment—State and outlook 2005. Luxembourg: Office for Official Publications of the European Communities.

Fan, Y.X., Tian, J., and Liu, X.M. 2011. Summary of the 43rd session of the Codex Committee on Food Additives. *China Health Standard Management* 2(2): 63–68 (in Chinese).

Fan, Z.Q. 2007. *An Investigation Report on Changes in Land Utilization in China in 2006.* Beijing, People's Republic of China: China Land Press (in Chinese).

FAO/WHO. 1997. *Codex Procedural Manual* (10th edition). ftp://ftp.fao.org/codex/Publications/ProcManuals/Manual_10e.pdf. (accessed April 6, 2014) (in Chinese).

Feder, G. 1980. Farm size, risk aversion and the adoption of new technology under uncertainty. *Oxford Economic Paper* 32: 263–283.

Feng, J.L. 2013. A discussion on the Chinese food safety management system in the super-ministry reform. *Science and Technology of Food Industry* 6: 1–7 (in Chinese).

Fernandez-Cornejo, J. 2004. The seed industry in U.S. agriculture: An exploration of data and information on crop seed markets, regulation, industry structure, and research and development. US Department of Agriculture, Economic Research Service, *Agriculture Information Bulletin* 786.

Firth, D. 1993. Bias reduction of maximum likelihood estimates. *Biometrika* 80: 27.

Food and Agriculture Organization. 2007. Review of agricultural water use per country (Rome). http://www.fao.org/nr/water/aquastat/water_use/index.stm (accessed July 1, 2011).

Food & Beverage Online. 2011. General review of food safety incidents in 2011. http://www.21food.cn/html/news/35/663093.htm (accessed June 1, 2012) (in Chinese).

Food Industry Net. 2011. Wal-Mart involved in counterfeit green pork scandal; consumers pay a higher price for psychological comfort. http://www.foodqs.cn/news/gnspzs01/2011102110125219.htm (accessed June 1, 2012) (in Chinese).

Foodmate Net. 2008. China will build a primary national standards system that meets the development needs. http://www.foodmate.net/news/guonei/2008/04/108851.html (accessed August 6, 2012) (in Chinese).

Foodmate Net. 2011. Work program for follow-up evaluation of standards for use of food additives and enzyme preparations for food industry application. http://www.foodmate.net/law/hebei/174128.html (accessed August 6, 2012) (in Chinese).

Foodmate Net. 2012. China National Center for Food Safety Risk Assessment runs open days for face-to-face communication of food safety standards. http://www.foodmate.net/news/guonei/2012/07/211091_4.html (accessed August 6, 2012) (in Chinese).

Fu, X.H. and Song, W.T. 2010. The influencing factors of farmers' purchase intention and behavior—The case of Sichuan Province. *Journal of Agrotechnical Economics* 6: 120–128 (in Chinese).

General Office of the Chinese State Council. 2013a. A notice on the issuance of key responsibilities, internal structure, and staffing prescription of China Food and Drug Administration. http://www.gov.cn/zwgk/2013-05/15/content_2403661.htm (accessed July 2, 2013) (in Chinese).

General Office of the Chinese State Council. 2013b. Instruction on the reform and improvement of local food and drug administration systems. http://www.gov.cn/zwgk/2013-04/18/content_2381534.htm (accessed July 12, 2013) (in Chinese).

Gratt, L.B. 1987. *Uncertainty in Risk Assessment, Risk Management and Decision Making.* New York: Plenum Press.

Guangxi Institute of Medical Scientific Information. 1980. Food poisoning analysis in Jiangsu Province during 1974–1976. *China Medical Abstracts: Epidemic Prevention (1979)*: 178 (in Chinese).

Guangzhou Daily. 2011. Ministry of Environmental Protection officials said that 10% of the cultivated land exceeded limits of heavy metals. Netease Discovery. http://discovery.163.com/11/1107/09/7I8DADC4000125LI.html (accessed October 10, 2012) (in Chinese).

Guler, G.O., Cakmak, Y.S., Dagli, Z., Aktumsek, A., and Ozparlak, H. 2010. Organochlorine pesticide residues in wheat from Konya region, Turkey. *Food and Chemical Toxicology* 48(5): 1218–1221.

Guo, Z.G. 2013. The secret of food safety information that cannot be told: Unable or unwilling. *Worker's Daily* (May 22): 3 (in Chinese).

Gut, L., Schilder, A., Isaacs, R., and Mcmanus, P. 2007. Managing the community of pests and beneficials. *Fruit Crop Ecology and Management*, ed. Joy Neumann Landis. Michigan: Michigan State University, pp. 34–75.

Han, Y.M. 2006. Research on food safety prediction and control model. PhD dissertation, Southeast University, Jiangsu Providence, People's Republic of China (in Chinese).

Health Department of Inner Mongolia Autonomous Region. 2012. A notice on completing the follow-up evaluation of national food safety standards. http://www.nmwst.gov.cn/html/ywlm/zhifajiandu/zhengcexinxi/201206/13-50231.html (accessed August 6, 2012) (in Chinese).

Health Inspection Bureau of Health Department of Hubei Province. 2012. Hebei Health Inspection Bureau holds a meeting to analyze the survey data of follow-up evaluation of national food safety standards implemented in Hebei. http://www.hbwsjd.gov.cn:8088/(S(edcgg4msv000yy2e2sgffc3q))/Item.aspx?id=15991 (accessed August 6, 2012) (in Chinese).

Health Standard Network. 2012a. National training meeting on the *General Rules for the Nutrition Labeling of Prepackaged Foods* held in Beijing. http://wsbzw.jdzx.net.cn/wsbzw/article/22/2012/3/2c909e8e35f6ca690136063c385d00b1.html (accessed August 6, 2012) (in Chinese).

Health Standard Network. 2012b. Seminar on the publicity and implementation of national food safety standards held in Beijing. http://wsbzw.jdzx.net.cn/wsbzw/article/22/2012/4/2c909e8e36e8260e0136f28277070036.html (accessed August 6, 2012) (in Chinese).

Heinze, G. and Schemper, M. 2002. A solution to the problem of separation in logistic regression. *Statistics in Medicine* 21: 2409–2419.

Hokkanen, M.T. and Hajek, A. 2003. *Environmental Impacts of Microbial Insecticides.* The Netherlands: Kluwer Academic Publishers.

Hoogeveen, H. and Oostendorp, R. 2003. On the use of cost-benefit analysis for the evaluation of farm household investments in natural resource conservation. *Environment and Development Economics* 8: 331–350.

Horowitz, J.K. and Lichtenberg, E. 1993. Insurance, moral hazard, and chemical use in agriculture. *American Journal of Agricultural Economics* 75: 926–935.

Hou, B. and Wu, L. 2010. Safety impact and farmer awareness of pesticide residues. *Food and Agricultural Immunology* 21: 191–200.

Hruska, A. and Corriols, M. 2002. The impact of training in integrated pest management among Nicaraguan maize famers: Increased net returns and reduced health risk. *International Journal of Occupation and Environmental Health* 8: 191–200.

Hu, H.Y. and Gan, X.P. 2009. Thinking on food safety problems caused by food additives. *Agricultural Technological Service* 11:138–140 (in Chinese).

ILSI. 1997. *A Simple Guide to Understanding and Applying the Hazard Analysis Critical Control Point Concept* (2nd edition). Brussels, Belgium: International Life Sciences Institute (ILSI) Europe Scientific Committed on Food Safety.

Information Office of the Ministry of Agriculture of China. 2007. Overview of agricultural product quality and safety in China. http://www.china.com.cn/news/2007-09/24/content_8940456.htm (accessed October 10, 2012) (in Chinese).

Information Office of the State Council of China. 2011. China's policies and actions to tackle climate change [EB/OL]. http://www.gov.cn/jrzg/2011-11/22/content_2000047.htm (accessed October 10, 2012) (in Chinese).

Information Office of the State Council of the PRC. 2007. Quality and safety of food in China. Website of the Central People's Government of the PRC (accessed November 10, 2012). http://www.gov.cn/jrzg/2007-08/17/content_719999.htm (in Chinese).

Isina, S. and Yildirim, I. 2007. Fruit-growers' perceptions on the harmful effects of pesticides and their reflection on practices: The case of Kemalpasa, Turkey. *Crop Protection* 26: 917–922.

Ji, W. 2012. On the necessity of multi-sectoral food safety regulation. *Chinese Public Administration* 2: 54–58 (in Chinese).

Jiang, H.Y. and Gong, S.Y. 2004. A review on research of lead pollution in tea. *Journal of Tea* 30(4): 210–212 (in Chinese).

Jiao, M.J. 2013. Improvement of the Chinese food safety management system: Reality and reflection. *People's Tribune* 5: 46–47 (in Chinese).

Jinghua News. 2011. The whole process of production and circulation of swill-cooked dirty oil in the Beijing-Tianjin-Hebei region. http://news.jinghua.cn/351/c/201111/18/n3540292.shtml (accessed June 1, 2012) (in Chinese).

Jolankai, P., Toth, Z., and Kismanyoky, T. 2008. The combined effect of N fertilization and pesticide treatments in winter wheat. *Cereal Research Communication* 36: 467–470.

Kerkaert, B., Mestdagh, F., Cucu, T., Shrestha, K., Van Camp, J., and De Meulenaer, B. 2012. The impact of photo-induced molecular changes of dairy proteins on their ACE-inhibitory peptides and activity. *Amino Acids* 43(2): 951–962.

Khwaja, M.A. 1990. Impact of pesticides on environment and health. *SDPI News and Research Bulletin* 8: 2.

Kishor, A. 2007. Pesticide use knowledge and practices: Gender differences in Nepal. *Environmental Research* 104: 305–311.

Kleter, G.A. and Marvin, H.J.P. 2009. Indicators of emerging hazards and risks to food safety. *Food and Chemical Toxicology* 47(5): 1022–1039.

Koleva, N.G. and Schneider, U.A. 2009. The impact of climate change on the external cost of pesticide application in US agriculture. *International Journal of Agricultural Sustainability* 7: 203–216.

Kong, X.Z., Pang, X.P., Ma, J.J. et al. 2005. *The Effect, Safety and Influencing Factors of Agricultural Technology Application.* Beijing, People's Republic of China: China Agriculture Press (in Chinese).

Kruger, D.J. and Polanski, S.P. 2011. Sex differences in mortality rates have increased in China following the single-child law. *Letters on Evolutionary Behavioral Science* 2: 1–4.

Laird, M., Lacey, L.A., and Davidson, E.W. 1990. *Safety of Microbial Insecticides.* Boca Raton, Florida: CRC Press.

Li, H.M., Gao, Y., and Feng, S.J. 2007. Strategies in enforcing food safety policies by exporting enterprises. *World Agriculture* 3: 1–4 (in Chinese).

Li, H.Y. and Zhang, Z. 2007. Risk identification and assessment of food safety risks during the 2008 Beijing Olympic Games. *Modern Preventive Medicine* 34(10): 1900–1901 (in Chinese).

Li, L. 2012. *China Development Report on Legal System NO.10 (2012).* Beijing, People's Republic of China: Social Sciences Academic Press (in Chinese).

Li, M.H., Fu, X.H., and Wu, X.M. 2007. The willingness to use pesticide safety and influenced factors—Based on 214 farmer households' investigation and analysis in Sichuan Guanghan. *Agricultural Technology Economy* 5: 99–104 (in Chinese).

Li, T. 2010. Sanitary conditions and management strategies of individual food processors. *Chinese Journal of Ethnomedicine and Ethnopharmacy* 19(006): 70 (in Chinese).

Li, X.M. 2009. China is the largest consumer of nitrogen fertilizers in the world: "The more fertilizers, the more yield" concept. *People's Daily Online.* http://scitech.people.com.cn/GB/8850811.html (accessed October 10, 2012) (in Chinese).

Li, Z.M. 2004. Research on the connotations of food security and evaluation indicator system. *Journal of Beijing Agricultural Vocation College* 18(1): 18–22 (in Chinese).

Li, Z.Q. and Li, Q.Q. 2007. Analysis of current consumption and effective demand of major foods in China. *Chinese Food and Nutrition* 1: 4 (in Chinese).

Lian, H.Y. 2013. Food safety information should be disclosed immediately. *Guangzhou Daily* (May 18): 2 (in Chinese).

Liao, H.J. 2013. Food safety information disclosure remains to be promoted. *China Food Safety News* (May 31): A02 (in Chinese).

Lin, H.C. and Li, X.W. 2009. Effects and countermeasures of agricultural pollution to the quality safety of agricultural products. *Ecological Economy* 9: 146–149 (in Chinese).

Liu, C.J., Men, W.J., Liu, Y.J., and Zhang, H. 2002. Soil contamination by pesticides and bioremediation of contaminated soils. *System Sciences and Comprehensive Studies in Agriculture* 18(4): 291–292, 297.

Liu, H.L., Li, Y.H., Li, L.Q. et al. 2006. Pollution and risk evaluation of heavy metals in soil and agro-products from an area in the Taihu Lake region. *Journal of Safety and Environment* 6(5): 60–63 (in Chinese).

Liu, J.W. 2012a. Analysis on China's food safety problems based on a signaling game model. *Special Zone Economy* 1: 303–304 (in Chinese).

Liu, P. 2010a. Food safety supervision in China: An empirical study based on institutional change and performance evaluation. *Journal of Public Management* 4: 63–77 (in Chinese).

Liu, S.C. 2010b. Organization F form of agricultural insurance in China. PhD dissertation, Wuhan University, Hubei, People's Republic of China (in Chinese).

Liu, X.H. and Yu, X.J. 2010. Study on characteristics and countermeasures of heavy metal pollution in Yangtze River Delta. *Ecological Economy* 10: 164–166 (in Chinese).

Liu, Y. 2012b. Yuejin Guo: It is recommended to establish a food safety information disclosure system. *China Industry and Commerce News* (March 13): A01 (in Chinese).

Lu, L.X., Zhou, D., Lu, C., and Zhou, Y.X. 2010. On the influencing factors of vegetables origin concentration in China: A study based on the structural equation model for the data from wholesalers in Shouguang County, Shandong. *Finance and Trade Economics* 6: 113–120 (in Chinese).

Lu, M. 2012. Microbial contamination of food and its prevention. *Food Guide* 2: 32–35 (in Chinese).

Luo, G. 1993. An explanation for the institutional reform plan of the State Council: At the first section of the Eighth National People's Congress on March 16, 1993. *Gazette of the State Council of the People's Republic of China* 10: 413–415 (in Chinese).

Luo, H. and Zhang, X.D. 2008. Analysis of food safety problems of refrigerated transport in terms of consumer demand characteristics. *China High Technology Enterprises* 1: 119–121 (in Chinese).

Lv, Z.M. and Yan, B. 2003. Procedural rationality of the economic law. *Science of Law* 1: 53–64 (in Chinese).

Ma, L.Q. 2012. Food safety information shall be disclosed pursuant to the law. *Economic Daily* (August 19): 5 (in Chinese).

Ma, X.S. 2013. A sampling inspection detects cadmium contamination in 44.4% of the rice samples in catering in Guangzhou. *Southern Metropolis Daily* (May 17): AA14 (in Chinese).

Mandal, S.M.A., Mishra, B.K., and Mishra, P.R. 2003. Efficacy and economics of some bio-pesticides in managing *Helicoverpa armigera* (Hübner) on chickpeas. *Annals of Plant Protection Science* 11: 26–30.

Meng, Q.Y. 2002. Litigation protection mechanism under the Chinese economic law: Reflections on and reconstruction of economic litigation in China. *Legal Forum* 2: 93–99 (in Chinese).

Michael, L., Morris, C., and Doss, R. 2001. How does gender affect the adoption of agricultural innovations? The case of improved maize technology in Ghana. *Agricultural Economics* 1: 27–39.

Ministry of Agriculture of China. 2011a. Conference on national uniform professional prevention and control of crop pests [EB/OL]. http://www.gov.cn/gzdt/2011-06/16/content_1885360.htm (accessed October 10, 2012) (in Chinese).

Ministry of Agriculture of China. 2011b. "Twelfth Five-Year" Plan for the development of quality and safety of agricultural products [EB/OL]. http://www.moa.gov.cn/zwllm/ghjh/201106/t20110616_2031099.htm (accessed October 10, 2012) (in Chinese).

Ministry of Agriculture of China. 2012a. China's pollution-free, green, organic, and geographically indicated agricultural products have entered a new stage, focusing on quality improvement and brand enhancement [EB/OL]. http://www.moa.gov.cn/zwllm/zwdt/201203/t20120329_2549706.htm (accessed October 10, 2012) (in Chinese).

Ministry of Agriculture of China. 2012b. Routine monitoring data of agricultural product quality and safety in 2011. http://www.aqsc.agri.gov.cn/zhxx/xwzx/201201/t20120118_87597.htm (accessed October 10, 2012) (in Chinese).

Ministry of Agriculture of China. 2009. *China Agricultural Statistical Yearbook.* Beijing, People's Republic of China: China Agriculture Press (in Chinese).

Ministry of Commerce of China. 2011. Export data of agricultural products of China in 2011. http://images.mofcom.gov.cn/wms/table/2011_12.pdf (accessed April 6, 2012) (in Chinese).

Ministry of Health of the PRC. 2011. *GB 2760-2011 National Food Safety Standard—Standards for Uses of Food Additives*. Beijing, People's Republic of China: China Standard Press (in Chinese).

Ministry of Public Security of the PRC. 2013. The Ministry of Public Security publishes ten typical food safety crimes. http://www.mps.gov.cn/n16/n1237/n1342/n803715/3679230.html (accessed February 3, 2013) (in Chinese).

Modern Agriculture of China. 2010. Constraints of development of agro-processing industry in China. http://www.caecn.org/detail.php?id=124 (accessed October 18, 2013) (in Chinese).

National Bureau of Statistics of China. 2010. *China Statistical Yearbook*. Beijing, People's Republic of China: China Statistics Press (in Chinese).

National Center for Health Inspection and Supervision. 2011. Kick-off meeting of follow-up evaluation project for national food safety standards held in Urumqi. http://www.jdzx.net.cn/article/40288ce4062bb7e101062bbbd0650002/2011/9/2c909e8c328b80f10132b45177c20029.html (accessed August 6, 2012) (in Chinese).

National Food Safety Information Center. 2006. China was selected as a new host country for two committees. http://www.fsi.gov.cn/news.view.jsp?id=8911 (accessed August 6, 2012) (in Chinese).

Ngow, A.V.F., Mbise, T.J., Ijani, A.S.M., London, L., and Ajayi, O.C. 2007. Smallholder vegetable farmers in Northern Tanzania: Pesticides use practices, perceptions, cost and health effects. *Crop Protection* 26: 1617–1624.

Ningbo Municipal Statistics Bureau. 2005. Four trends and five problems in agricultural development of Zhejiang Province. http://www.nbstats.gov.cn/read/20051009/20721.aspx (accessed October 18, 2013) (in Chinese).

Padgitt, M.D., Newton, R.P., and Sandretto, C. 2000. Production practices for major crops in US agriculture, 1990–1997. Working paper, USDA Economic Research Service.

Palikhe, B.R. 2001. *Pesticide pollution management in Nepal: In harmony with nature*. Kathmandu, Nepal: Ministry of Agriculture and Cooperatives.

Pan, Y.H. and Zhang, N.N. 2011. Research on the entry of multinational seed industry to seed purchase and using behavior of vegetable farmers: Evidence from Shouguang in Shandong Province. *Issues in Agricultural Economy* 8: 10–18 (in Chinese).

Peng, W.P. 2002. Farmer economics: A new development in development economics. *Foreign Economies and Management* 2: 2–6 (in Chinese).

People's Daily Online. 2013. Supreme law: Food crimes pose a threat on national prosperity. http://365jia.cn/news/2013-05-03/9E29C1F2E1A94CA8.html (accessed June 21, 2013) (in Chinese).

Qi, J.K. 2011. Transformation of the Chinese food safety risk regulation model. *Chinese Journal of Law* 1: 33–49 (in Chinese).

Qianlong News Network. 2012. Training meeting on national food safety standards held in Beijing by the China Dairy Industry Association. http://www.qianlongnews.com/index.php/cms/item-view-id-62741.shtml (accessed August 6, 2012) (in Chinese).

Rizwan, S., Rizwan, S., Ahmad, I. et al. 2005. Advance effect of pesticides on reproduction hormones of women cotton pickers. *Pakistan Journal of Biological Sciences* 8: 1588–1591.

Rother, H.A. 2005. Risk perception, risk communication, and the effectiveness of pesticide labels in communicating hazards to South African farm workers. PhD dissertation, Michigan State University, Michigan.

Sanzidur, R. 2003. Farm-level pesticide use in Bangladesh: Determinants and awareness. *Agriculture, Ecosystems & Environment* 95: 241–252.

Sarig, Y. et al. 2003. Traceability of food products. *CIGR Journal of Scientific Research and Developments* 5(12): 54–65.

Schultz, T.W. 1964. *Transforming Traditional Agriculture*. New Haven, CT: Yale University Press.

Shang, Y.B. and Tang, H.G. 2007. *Puffed Food Processing Technology*. Beijing, People's Republic of China: Chemical Industry Press (in Chinese).

Shanghai Government Website. 2011. CNCA will build a database for "organic food" certification mark. http://www.shanghai.gov.cn/shanghai/node2314/node2315/node17239/node27048/u21ai553697.html (accessed June 1, 2012) (in Chinese).

Shenton, L. and Wallington, P. 1962. The bias of moment estimators with an application to the negative binomial distribution. *Biometrika* 49: 193.

Shi, H.G. 2006. Investigation and analysis of the use of food additives by rural food enterprises in Jiaxing City. *China Preventive Medicine* 6: 548–550 (in Chinese).

Shi, J.P. 2010. *Food Safety Risk Assessment*. Beijing, People's Republic of China: China Agricultural University Press (in Chinese).

Sina. 2011. NPC Standing Committee listens to the inspection report on the implementation of Food Safety Law. http://news.sina.com.cn/c/2011-06-29/175422728330.shtml (accessed 20 March, 2012) (in Chinese).

Sina Finance. 2011. The normal use of watermelon swelling agent yields no side effects. http://finance.sina.com.cn/roll/20110518/02379857697.shtml (accessed June 1, 2012) (in Chinese).

Song, W.C., Shan, W.L., and Ye, J.M. 2010. Summary and debate focus of the 41st Session of the Codex Committee on Pesticide Residues. *World Agriculture* 2: 42–44 (in Chinese).

Spicer, J. 2005. *Making Sense of Multivariate Data Analysis*. Thousands Oak, CA: Sage Publications Inc.

Sule, I. and Ismet, Y. 2007. Fruit-growers' perceptions on the harmful effects of pesticides and their reflection on practices: The case of Kemalpasa, Turkey. *Crop Protection*, 26: 917–922.

Sun, R.Z. 2011a. Combating against swill-cooked dirty oil requires improvement of the legal system. Sina Finance. http://finance.sina.com.cn/review/mspl/20110919/000010498822.shtml (accessed June 1, 2012) (in Chinese).

Sun, X.Y. 2008. Agricultural product quality security problem causes and management. PhD dissertation, Southwest University of Finance and Economics, Sichuan, People's Republic of China (in Chinese).

Sun, X.Z. 2011b. Nanyang: Anyone planting tainted Chinese chives should be held responsible. Nanyang Reporter Station. http://www.radiohenan.com/Article/news/shts/2011/04/01/67294.htm (accessed June 1, 2012) (in Chinese).

Sun, Z.H. 2011c. A speech in the forum on construction and development policy of farmers' professional cooperatives. http://www.cfc.agri.gov.cn/cfc/html/78/2011/20110920130130968989502/20110920130130968989502_.html (accessed April 6, 2014) (in Chinese).

Tang, S., Tang, G., and Cheke, R.A. 2010a. Optimum timing for integrated pest management: Modelling rates of pesticide application and natural enemy releases. *Journal of Theoretical Biology* 264: 623–638.

Tang, X.C. 2005. A design for evaluation indexes of food safety early warning system. *Science and Technology of Food Industry* 26(11): 152–155 (in Chinese).

Tang, Z.G., Wen, C., Wang, J.F. et al. 2010b. Status and control technology of heavy metal pollution in animal food. *Livestock Environment and Ecology Symposium* 235–240 (in Chinese).

Teng, W., Liu, Q., Li, Q., and Liu, Y.B. 2010. *Hazards of Heavy Metal Pollution on Agricultural Products and Risk Assessment.* Beijing, People's Republic of China: Chemical Industry Press (in Chinese).

Tu, Y.Q. and Xu, J. 2012. Path selection for food safety regulation in China. *Law Review* 3: 95–101 (in Chinese).

Valeeva, N.I., Meuwissen, M.P.M., and Huirne, R.B.M. 2004. Economics of food safety in chains: A review of general principles. *Wageningen Journal of Life Sciences* 51(4): 369–390.

Waichman, A.V. and Eve, E. 2007. Do farmers understand the information displayed on pesticide product labels? A key question to reduce pesticides exposure and risk of poisoning in the Brazilian Amazon. *Crop Protection* 26: 576–583.

Wang, H.S. and Xu, X. 2004. Micro-behavior and safety of agri-products: An analysis of rural household' production and residents' consumption. *Journal of Nanjing Agricultural University Social Sciences Edition* 1: 23–28 (in Chinese).

Wang, H.Y. and Wang, J. 2006. Pollution and control of agricultural non-point source on water environment. *Environment Science and Technology* 4: 53–55 (in Chinese).

Wang, L.H. 2004. General situation and management of food additive producers in Beijing. *Food Science and Technology* 9: 46–48 (in Chinese).

Wang, Y. 2008. Safety of Chinese food export since accession to the WTO and its countermeasures. PhD dissertation, Hefei University of Technology, Anhui, People's Republic of China (in Chinese).

Website of the Central People's Government of the PRC. 2010. Measures for administration of disclosure of agricultural product quality and safety information (for trial implementation). http://www.moa.gov.cn/govpublic/ncpzlaq/201011/t20101111_1698591.htm (accessed September 21, 2012) (in Chinese).

Website of the Central People's Government of the PRC. 2012. A notice on the issuance of the Five-Year Plan on the food safety management system from the General Office of the State Council. http://www.gov.cn/zwgk/2012-07/21/content_2188309.htm (accessed June 29, 2012) (in Chinese).

Website of the Central People's Government of the PRC. 2013a. A notice on the issuance of arrangement of key work on food safety in 2013 from the General Office of the State Council. http://www.gov.cn/zwgk/2013-04/16/content_2378952.htm (accessed April 8, 2013) (in Chinese).

Website of the Central People's Government of the PRC. 2013b. 2012 annual report on government information disclosure of the State Food and Drug Administration. http://www.sda.gov.cn/WS01/CL0633/79476.html (accessed March 30, 2013) (in Chinese).

Website of the Central People's Government of the PRC. 2013c. 2012 annual report on government information disclosure of the Ministry of Health. http://www.gov.cn/gzdt/2013-02/25/content_2339497.htm (accessed January 26, 2013) (in Chinese).

Website of the Central People's Government of the PRC. 2013d. A notice on the issuance of key responsibilities, internal structure, and staffing of the China Food and Drug Administration. http://www.gov.cn/zwgk/2013-05/15/content_2403661.htm (accessed March 27, 2013) (in Chinese).

Website of Zhongshan Municipal Bureau of Quality and Technical Supervision. 2008. Labeling methods for 28 product categories and license application units. http://www.zsqts.gov.cn/FileDownloadHandle?fileDownloadId=522 (accessed 13 January, 2013) (in Chinese).

Wen, X.W. and Liu, M.L. 2012. Cause, dilemma and supervision of food safety from the year 2002 to 2011. *Reform* 9: 37–42 (in Chinese).

Wu, J.W. and Lin, S.Q. 2005. Use and management of food additives in Fujian Province. *Strait Journal of Preventive Medicine* 3: 57–59 (in Chinese).

Wu, L.H., Hou, B., and Gao, S.R. 2011. Scattered farmers' awareness of pesticide residue and its main influencing factors: An analysis based on structural equation model. *Chinese Rural Economy* 3: 35–48 (in Chinese).

Wu, L.H. and Qian, H. 2012. *China Development Report on Food Safety*. Beijing, People's Republic of China: Peking University Press (in Chinese).

Wu, L.H. and Xu, L.Q. 2009. *International Food Trade*. Beijing, People's Republic of China: China Light Industry Press (in Chinese).

Wu, L.L., Zhang, X.L., and Shan, L.J. 2011. The effect of economic and social characteristics of pesticide sprayers on their application behaviors: The case of Hebei Province. *Journal of Dialectics of Nature* 6: 23–29 (in Chinese).

Xing, M.H., Zhang, J.B., and Huang, G.T. 2009. Environmental awareness of farmers not participating in circular agriculture and its influencing factors: A case of Shanxi and Hubei Province. *Chinese Rural Economy* 4: 72–79 (in Chinese).

Xinhuanet. 2012. The Ministry of Health refuses to disclose the decision-making process of the new national raw milk standards and says it may affect stability. http://health.people.com.cn/n/2012/1024/c14739-19374063-5.html (accessed June 25, 2013) (in Chinese).

Xinhuanet. 2013a. 53.1% of consumers consider agricultural problems to be serious. http://www.bj.xinhuanet.com/bgt/2013-06/18/c_116188047.htm (accessed July 5, 2013) (in Chinese).

Xinhuanet. 2013b. The State Council determined the reform timeline of local food and drug administrative agencies. http://news.xinhuanet.com/politics/2013-04/18/c_115445397.htm (accessed July 5, 2013) (in Chinese).

Xinhuanet. 2013c. Work report of the Supreme People's Procuratorate (summary). http://news.xinhuanet.com/2013lh/2013-03/10/c_114970512.htm (accessed April 23, 2013) (in Chinese).

Xu, W.G. 1992. Overlap of regulations in the Food Hygiene Law. *Chinese Rural Health Service Administration* 5: 43 (in Chinese).

Yan, P.M., Xue, W.T., Zhang, H., Hu, X.P., and Tan, L.P. 2006. Study on nitrite formation in vegetables stored in varied storage conditions. *Food Science* 6: 242–247 (in Chinese).

Yang, H. and Zhang, Y.M. 2011. Food safety in enterprises providing nutritious meals for students in Chaoyang District, Beijing. *China Preventive Medicine* 12: 889–890 (in Chinese).

Yang, J., Gao, S.R., and Wu, L.H. 2010. The scattered farmers' behaviors of pesticide application and the influencing factors. *Heilongjiang Agricultural Sciences* 1: 45–48 (in Chinese).

Yang, L.K. and Xu, G.T. 1988. The Chinese food industry has been developing rapidly and ranks third in the industrial sector this year. *People's Daily* (November 29) (in Chinese).

Yin, X.L. and Yang, Z. 2011. Chinese milk standards in line with international practice. *Legal Evening News.* http://www.fawan.com/Article/jj/jd/2011/06/29/130013121591.html (accessed June 1, 2012) (in Chinese).

Yu, H.R. 2012. 15,000 food safety managers will be trained this year. *Shenzhen Special Zone Daily.* http://sztqb.sznews.com/html/2012-02/08/content_1919517.htm (accessed January10, 2013) (in Chinese).

Yue, N. 2010. Studies on the scientific technology support system in China's food import and export trade. PhD dissertation, Jiangnan University, Jiangsu, People's Republic of China (in Chinese).

Yunnan Health Inspection Information Network. 2011. A notice on conducting follow-up evaluation of national food safety standards from the Center for Health Inspection and Supervision. http://www.ynwsjd.cn/Item/12470.aspx (accessed August 6, 2012) (in Chinese).

Zeng, X.G., Li, Y., Niu, X.J., and Wang, S.F. 2002. The role of chemical pesticides in agricultural pest control and scientific evaluation. *Pesticide Science and Administration* 23(6): 30–31.

Zhang, L.J. 2011a. Situation and prospect of agricultural product quality and safety in China. Xinhuanet. http://news.xinhuanet.com/world/2011-09/14/c_122034007.htm (accessed October 10, 2012) (in Chinese).

Zhang, Q.Q., Chen, Z.X., and Wu, L.H. 2012. Investigation and analysis of use of food additives by producers and processors. *Food and Machinery* 2: 229–232 (in Chinese).

Zhang, T.J. 2010. Food safety in foreign trade of China. PhD dissertation, Wuhan University of Technology, Hubei, People's Republic of China (in Chinese).

Zhang, Y. 2001. Heavy metal pollution of soil and agricultural products in Shenyang suburbs: Current situation. *Chinese Journal of Soil Science* 32(4): 11–14 (in Chinese).

Zhang, Y. 2012. Define tasks and strengthen coordination to improve food safety information disclosure. *China Food Newspaper* (May 26): A01 (in Chinese).

Zhang, Y.H. 2011b. Development of cold chain logistics of food in Henan Province. *China Business and Trade* 13: 68–70 (in Chinese).

Zhang, Y.H., Ma, J.J., Kong, X.Z., and Zhu, Y. 2004. The influencing factors of pollution-free and green pesticides: An empirical analysis in 15 counties and cities in Shanxi and Shaanxi Province. *Chinese Rural Economy* 1: 41–49 (in Chinese).

Zhao, G.P. 2002. Sanitary conditions and management of small-scale food processors. *Modern Preventive Medicine* 29: 793–794 (in Chinese).

Zhao, W. 2009. Management strategies of bulk food in distribution chain. *China Practical Medical* 4: 248–249 (in Chinese).

Zheng, D.M. 2006. Analysis of competitiveness of biopesticide industry. *Journal of Fujian Agriculture and Forestry University Philosophy and Social Sciences* 3: 43–47 (in Chinese).

Zheng, F.T. 2013. What is missing in the new food safety regulatory bodies. *China Animal Industry* 8: 30 (in Chinese).

Zhou, J.H. 2006. Vegetable farmers' quality safety control behavior and its influencing factors: The case of Zhejiang Province. *Chinese Rural Economy* 11: 25–34 (in Chinese).

Zhou, J.H. and Hu, J.F. 2009. Quality and safety management behavior of vegetable processing enterprises and its influencing factors: The case of Zhejiang. *Chinese Rural Economy* 3: 45–56 (in Chinese).

Zhou, N.Y., Pan, J.R., and Wang, M. 2009. Study on the mathematical model for comprehensive evaluation of food safety. *Chinese Journal of Food Hygiene* 21(3): 198–202 (in Chinese).

Zhou, X.M. 2013. Man-made pollution in food safety in China. *Price: Theory and Practice* 1: 45–46 (in Chinese).

Zhou, Y.H. et al. 2008. *Food Safety and Regulation Issues.* Beijing, People's Republic of China: Economic and Management Publishing House (in Chinese).

Zou, J. and Yu, L.Y. 2011. Re-investigation of organic food chaos. *Oriental Morning Post.* http://www.dfdaily.com/html/3/2011/11/4/691323.shtml (accessed June 1, 2012) (in Chinese).

Zou, L.H. 2005. A study of mechanism of the early warning on food safety crisis. PhD dissertation, Tsinghua University, Beijing, People's Republic of China (in Chinese).

Zou, Z.F. 2010. *Guidance for Detection of Food Additives.* Beijing, People's Republic of China: China Standard Press (in Chinese).

Index